DISCARD

SEP 1 4 2010

Serious Game Design and Development:
Technologies for Training and Learning

Jan Cannon-Bowers
University of Central Florida, USA

Clint Bowers
University of Central Florida, USA

INFORMATION SCIENCE REFERENCE

Hershey · New York

Director of Editorial Content:	Kristin Klinger
Director of Book Publications:	Julia Mosemann
Acquisitions Editor:	Lindsay Johnston
Development Editor:	Christine Bufton
Typesetter:	Devvin Earnest
Quality control:	Jamie Snavely
Cover Design:	Lisa Tosheff
Printed at:	Yurchak Printing Inc.

Published in the United States of America by
Information Science Reference (an imprint of IGI Global)
701 E. Chocolate Avenue
Hershey PA 17033
Tel: 717-533-8845
Fax: 717-533-8661
E-mail: cust@igi-global.com
Web site: http://www.igi-global.com/reference

Copyright © 2010 by IGI Global. All rights reserved. No part of this publication may be reproduced, stored or distributed in any form or by any means, electronic or mechanical, including photocopying, without written permission from the publisher.
   Product or company names used in this set are for identification purposes only. Inclusion of the names of the products or companies does not indicate a claim of ownership by IGI Global of the trademark or registered trademark.

Library of Congress Cataloging-in-Publication Data

Serious game design and development : technologies for training and learning / Janis Cannon-Bowers and Clint Bowers, editors.
    p. cm.
  Includes bibliographical references and index.
  Summary: "With an increasing use of vido games in various disciplines within the scientific community, this book seeks to understand the nature of effective games and to provide guidance for how best to harness the power of gaming technology to successfully accomplish a more serious goal"--Provided by publisher.
  ISBN 978-1-61520-739-8 (hardcover) -- ISBN 978-1-61520-740-4 (ebook) 1. Video games--Design. 2. Video games industry--Technological innovations. 3. Game theory. I. Cannon-Bowers, Janis A. II. Bowers, Clint A.   GV1469.3.S48 2010
  794.8--dc22
                                                            2009050068

British Cataloguing in Publication Data
A Cataloguing in Publication record for this book is available from the British Library.

All work contributed to this book is new, previously-unpublished material. The views expressed in this book are those of the authors, but not necessarily of the publisher.

Editorial Advisory Board

Gil Muniz, *Uniformed Services University of the Health Sciences, USA*
Perry McDowell, *Navy Postgraduate School, Canada*
Denise Nicholson, *ACTIVE Laboratory, UCF, USA*
Ray Perez, *Office of Naval Research, USA*
Doug Watley, *BreakAway Ltd., USA*

List of Reviewers

Lucas Blair, *RETRO Laboratory, UCF, USA*
Sae Schatz, *ACTIVE laboratory, UCF, USA*
Janan Smither, *Dept. of Psychology, UCF, USA*
Peter Smith, *ADL Co-Lab, USA*
Rachel Joyce, *RETRO Laboratory, UCF, USA*
Denise Nicholson, *ACTIVE Laboratory, UCF, USA*
Steve Fiore, *Department of Philosophy, UCF, USA*
Rudy McDaniel, *Department of Digital Media, UCF, USA*
Florian Jentsch, *Dept. of Psychology, UCF, USA*
Bob Kenny, *Dept. of Digital Media, UCF, USA*

Table of Contents

Foreword xiv

Preface xvii

Section 1
Design Principles for Serious Games

Chapter 1
Mini-Games with Major Impacts 1
    Peter A. Smith, Joint ADL Co-Lab, USA
    Alicia Sanchez, Defense Acquisition University, USA

Chapter 2
Serious Storytelling: Narrative Considerations for Serious Games Researchers
and Developers 13
    Rudy McDaniel, University of Central Florida, USA
    Stephen M. Fiore, University of Central Florida, USA
    Denise Nicholson, University of Central Florida, USA

Chapter 3
An Adventure in Usability: Discovering Usability Where it was not Expected 31
    Holly Blasko-Drabik, University of Central Florida, USA
    Tim Smoker, University of Central Florida, USA
    Carrie E. Murphy, University of Central Florida, USA

Chapter 4
Development of Game-Based Training Systems: Lessons Learned in an Inter-Disciplinary
Field in the Making 47
    Talib Hussain, BBN Technologies, USA
    Wallace Feurzeig, BBN Technologies, USA
    Jan Cannon-Bowers, University of Central Florida, USA
    Susan Coleman, Intelligent Decision Systems, Inc., USA
    Alan Koenig, National Center for Research on Evaluation, Standards and Student
        Testing (CRESST), USA
    John Lee, National Center for Research on Evaluation, Standards and Student
        Testing (CRESST), USA
    Ellen Menaker, Intelligent Decision Systems, Inc., USA
    Kerry Moffitt, BBN Technologies, USA
    Curtiss Murphy, Alion Science and Technology, AMSTO Operation, USA
    Kelly Pounds, i.d.e.a.s. Learning, USA
    Bruce Roberts, BBN Technologies, USA
    Jason Seip, Firewater Games LLC, USA
    Vance Souders, Firewater Games LLC, USA
    Richard Wainess, National Center for Research on Evaluation, Standards and Student
        Testing (CRESST), USA

Chapter 5
DAU CardSim: Paper Prototyping an Acquisitions Card Game 81
    David Metcalf, University of Central Florida, USA
    Sara Raasch, 42 Entertainment, USA
    Clarissa Graffeo, University of Central Florida, USA

Chapter 6
Kinesthetic Communication for Learning in Immersive Worlds 102
    Christopher Ault, The College of New Jersey, USA
    Ann Warner-Ault, The College of New Jersey, USA
    Ursula Wolz, The College of New Jersey, USA
    Teresa Marrin Nakra, The College of New Jersey, USA

Section 2
Applications of Serious Games

Chapter 7
How Games and Simulations can Help Meet America's Challenges in Science
Mathematics and Technology Education 117
    Henry Kelly, Federation of American Scientists, USA

Chapter 8
Games for Peace: Empirical Investigations with PeaceMaker 134
    Cleotilde Gonzalez, Carnegie Mellon University, USA
    Lisa Czlonka, Carnegie Mellon University, USA

Chapter 9
Play's the Thing: A Wager on Healthy Aging 150
    Mihai Nadin, antÉ – Institute for Research in Anticipatory Systems
       University of Texas at Dallas, USA

Chapter 10
Re-Purposing a Recreational Video Game as a Serious Game for Second
Language Acquisition 178
    Yolanda A. Rankin, IBM Almaden Research Center, USA
    Marcus W. Shute, Clark Atlanta University, USA

Section 3
Games in Healthcare

Chapter 11
Application of Behavioral Theory in Computer Game Design for Health Behavior Change 196
    Ross Shegog, UT-School of Public Health, USA

Chapter 12
Avatars and Diagnosis: Delivering Medical Curricula in Virtual Space 233
    Claudia L. McDonald, Texas A&M University-Corpus Christi, USA

Chapter 13
Using Serious Games for Mental Health Education 246
    Anya Andrews, Novonics Corporation, Training Technology Lab (TTL), USA
    Rachel Joyce, University of Central Florida, USA
    Clint Bowers, University of Central Florida, USA

Chapter 14
Pervasive Health Games 260
    Martin Knöll, University of Stuttgart, Germany

Chapter 15
Influencing Physical Activity and Healthy Behaviors in College Students: Lessons
from an Alternate Reality Game 270
    Jeanne D. Johnston, Indiana University, USA
    Lee Sheldon, Indiana University, USA
    Anne P. Massey, Indiana University, USA

Section 4
The Way Ahead: The Future of Serious Games

Chapter 16
Establishing a Science of Game Based Learning 290
Alicia Sanchez, Defense Acquisition University, USA
Jan Cannon-Bowers, University of Central Florida, USA
Clint Bowers, University of Central Florida, USA

Chapter 17
The Way Ahead in Serious Games 305
Jan Cannon-Bowers, University of Central Florida, USA

Compilation of References 311

About the Contributors 341

Index 352

Detailed Table of Contents

Foreword xiv

Preface xvii

Section 1
Design Principles for Serious Games

This section provides several different perspectives on designing and developing serious games. Each chapter offers a design principle or strategy that can be employed to enhance the effectiveness of serious games. Several also include lessons learned drawn from specific serious game development efforts.

Chapter 1
Mini-Games with Major Impacts 1
    Peter A. Smith, Joint ADL Co-Lab, USA
    Alicia Sanchez, Defense Acquisition University, USA

The authors describe a strategy for developing mini games that can be embedded in game-based training. They also present descriptions of several case studies that used mini-games as part of the learning strategy.

Chapter 2
Serious Storytelling: Narrative Considerations for Serious Games Researchers
and Developers 13
    Rudy McDaniel, University of Central Florida, USA
    Stephen M. Fiore, University of Central Florida, USA
    Denise Nicholson, University of Central Florida, USA

This chapter discusses the importance of narrative in serious games. These authors contend that narrative aids can help in game design in several ways, including: increasing the player's motivation to remain in the game; stories can embed learning objectives; narrative can tie together elements in the game into a coherent whole.

Chapter 3
An Adventure in Usability: Discovering Usability Where it was not Expected 31
Holly Blasko-Drabik, University of Central Florida, USA
Tim Smoker, University of Central Florida, USA
Carrie E. Murphy, University of Central Florida, USA

This chapter describes the goals of usability and how it is traditionally performed using two popular methods. It goes on to discuss appropriate usability measures for serious games.

Chapter 4
Development of Game-Based Training Systems: Lessons Learned in an Inter-Disciplinary Field in the Making 47
Talib Hussain, BBN Technologies, USA
Wallace Feurzeig, BBN Technologies, USA
Jan Cannon-Bowers, University of Central Florida, USA
Susan Coleman, Intelligent Decision Systems, Inc., USA
Alan Koenig, National Center for Research on Evaluation, Standards and Student Testing (CRESST), USA
John Lee, National Center for Research on Evaluation, Standards and Student Testing (CRESST), USA
Ellen Menaker, Intelligent Decision Systems, Inc., USA
Kerry Moffitt, BBN Technologies, USA
Curtiss Murphy, Alion Science and Technology, AMSTO Operation, USA
Kelly Pounds, i.d.e.a.s. Learning, USA
Bruce Roberts, BBN Technologies, USA
Jason Seip, Firewater Games LLC, USA
Vance Souders, Firewater Games LLC, USA
Richard Wainess, National Center for Research on Evaluation, Standards and Student Testing (CRESST), USA

This chapter describes a recent experience developing a serious game for U.S. Navy recruits to describe a multi-disciplinary approach to serious game design. They describe their process in terms of the selection of training requirements, the domain and the gaming platform; knowledge acquisition; story development; game design; initial instructional design; assessment strategy; software development; introductory video; and review, refinement and testing.

Chapter 5
DAU CardSim: Paper Prototyping an Acquisitions Card Game 81
David Metcalf, University of Central Florida, USA
Sara Raasch, 42 Entertainment, USA
Clarissa Graffeo, University of Central Florida, USA

This chapter describes the development of a multiplayer card game that was first developed as a paper prototype. The chapter provides a post-mortem of the iterative design process that included development of varying levels of simple prototypes for initial design and playtesting, followed by evaluation of game balance and refinement. They also cover the process they employed to digitize the game, and expand the game to cover additional learning objectives.

Chapter 6
Kinesthetic Communication for Learning in Immersive Worlds 102
Christopher Ault, The College of New Jersey, USA
Ann Warner-Ault, The College of New Jersey, USA
Ursula Wolz, The College of New Jersey, USA
Teresa Marrin Nakra, The College of New Jersey, USA

This chapter discusses a game design architecture that exploits the pedagogical potential of a rich graphical environment using a kinesthetic interface. The authors conclude by describing directions for future testing and application of the kinesthetic input devices in serious games.

Section 2
Applications of Serious Games

Our conception of Serious Games is the use of games for any non-entertainment purpose, although the preponderance of attention has been given to educational or learning games. In this section, we have included several chapters that are not strictly educational in nature to highlight the fact that other applications are possible. That said, we believe that the potential application of games to learning (across settings and age groups) is vast and only beginning to be tapped.

Chapter 7
How Games and Simulations can Help Meet America's Challenges in Science
Mathematics and Technology Education 117
Henry Kelly, Federation of American Scientists, USA

The author addresses three key issues in educational game design: (1) designing the course of instruction so that it is both rigorously correct and constantly engaging, (2) ensuring that the system adapts to the background and interests of individual learners, and (3) evaluating the expertise of learners in ways that make sense to them and to future employers.

Chapter 8
Games for Peace: Empirical Investigations with PeaceMaker 134
Cleotilde Gonzalez, Carnegie Mellon University, USA
Lisa Czlonka, Carnegie Mellon University, USA

This chapter describes the use of a video game to conduct empirical investigations designed to build theoretical models of socio-psychological variables that influence dynamic decision making. Specifically, an investigation on decision making in a dynamic and complex situation, the solution of international conflict and the achievement of peace, using PeaceMaker, a popular video game, is presented.

Chapter 9
Play's the Thing: A Wager on Healthy Aging 150
    Mihai Nadin, antÉ – Institute for Research in Anticipatory Systems
        University of Texas at Dallas, USA

This chapter centers on the hypothesis that the aging process results in diminished adaptive abilities resulting from decreased anticipatory performance. To mitigate the consequences of reduced anticipatory performance, the addresses brain plasticity through game play.

Chapter 10
Re-Purposing a Recreational Video Game as a Serious Game for Second
Language Acquisition 178
    Yolanda A. Rankin, IBM Almaden Research Center, USA
    Marcus W. Shute, Clark Atlanta University, USA

The authors report their efforts to re-purpose a recreational game as a serious game to promote learning in the context of Second Language Acquisition. They outline the process of game transformation, which leverages the entertainment value and readily accessible developer tools of the game.

Section 3
Games in Healthcare

Given the number of high quality proposals we received in the healthcare area, we decided to create a separate section to highlight this important area. The chapters in this section offer a sampling of the types of Serious Games being developed in this area. These include: games being used in the therapeutic process, games to promote healthy behaviors, games to train healthcare professionals and pervasive health games. These applications, as well as others related to healthcare, have the potential to play an important role in the future of healthcare in the U.S. and across the world.

Chapter 11
Application of Behavioral Theory in Computer Game Design for Health Behavior Change 196
    Ross Shegog, UT-School of Public Health, USA

The chapter introduces serious game developers to processes, theories, and models that are crucial to the development of interventions to change health behavior, and describes how these might be applied by the serious games community.

Chapter 12
Avatars and Diagnosis: Delivering Medical Curricula in Virtual Space 233
    Claudia L. McDonald, Texas A&M University-Corpus Christi, USA

The author describes Pulse!! The Virtual Clinical Learning Lab—a project designed to explore the use of games in health care by developing a reliable and valid learning platform for delivering medical curricula in virtual space. She uses the Pulse!! example to describe lessons learned in the general area of collaboration, including issues such as funding, technology and evaluation.

Chapter 13
Using Serious Games for Mental Health Education 246
    Anya Andrews, Novonics Corporation, Training Technology Lab (TTL), USA
    Rachel Joyce, University of Central Florida, USA
    Clint Bowers, University of Central Florida, USA

The chapter addresses the mental health training and education needs of modern "at risk" populations and discuss the potential of serious games as effective interventions for addressing those needs.

Chapter 14
Pervasive Health Games 260
    Martin Knöll, University of Stuttgart, Germany

The author describes the potentials of serious game applications in a health context to improve user's motivation, education and therapy compliance. He focuses on "Pervasive Health Games", which combine pervasive computing technologies with serious game design strategies.

Chapter 15
Influencing Physical Activity and Healthy Behaviors in College Students: Lessons
from an Alternate Reality Game 270
    Jeanne D. Johnston, Indiana University, USA
    Lee Sheldon, Indiana University, USA
    Anne P. Massey, Indiana University, USA

The authors investigated the effectiveness of a prototype Alternate Reality Game – called The Skeleton Chase – in influencing physical activity and wellness of college-age students.

Section 4
The Way Ahead: The Future of Serious Games

This section includes chapters that focus on looking toward the future of serious games. Specifically, it addresses how to establish a science of serious game design that is meant to stimulate research and applications. In addition, it includes a commentary on the way ahead in Serious Games.

Chapter 16
Establishing a Science of Game Based Learning 290
    Alicia Sanchez, Defense Acquisition University, USA
    Jan Cannon-Bowers, University of Central Florida, USA
    Clint Bowers, University of Central Florida, USA

The authors offer a simple framework for organizing variables important in the learning process and then discuss findings from psychology and education as a basis to formulate a research agenda for game-based training. The goal of the framework is to stimulate researchers to conduct systematic, appropriately controlled experiments that will provide insight into how various game features affect motivation and learning.

Chapter 17
The Way Ahead in Serious Games 305
Jan Cannon-Bowers, University of Central Florida, USA

The author summarizes the major themes that emerge from the previous chapters and offers some observations and presents suggestions for the way ahead in Serious Games and their application to important societal challenges.

Compilation of References 311

About the Contributors 341

Index 352

Foreword: Does Game Technology Matter?

Among the ruins of ancient Egypt there are multiple references to games that were popular among the Pharaohs. The remains and images of the game of Senet date back to 3,000BC. This board game contains features similar to modern checkers and a method of play reminiscent of a horse race around the board. Though primarily a game for entertainment, it was also used as a mystic tool to foretell the future. Egyptians believed that the square that a player's piece ended on contained special significance about what would happen to the person in the future. Though we would consider this superstition, the players at that time took the results as guidance on decisions about commerce, farming, religion, or family.

Around 1,400BC the game of Mancala emerged in Africa. It was a tool used to account for livestock and crops, and a form of entertainment. Tribesmen used the board and stones to negotiate the trade of goods, and perhaps to gamble for a better exchange. But they also passed the time in the fields playing a version of Mancala that had no economic consequences, but was purely a form of entertainment.

In 1956, Charles Roberts developed the components of the modern board wargame as a tool to help him prepare for his commissioning in the U.S. Army. But by 1958 he realized the commercial value of this wargame and created the Avalon Hill game company to market it to thousands of avid "armchair generals" who were eager to test and develop their own tactical military skills, but for entertainment. For the next four decades Avalon Hill and several competitors created wargames for both entertainment and military training.

Were these games primarily and initially entertainment or serious tools for guiding life decisions? There was really no hard division between the two purposes. There is no law of nature that says tools for education and training cannot be enjoyable to use, or that such tools cannot be inspired by or created from applications that were initially entertainment. The dual nature of games has been with us for at least 5,000 years. Today we may have replaced dice made from sheep knucklebones for computerized, pseudo-random number generation algorithms, but we continue to look to the results of game play for insight into important problems in our lives. Now we place our faith in the accuracy of mathematical and logical algorithms rather than the mystical forces influencing the roll of the die, but we continue to construct games that can challenge our thinking and guide us to a better understanding of the world.

WHAT IS A GAME?

What makes some activities and tools into games, while others are considered completely serious tools? In his 1970 book entitled *Serious Games*, Clark Abt defined a game with these words, "reduced to its formal essence, a game is an activity among two or more independent decision-makers seeking to achieve their objectives in some limiting context. A more conventional definition would say that a game is a context with rules among adversaries trying to win objectives." In a 2005 issue of *IEEE Com-*

puter, Mike Zyda defined a game as, "a physical or mental contest, played according to specific rules, with the goal of amusing or rewarding the participant." He went on to suggest that a serious game was, "a mental contest, played with a computer in accordance with specific rules that uses entertainment to further government or corporate training, education, health, public policy, and strategic communication objectives." Zyda explicitly points to the desirable goal of using "entertainment" to further the goals of the organization, to harness entertainment, fun, engagement, challenge, and trail-and-error to get people to learn more or to learn faster.

Academics like Andrew Hargadon at University of Southern California explore the difficulties involved in adopting tools and practices from other industries. There is a psychological, social, and professional barrier that keeps people from accepting ideas that were "not invented here." The barrier between "serious business" and "frivolous entertainment" is even higher, wider, and deeper than those between industrial professions. Industries may adopt new computers, networks, materials, and energy sources. But reaching into the entertainment industry for something that can improve effectiveness is considered quite a daring and questionable move.

GAME TECHNOLOGY

Games have created and introduced new technologies for centuries. Ancient games offered numbered throwing sticks, the predecessors to dice and random number generators, as a means of making decisions with limited information. Board wargames of the 1950's introduced the hexagonal tessellation of terrain, a concept that is still used in cellular communications models as an approximation to the circular area covered by a tower. Charles Roberts introduced the combat results table as a means of enriching the military results from the throw of a die. Today all military models use extensive algorithms to make decisions, but often retain a random number generator as a nondeterministic influence in those algorithms.

Currently it is difficult to determine whether computer hardware and software technologies are "game technologies" or "serious technologies". Graphics cards, network cards, and multi-core chips are all essential for the play of the latest computer games. But should they be tagged as serious or entertainment technologies? Does it matter? Does it help?

Recently the gaming industry has been the source of some of the best software technologies on the market. The 3D scene generators or game engines are far superior in performance and features to competing applications created in serious industries and academia. Game companies have adopted the principles of man-machine interfaces and effective graphical user interfaces to create complex applications for which no user's manual is required. But similar interfaces in serious industries can be so complex that multi-day courses are required to learn to use them. Games have isolated the most essential physics and human behavior features such that they can be incorporated into an application that can run on a consumer PC. They are certainly not the highest fidelity models of physics or artificial intelligence, but they are the most accessible and among most useful. Multiplayer games have advanced networking protocols and libraries so that players can join the virtual world from anyplace on the planet. But what serious industry applications provide this type of ad hoc collaboration?

The financial incentives and the personal energy that drive the creation of new technologies in the game industry have led to technologies that are just too valuable to be excluded from other serious industrial applications. All industries have got to take these technologies seriously or risk being passed by competitors who will use them.

DOES GAME TECHNOLOGY MATTER?

Game technologies have been adopted for military training, medical education, emergency management, city planning, spacecraft engineering, architectural design, religious proselyzation, political communication, movie making, and advertising – to name a few. These are far from being the dominant applications in any of these fields. But they gain ground every year as young game players become serious business people and as older business people become more avid game players. The barriers are falling. Each year more people are able to peer through the science fiction veneer of a space game and see the powerful computer science beneath. They understand the advantages of putting this technology to use, and doing so before a competitor does the same. In his 2003 *Harvard Business Review* article entitled "IT Doesn't Matter", Nicholas Carr shook up the business and the IT worlds with his observation that IT initially provided a competitive advantage. But after mass adoption, all industries had harnessed its power, and IT became as essential to modern business as electricity had been to the industrial revolution. It had transcended its own uniqueness and become essential. If game technology is as successful, it will lose its niche status to become an essential part of running an effective and profitable business.

Roger Smith

REFERENCES

Abt, C. (1970). *Serious games*. New York: The Viking Press.

Beck, J.C. and Wade, M. (2004). *Got game: How the gamer generation is reshaping business forever*. Boston, MA: Harvard Business School Press.

Carr, N. (May 2003). "IT doesn't matter". *Harvard Business Review*.

Michael, D and Chen, S. (2005). *Serious games: Games that educate, train, and inform*. New York: Thompson Publishing.

Orbanes, P.E. (2004). *The Game makers: The Story of Parker Brothers*. Boston: Harvard Business School Press.

Perla, P. (1990). *The Art of wargaming*. Naval Institute Press.

Smith, R. (January 2006). "Technology disruption in the simulation industry". *Journal of Defense Modeling and Simulation*.

Zyda, M. (September 2005). "From visual simulation to virtual reality to games". *IEEE Computer*.

Preface

As many have observed, the use of video game techniques and technologies for purposes other than purely entertainment has gained attention in recent years. So called serious games—those that have a non-entertainment purpose—are beginning to be developed in a variety of settings, including healthcare, education, and workplace learning. Despite the popularity of serious games, however, there are only now beginning to be rigorous attempts to guide application of the technologies, and evaluation of their ability to meet their intended goals. The purpose of this volume is to provide a cross section of the work being done in this burgeoning area.

The volume is organized around three themes: Design Principles for Serious Games, Applications of Serious Games, Games in Healthcare, The Way Ahead: A Roadmap for the Future of Serious Games. We should note that we did not necessarily intend to pull Healthcare out as a separate section, but we received so many quality chapter proposals in this area that we decided to group them together. This may be a function of the funding available to study health-related games (e.g., Robert Woods Johnson Foundation's Games for Health program) or attention being given to this area (e.g., the annual Games for Health Conference and Healthcare reform in general). In any case, much good work is taking place in this sector and will hopefully transfer over to other application areas.

The following sections describe the major themes of the book, along with a description of the chapters that fall within them.

Section 1: Design Principles for Serious Games

This section provides several different perspectives on designing and developing serious games. Each chapter offers a design principle or strategy that can be employed to enhance the effectiveness of serious games. Several also include lessons learned drawn from specific serious game development efforts.

In the chapter entitled "*Mini-Games with Major Impacts*," Smith and Sanchez describe a strategy for developing mini games that can be embedded in game-based training. These authors address how mini-games can be used for conceptual or procedural knowledge and provide theoretical arguments from: Cognitive Learning Theory, Social Cognitive Theory, and Motivation. They also present descriptions of several case studies that used mini-games as part of the learning strategy. Smith & Sanchez conclude that mini-games have become sophisticated enough to be included in serious games.

McDaniel, Fiore, and Nicholson then discuss the importance of narrative in serious games in their chapter, "*Serious Storytelling: Narrative Considerations for Serious Games Researchers and Developers*." Specifically, they highlight the congruence between the game's story and its learning content as a mechanism to enhance the player's immersion in the game. These authors contend that narrative aids can help in game design in several ways, including: increasing the player's motivation to remain in the game, stories can embed learning objectives, and narrative can tie together elements in the game into a coherent whole. They go on to cover selected narratological principles, interactive narratology, and then present a preliminary narrative taxonomy to guide research and development. They conclude with implications for the field.

In the chapter by Blasko-Drabik, Smoker, and Murphy, "*An Adventure in Usability: Discovering Usability Where it was not Expected,*" these authors define usability as it is employed in software design. As with other software applications, it is important to establish the usability of a serious game to ensure that poor interface design does not interfere with learning. These authors describe the goals of usability and how it is traditionally performed using two popular methods. They go on to discuss appropriate usability measures for serious games. They compare two major methods and then conclude with a description of how usability analyses can be used to improve game design.

Next, Hussain and colleagues use a recent experience developing a serious game for U.S. Navy recruits to describe a multi-disciplinary approach to serious game design. In the chapter entitled, "*Development of Game-Based Training Systems: Lessons Learned in an Inter-Disciplinary Field in the Making*", these authors begin with a number of theoretical justifications for using games in learning, and then describe the process they employed in developing the serious game. Specifically, they describe their process in terms of the selection of training requirements, the domain and the gaming platform; knowledge acquisition; story development; game design; initial instructional design; assessment strategy; software development; introductory video; and review, refinement and testing. In each of the sections, they identify a number of tensions that need to be resolved as the game is being developed. They go on to provide lessons learned by describing how each of the tensions was resolved. These lessons learned can be of use to future serious game designers.

In the chapter entitled, "*DAU CardSim: Paper Prototyping an Acquisitions Card Game*", Metcalf, Raasch, and Graffeo describe development of a multiplayer card game that was first developed as a paper prototype. The game, a multiplayer scenario-based card game, was designed to teach skills associated with Department of Defense acquisition procedures and teamwork. The chapter provides a post-mortem of the iterative design process that included development of varying levels of simple prototypes for initial design and playtesting, followed by evaluation of game balance and refinement. They also cover the process they employed to digitize the game, and expand the game to cover additional learning objectives. Finally, they provide a series of lessons learned as they relate to paper prototyping as a design strategy.

The final chapter in this section, "*Kinesthetic Communication for Learning in Immersive Worlds*", by Ault, Warner-Ault, Wolz, and Nakra, posits a game design architecture that exploits the pedagogical potential of a rich graphical environment using a kinesthetic interface (such as the one used by the Nintendo Wii). They explain that their approach is grounded in the game's content so that genuine learning can occur in context. Furthermore, the kinesthetic interface is consistent with research showing that movement-based methods are more effective in language learning than more traditional methods. The authors conclude by describing directions for future testing and application of the kinesthetic input devices in serious games.

Section 2: Applications of Serious Games

As noted, our conception of Serious Games is the use of games for any non-entertainment purpose, although the preponderance of attention has been given to educational or learning games. In this section, we have included several chapters that are not strictly educational in nature to highlight the fact that other applications are possible. That said, we believe that the potential application of games to learning (across settings and age groups) is vast and only beginning to be tapped.

To begin this section, Kelly provides compelling statistics showing that the quality of education in the U.S. is in dire need of improvement in his chapter, "*How Games and Simulations can Help Meet America's Challenges in Science Mathematics and Technology Education.*" Fortunately, he contends that modern technology has the potential to make learning more productive, more engaging, and more closely tailored to the interests and backgrounds of individual learners. According to Kelly, computer games provide a particularly good example of what can be achieved because they often require players to master complex skills to advance in the game. He goes on to address three key issues in educational game design: (1) designing the course of instruction so that it is both rigorously correct and constantly engaging, (2) ensuring that the system adapts to the background and interests of individual learners, and (3) evaluating the expertise of learners in ways that make sense to them and to future employers, using a game called "Immune Attack" as his example.

In the next chapter, "*Games for Peace: Empirical Investigations with PeaceMaker,*" Gonzalez and Czlonka provide a example of using a video game to conduct empirical investigations designed to build theoretical models of socio- psychological variables that influence dynamic decision making. Specifically, they present an investigation on decision making in a dynamic and complex situation, the solution of international conflict and the achievement of peace, using PeaceMaker, a popular video game. PeaceMaker represents the historical conditions of the Israeli-Palestinian conflict and provides players with an opportunity to resolve the conflict. Students in an Arab-Israeli history course played perspectives of the Israeli and Palestinian leaders at the beginning and end of the semester. Student actions were recorded and analyzed along with information about their personality, religious, political affiliation, trust attitude, and number of gaming hours per week. The authors offer several conclusions regarding the manner in which these variables affect conflict resolution, hence the game served as a mechanism to better understand the phenomenon of interest. Many other applications of this approach to sutdy human behavior in complex systems seem obvious.

Nadin begins the next chapter, "*Play's the Thing: A Wager on Healthy Aging,*" with the hypothesis that the aging process results in diminished adaptive abilities resulting from decreased anticipatory performance. To mitigate the consequences of reduced anticipatory performance, he addresses brain plasticity through game play. Since anticipation is expressed in action, the games conceived, designed, and produced for triggering brain plasticity need to engage the sensory, cognitive, and motoric aspects of performance. Nadin offers a rich theoretical foundation upon which to design and validate such games.

A popular notion among those developing serious games is that entertainment games can be repurposed to accomplish serious objectives. In their chapter, "*Re-Purposing a Recreational Video Game as a Serious Game for Second Language Acquisition,*" Rankin and Shute describe efforts to re-purpose the recreational Massively Multiplayer Online Role Playing Game (MMORPG) EverQuest® II as a serious game to promote learning in the context of Second Language Acquisition (SLA). They outline the process of game transformation, which leverages the entertainment value and readily accessible developer tools of the game. They identify the affordances attributed to MMORPGs and then evaluate the impact of gameplay experiences on SLA. Promising results are described.

Section 3: Games in Healthcare

Given the number of high quality proposals we received in the healthcare area, we decided to create a separate section to highlight this important area. The chapters in this section offer a sampling of the types of Serious Games being developed in this area. These include: games being used in the therapeutic process, games to promote healthy behaviors, games to train healthcare professionals, and pervasive health games. These applications, as well as others related to healthcare, have the potential to play an important role in the future of healthcare in the U.S. and across the world.

In the introductory chapter in this section, "*Application of Behavioral Theory in Computer Game Design for Health Behavior Change*," Shegog provides an excellent overview of behavioral theories and how they might be used to promote health behaviors. The chapter introduces serious game developers to processes, theories, and models that are crucial to the development of interventions to change health behavior, and describes how these might be applied by the serious games community. Shegog goes on to describe the protocols, theories, and models that have informed the development of interventions in health behavior change and reviews them in terms of their potential contribution to serious game design, implementation, and evaluation. The author describes a serious game application aimed at cognitive-based gaming in adolescents to exemplify this.

Next, McDonald asserts that virtual-world technologies have advanced to the point where they can be considered as a viable method for delivering medical curricula effectively and safely. In her chapter entitled "*Avatars and Diagnosis: Delivering Medical Curricula in Virtual Space*," she contends further that research must establish that such systems are reliable and valid tools for delivering medical curricula; otherwise, they are of no use to the medical community, regardless of their technical sophistication. McDonald then describes Pulse!! The Virtual Clinical Learning Lab—a project designed to explore these issues by developing a reliable and valid learning platform for delivering medical curricula in virtual space. She uses the Pulse!! example to describe lessons learned in the general area of collaboration, including issues such as funding, technology and evaluation. She concludes with a discussion of what lies ahead for the Pulse!! research and development project.

In the chapter by Andrews, Joyce, and Bowers, called "*Using Serious Games for Mental Health Education*," these authors address the mental health training and education needs of modern "at risk" populations and discuss the potential of serious games as effective interventions for addressing those needs. These authors pay particular attention to the importance of prevention training and ways in which serious games can be designed to facilitate the prevention process. They focus specifically on interventions targeted at the development of appropriate coping skills associated with certain sets of mental health risks. Within the chapter, the authors describe several specific mental health-related serious game efforts and discuss design considerations for effective serious games.

Knöll then discusses the potentials of serious game applications in a health context to improve user's motivation, education, and therapy compliance. He focuses on "*Pervasive Health Games*," which combine pervasive computing technologies with serious game design strategies. They represent a new instantiation of gameplay essentially using the user's environment as the play space, and therefore extending into their everyday life. Knöll presents the new typology of PHG as an interdisciplinary field, consisting of health care, psychology, game design, sports science, and urban research. A brief introduction to the theme is illustrated with a conceptual "showcase," a pervasive game for young diabetics.

Capitalizing on the trend toward developing games for physical activity (so called, "exergaming"), Johnston, Sheldon, and Massey describe a game designed to influence physical activity and wellness in the college-age population. In their chapter entitled "*Influencing Physical Activity and Healthy Behaviors in College Students: Lessons from an Alternate Reality Game*," these authors describe how they were motivated to develop the game based on statistics showing that in the transition to college individual demonstrate an alarming decrease in physical activity. Simultaneously, a significant weight gain during early college years has been shown to increase the risk of obesity and associated diseases later in life such as diabetes and coronary heart disease. In this study, the authors investigated the effectiveness of a prototype Alternate Reality Game (ARG) – called *The Skeleton Chase* – in influencing physical activity and wellness of college-age students. A growing game genre, an ARG is an interactive narrative that uses the real world as a platform, often involving multiple media (e.g., game-related web sites, game-related blogs, public web sites, search engines, text/voice messages, video, etc.) to reveal a story. They provide preliminary findings on the effectiveness of the game as well as lesson learned to guide future efforts.

Section 4: The Way Ahead: A Roadmap for the Future of Serious Games

In the final section, we included chapters that focus on looking toward the future of serious games.

First, in the chapter entitled *"Establishing a Science of Game Based Learning,"* Sanchez, Cannon-Bowers, and Bowers offer a simple framework for organizing variables important in the learning process and then discuss findings from psychology and education as a basis to formulate a research agenda for game-based training. These include: characteristics of the user, pedagogical features embedded in the game, and game design features. These can all affect the user's motivation to interact with the game, and in turn, influence learning, while some of the features may also exert a direct impact on learning. The authors' purpose in presenting this framework is to stimulate researchers to conduct systematic, appropriately controlled experiments that will provide insight into how various game features affect motivation and learning. According to these authors, by following theoretically-based roadmap, a true science of educational games can be formed.

In the final chapter, *"The Way Ahead in Serious Games,"* Cannon-Bowers attempts to summarize some of the major themes found throughout the volume. She offers some observations and presents suggestions for the way ahead in serious games and their application to important societal challenges.

Overall, we are moved to comment that serious games hold great promise as a means to reach and affect large numbers of people in a positive way. Capitalizing on the popularity of video games, along with emerging digital technologies and more accessible delivery methods, those seeking to affect positive change in the future may find that serious games are a useful mechanism to both study and influence human behavior. We believe that efforts to investigate serious games and their impact in scientifically valid and rigorous ways must continue if this potential is to be reached.

Jan Cannon-Bowers & Clint Bowers
Orlando, Florida
July, 2009

Section 1
Design Principles for Serious Games

Chapter 1
Mini-Games with Major Impacts

Peter A. Smith
Joint ADL Co-Lab, USA

Alicia Sanchez
Defense Acquisition University, USA

ABSTRACT

The concept of mini-games has long been associated with small uninspired games found in conventional Computer Based Training (CBT). They have traditionally been made up of simple quizzes or matching games that have done little to engage the players in the learning event. This, however, is no longer the case. With advances in mini-game design paradigms, mini-games have become an effective means to engage learners with a specific learning objective both standalone and in the context of a greater training application. This work will explore educational and training mini-game development within Defense Acquisition University (DAU), National Science Foundation (NSF), and others.

INTRODUCTION

Mini-games, those simple little downloadable games that are commonly found in conventional web-based training courses, should no longer be considered as nothing more than a distraction breaking up the content from the inevitable test that will be presented on the next slide. Mini-games have come into their own as a legitimate form of training and education through games.

Mini-Games commonly reside on the opposite side of the gaming spectrum from conventional games. They are usually small games that are easy to learn, hard to master. Think of "Tetris" as a good example of a Mini-Game. Anyone can play "Tetris" but it is hard to be very good at "Tetris." While conventional games might take days or weeks to play, Mini-Games are often played for under an hour.

Educational Mini-Games follow the same philosophy while containing a single learning goal. A Mini-Game could, for example, teach vector addition. It would not go further to include positive and negative acceleration, but provide a concentrated experience for only the one learning objective. The design of mini-games has matured from simple matching games, and quizzes to allow for real interaction with training concepts in a meaningful way.

Using Mini-Games for Procedural and Conceptual Learning Objectives

Mixed results have been generated on the use of games and simulations in the classroom. A study by Randel, Morris, Wetzel & Whitehill (1992) examined 68 studies that used games and simulations in the classroom to enhance learning. Finding indicated that of the 68 studies in which games and simulations were considered, 22 of them enhanced student performance. Twelve of the studies also indicated that students were more interested in games and simulations than traditional classroom instruction. Thirty-eight of the studies had no impact on student performance, however, making the implementation of games and simulations into classrooms a risky notion. Ricci, Salas & Cannon-Bowers (1996) supported these findings by explaining that although games could stimulate more interest than traditional classroom based instruction, they might not provide any additional value to the education.

Over the last several years, the concept of using serious games for teaching and training has gained a considerable amount of popular support in a wide array of fields. Unfortunately, the potential benefits of the use of games in education and training has been relegated to the use of large and often very expensive game systems, designed to target entire learning systems or to serve as capstone and cumulative experiences. There has been little to no attention paid to the use of mini-games in order to target both part task training and smaller learning objectives.

Taking their cues from the casual gaming market, mini-games are essentially small games that distil a complex learning concept into a small extremely targeted amount of game play. Mini-games have the potential to reinforce a single or small group of learning objectives by providing bite sized, replayable, engaging, and motivating learning experiences.

Often education and training systems as a whole are designed to provide a student with both core knowledge and the application of that knowledge. While learning systems as a whole are usually targeted towards a performance oriented outcome, creating meaningful relationships between the concepts required to achieve those outcomes and practicing the concepts learned within context can both be achieved through the use of single serving game applications.

Mini-Games for Conceptual Information

Mini-Games that are used to provide conceptual information often rely on the retention of information. A good example of this type of game is the common children's game "Memory." In "Memory" the player has a field of cards laid out in front of them face down. They first flip a card over revealing its value and then flip another card hoping to find the match of the previous card. If a match is found they remove the card from the group. If no match is found they try again, until all cards are removed from the group. This game requires the player to utilize memorization to complete the game. The intended result of these games if for the player to memorize the concepts contained on the cards.

Mini-Games for Procedural Information

Procedural focused mini-games are a newly formed incarnation of the mini-game genre. They have become a staple of the Party Game genre of entertainment games and are much more complex than their Conceptual counterparts but still maintain the easy to pick up and play, targeted information delivery, of the mini-game paradigm. These mini-games provide the player with a situation in which they can apply a concept. In order for these games to work the player must create a meaningful relationship between nuggets of conceptual information.

Theories Supporting the Use of Mini-Games

Before delving into case studies in which mini-games have been used, it is important to understand some of the theory behind why mini-games should be considered a useful training paradigm. The most pertinent of theories to this discussion are Cognitive Learning Theory, Social Cognitive Theory, and Motivation.

Cognitive Learning Theory

Cognitive learning theories focus on how humans acquire, process, store, and retrieve knowledge; and how the environment affects their learning. With origins in philosophy stemming from Plato and Descartes: cognitive psychology has evolved through the decades into strategies used today that incorporate the new environment we live in. In 1986, Bell-Gredler reviewed cognitive theories and synthesized their findings. Beginning with Gestalt, cognition was defined as the human process of organizing stimuli that gave it meaning. Gestalt theorized that when stimuli were introduced to humans, they would organize those stimuli cognitively and that stimuli could only be utilized when the purpose of the stimuli was understood. He argued that how an individual initially perceives an object could determine their application of that object. This gave way to the idea of frameworks within human cognition and the relationships between them.

According to Bell-Gredler, Frederic C. Bartlett developed the idea of schemata in the 1930's. Schemata are the frameworks in which new stimuli or information can be stored. Barlett's (1958) research indicated that gaps in schemata were filled in using expectations until confirmation could be reached through the acquisition of new stimuli. This was evidenced in an experiment conducted in which successive patterns were shown to individuals who were able to predict the final display without seeing it.

The storage framework, schemata, served as structures in which new information could be assimilated and processed. New information or stimuli were encoded during the assimilation process into existing schemata. Understanding came from the ability to make relationships with new information and evolving schemata. Baron & Byrne (1977) offered further insight on the process of assimilating new information by theorizing that the encoding process involved changing the new information in order to fit it into an individual's existing schemata, changing or distorting it based on that individual's perceptions, interests, and motivations.

The cognitive constructivist work of Bruner (1966) provided a unifying understanding of human cognition as an active process. This active process incorporated new information into existing knowledge. When learning activities were relevant and engaging, students could construct their own understanding of the information based on their prior knowledge; therefore each individual would understand things slightly differently. His approach to education was to allow students to make connections between new information and their existing knowledge themselves, continually adding to the existing knowledge structures. Key to constructivism were three components of effective learning: anchored or situated learning; cognitive apprenticeships, and social negotiation of knowledge (Asynchronous Learning, 1997).

Craik & Lockhart (1972) developed a framework involving levels of processing that was intended to explain how information was stored. Within this framework, stimuli were processed simultaneously within multiple stages including sensory, working, and long-term memory. Attention and existing knowledge provided the basis for the depth of processing. Stimuli that received attention or were identified as related to previous knowledge would be processed more deeply and therefore more durable as memories.

Several facets of memory and information storage were filtered into 3 key known components to memory; short-term memory (STM), working memory (WM), and long-term memory (LTM). Incoming stimuli were first held in a buffer that had unlimited capacity prior to assimilation. This buffer would hold information, but dispose of it quickly if an individual's attention on the information did not transfer it into STM. Short term memory could hold approximated seven pieces of information at a time for a short period of time, approximately 15-30 seconds. This information was active and readily accessible and usually included sensory input information and items retrieved from LTM (Miller, 1956). Information needed for a specific purpose would be transferred from the buffer into WM, where it could be held temporarily and manipulated (Baddeley, 1986; 2000). Long term memory held an unlimited storage capacity and information could be held there indefinitely. Information held in LTM was organized in a meaningful way (i.e., frameworks and schemata) and was available for recall based on need (Bower, 1975).

In Bell-Gredler's 1986 review of cognitive theory, two types of LTM were discussed: semantic and episodic. Semantic memory was information from the environment that was received directly while episodic knowledge was based on an individual's experiences. These two types of memory could be readily decoded and made available for further processing, or could be modified and expanded by encoding of new information.

Based on these findings, Bell-Gredler also discussed two theorists who made further classifications on how knowledge was prioritized and encoded. Edward Tolman put forth the idea of purposive behaviorism in which learning specific information was related to the need of that information in meeting a goal. This indicated that behavior and learning were goal oriented and involved the fulfillment of an individual's expectations in order to remain in their schemata. Kurt Lewin theorized that motivation played a large role in learning, suggesting that an individual's motivation to learn would predict their learning, or in essence, people only learn what and when they want to.

Social Cognitive Theory

Social Cognitive Theory considers an individual to be constantly affected by influences from behavioral, cognitive, and environmental forces. When applied to a learning context, Social Cognitive Theory suggests influences regarding, for example, an individual's performance, their learning, and the strategy of teaching might influence an individual's experience. It is generally believed that individual behavior can be predicted by past experiences regarding success and failure at a given task. People who have had a positive experience with something are more likely to do it again, while people who have had negative experiences are less likely to do something again (Bandura, 1997). On a more basic level, a person's expectations regarding an outcome might affect their willingness to invest effort into a task. These expectations might be based on a person's beliefs regarding their own ability to be successful at this task, also known as self-efficacy.

Motivation

Motivation or the driving factor behind a behavior is often separated into two subsections: intrinsic or extrinsic. Intrinsic motivation is defined as the desire to engage in a behavior for no other reason than enjoyment, while extrinsic motivation has been defined as the desire to engage in a behavior due to an external force, such as a reward or penalty (Berlyne, 1960; White, 1959). Social Cognitive Theory considers motivation to be a product of self-efficacy and as such a measure of the effort that is exerted on a task such as learning

(Zimmerman, 2000). For example, a student with a high low self-efficacy might have lower extrinsic motivation for pleasing and lower intrinsic motivation because they view their chances of succeeding as low.

In learning tasks, these two motivations are not two opposing forces as was originally proposed by Harter (1981). While researching motivation to read, Harter used the two scales to determine explicitly if their motivation was due to intrinsic motivational factors such as enjoyment or extrinsic motivational factors like pleasing the teacher. Later, researchers Lepper, Corpus, & Iyengar (2005) found that these two types of motivation could exist simultaneously and increase learning motivation. They also extended the use of this scale to diverse populations and varying age groups to address issues of generalizability of their metric.

Motivation becomes increasingly important when retention and depth of learning are considered. Hatano & Inagaki (1987, in Brown 1988) in a recipe for making sashimi uncovered levels of mastery ranging from the ability to follow the recipe, or low level mastery to the ability to understand the relationships between the steps and to understand why the recipe worked, or high level mastery. They believed that interactive learning in the question answer format would increase depth of processing as they believed the ability to ask questions would lead to increased comprehension. According to theories of memory, deep meaningful learning that can be applied and transferred requires effort and this effort could be a result of motivation.

Summary of Background Theory

In summary, cognitive theories articulate how information is stored and how understanding of information develops through relationships with existing information. The motivation for storing and/or understanding information is also an important construct for teaching strategies. This viewpoint provides valuable insight into instructional design and the process of teaching and provides substance to the assertion that the use of mini-games provides a significant opportunity to enhance learning by:

- Providing learners with smaller lower risk opportunities to succeed or fail
- Providing fun but short interactions that focus on subsections of larger learning goals
- Increasing a student's motivation to both learn and succeed by providing opportunities for students to accelerate their learning

Case 1 Lunar Quest

Seymour and Hewett showed that approximately 50% or prospective engineers leave the discipline, regardless of GPA (1997). It is believed that this is caused by student unwillingness to endure the unpalatable pedagogical experience provided by engineering classes. Lunar Quest was developed to help provide not only a learning experience, but also a venue in which an aspiring engineer can envision them self in the role of an engineer.

Lunar Quest provides an engaging multiplayer learning environment in the form of a Retro-Future Moon Colony in an alternate future in which the United States continued the race to moon and is now building their first moon base. The world of Lunar Quest is managed by a large bureaucracy, the Lunar Colonization Authority, in which the player is enlisted. The player is cast as a physicist, fresh out of the Lunar Colonization Academy, tasked with repairing a series of problems (caused by a misunderstanding of physics) impeding the success of the colony as a whole. The player solves the problems to save the colony from certain disrepair and becomes the hero while at the same time learning valuable physics information that can be directly applied to the player's real world physics classes.

Lunar Quest is at its heart a Massively Multiplayer Online Game (MMOG). MMOGs are defined by their ability to provide a multiplayer gaming experience. The Multiverse Server, the technology behind Lunar Quest, is capable of managing over 1000 players per server. Commercial MMOGs commonly are capable of hosting many times this number through careful server management. The use of a MMOG environment was beneficial because it provided opportunities for customizable characters, public recognition, a social environment, and an expansive fantasy world which are thought to help lead to improved learning outcomes (Smith, Bowers, & Cannon-Bowers, 2008).

It was quickly discovered, however, that the MMOG environment had limitations that made teaching physics more difficult than the single player environment. It is difficult to model realistic physics in a MMOG due to network latency issues; it is difficult to insure an identical learning experience between players due to the ability for players to assist each other; and it is difficult to grade a player on a deeper level than pass/fail due to the stringent questing rules found in an MMOG. This is why the design team incorporated mini-games into Lunar Quest to deliver the learning content.

Mini-games provide a single player learning experience in the otherwise multiplayer world of Lunar Quest providing a hybrid environment that takes advantage of the desired benefits of both technologies. Mini-games can easily replicate simple physics on the player's computer without concerning itself with syncing objects across a network. Mini-games are single player which allows the educational content to be identical for each player while stopping players from cheating off of other player's accomplishments. Mini-games usually contain a score which allows players to be ranked on skill and not just on their ability to complete the game. This essentially removed the barriers previously created for using MMOG technology.

Lunar Quest is unique in the educational MMOG space in that mini-games are used to deliver the learning content. This also illustrates one of the greatest benefits of using mini-games, they are small enough to be embedded within other more complex games that cover a greater breadth of information than could be delivered through a mini-game platform.

The mini-games in Lunar Quest cover content both conceptual and procedural information. Each topic covered in the game is instantiated by two mini-games. The first is what was called the training game. This would cover the conceptual information for the player. The player would be exposed to the core information and asked simple quiz questions about it. After these games were played the player's quest would bring them to the procedural game. This game would give them an opportunity to apply the information within the context of the game's fantasy world. The combination of using both types of information constructs provide the player's with meaningful and useful information that can be used in their physics classes.

Case 2 Virtual Field Trip

The Virtual Field Trip project was established upon the firm belief that digital media can be an important tool to reduce the amount of time teachers spend trying to introduce students to real-world concepts. As students' family life and environments continue to change, many of them are now lacking in the real-world experiences that normally would be supplied by travel and tutelage from older family members. The establishment of standardized testing within lower-level schools has revealed that mush of the missing experiences is translating into poor scores in reading comprehension. Virtual reality simulation technologies can go a long way to fill the missing experience opportunities of these students.

Virtual Field Trips (VFTs) should reduce the time spent developing reading comprehension by populating general knowledge of a child's world. The VFT sought to surpass existing games by introducing several new aspects.

- Provide sufficient proof that a teacher or administrator can justify this technology purchase by documenting learning gains
- Meet or exceed the caliber of quality that is standard for the industry of educational games.
- Increase educational value above a traditional field trip, showing times and locations that normally would not be able to be shown together.
- Engage the child to make them more investigative, spawning extracurricular learning.
- Provide an immersive environment around vocabulary items by providing proper social, intellectual and physical context in the environment important to the development of connotative knowledge.
- Employ the use of mini-games as opportunities for students to create meaningful cognitive references with vocabulary

Given the importance of vocabulary knowledge in other developmental processes such as reading ability and comprehension, and the need for identifying and testing tools prior to the development of reading problems that have the potential to increase and/or facilitate vocabulary acquisition; this research program evaluated the utility of simulated learning experiences and mini-games in early vocabulary acquisition and retention.

Previous research findings have all reached similar conclusions regarding the importance of vocabulary acquisition; that it is a critical component reading comprehension. Neuman (2005) theorized that vocabulary development was an integral part of school readiness, a reference to the motivational behaviors and the common knowledge and experiences that are necessary for children to enter into school meaningfully. Students who exhibited school readiness were more likely to be successful in school and to have more productive and happy lives.

Acquisition of vocabulary impacts reading in several ways. It is not enough to recognize and be able to identify a word, the words meaning must be understood in order to make that word a tool. Notably, Stahl (1983) categorized word knowledge into three levels: association, comprehension, and generation. These three levels describe the depth of processing of vocabulary words. Word knowledge need not pass through these levels as if they were stages, but each represents an increasing depth of knowledge regarding the word. Association knowledge is characterized by the ability to hold a single definition for a word or to understand it in a single context. Comprehension involves a more generalized understanding of the word characterized by the ability to categorize a word, understand its use in a sentence and understands similar and dissimilar words and their relationships. Finally, generation is the ability to use the word without cues by creating sentences with the word and appropriately defining the word without clues.

Beck & McKeown (1991) also concluded that vocabulary knowledge included levels related to the ability to store, use, and recall the word and that vocabulary development goaled instruction could create greater understanding of words if strategies related to the depth of word knowledge were employed. Specifically, the levels of understanding could help determine the learning strategies to be employed.

Motivation could be an important factor in an individual's acquisition of vocabulary. Ediger (2001) found that extrinsic motivation, testing in particular, could be a large force in motivation to read. Individuals learning plans aimed at increasing intrinsic motivation and teacher based extrinsic motivation were identified as the best combination in motivating children to read.

Sweet & Gurthrie's (1996) introspect on motivation to read related intrinsic motivation to long-term literacy. They speculated that intrinsic motivation demonstrated that enhanced long term learning commitments such as spending time searching for books, reading, and learning while extrinsically motivated students had short term behaviors that controlled behavior for reasons such as competition. Extrinsic behaviors were linked to work-avoidance and minimized the importance of positive behaviors. Cameron & Pierce (1994) additionally found that when extrinsic rewards were attached to learning objectives, intrinsic motivations decreased in their meta-analysis of 150 related studies.

This experiment was designed to determine if a virtual experiences and mini-games could increase vocabulary acquisition in second graders when compared to similar content delivered via a story being read aloud. Students using the SLE, the VFT saw pictures and videos associated with words; they saw them in print and in a context in which the word made sense. They also had the opportunity to experience the words in the frame of a field trip. They interacted with words and concepts in accurate and interesting ways such as through flash based games.

Mini-Game 1 Squirrel Game

This provides a competitive game with scoring in a short interactive lesson about squirrels collecting enough food to survive during winter when food is not readily available.

This game opens with two squirrels in a sort of race to gather acorns before the winter. The user clicks on the acorns and their squirrel hops over and gathers that acorn, then the user clicks a hole in their tree and the squirrel stores the acorn away. There is a second squirrel that also gathers acorns. This goes on until all the nuts are gone, then both squirrels go into their holes. A short animation of the leaves falling and snow falling, while the squirrels sleep, and then they each come out of their holes, either skinny and a little sad or healthy and happy, depending on how many nuts they gathered before the winter.

Mini-Game 2 Tracks Matching

This game allows students to match animal tracks with the animals that made those tracks. Clicking on the animal tracks in the VR scene should activate a matching game. There is a set of animal tracks on the left side of the screen (horse, dog, bird, cow, fox), and a set of animals on the right side of the screen. Click a track and drag it onto an animal to attempt a match. When a match is made, the name of the animal in text should pop out of the animal's picture, and be pronounced in audio before fading away. The animal should move around to signify the match as well. If the match was not made, the tracks graphic should snap back to its original position, and the friend's voice should say, "Wait, that's not it. Let's look again."

Mini-Game 3 Lizard Food

In this mini-game, students maneuver a lizard around a stump in order to eat as many bugs as possible. A close up of the log with a bunch of holes is shown with bugs all over it and a lizard. Insects are crawling around going in and out of the holes. Chasing an insect with the lizard causes it make a little noise and move to a different part of the log. If you catch the bug, then the lizard eats it with a little slurp. This is a sort of score-free game meant to be more of an interactive activity.

Mini-Game 4 Matching Insects

This is a matching game to improve student identification skills so that the insects become more than general "bugs." Included are ants, bees, butterfly, and dragon flies. The game should be found on an old log. The game opens with a close-up of a tree. The insect names are placed on a tree spaced around. Insects, (Bee, dragonfly, ant, lovebug, butterfly) are on the right. The student is instructed to drag the insect to its name. They receive encouragement upon unsuccessful attempts and positive feedback on successful attempts. The insect remains on the tree after they are matched, and their name is spoken aloud and the written word is flashed.

Results indicated that the VFT's rich experiential learning environments filled with contextually appropriate and semantic cues did increase the breadth of knowledge for vocabulary words as demonstrated on a writing exercise within the deeper level of understanding required for a word to be used within a writing sample. In summary, this indicates an increase in words known at that deeper level. Therefore, it could be concluded that students who used VFTs learned words more deeply when this learning was surrounded by contextually appropriate semantic information.

Case 3 Business Rat Race

Defense Acquisition University (DAU) is positioned as the Department of Defense's premier award winning corporate university. Servicing the Acquisition, Techology, and Logistics Worforces, DAU services of workforce of over 300,000 military and civilian professionals. While its students are comprised of 15 career fields, major educational concentration areas center around General Acquisition, Business Contracting and Finance, Logistics, Systems Engineering, Contracting, and Program Management. DAU has over 100 courses with a remarkable 70% of those courses being offered in distance learning or blended paradigms. Recently within their e-Learning Technologies Center, DAU has transformed its approach from traditional slide based CBTs to a new approach that includes the use of context and relevance centered games designed to foster motivation and increase content retention amongst its students.

This project focuses on the Business, Contracting and Finance (BCF) career field. Specifically, a low level course that serves as a required course for all students in the career field was selected to include a mini-game based intervention. While the course selected, BCF 103 represented a high performing course, a content analysis of the course indicated that this course transmitted primarily conceptual knowledge often including a heavy emphasis on vocabulary memorization while providing little use case information or context. It was hypothesized that by including mini-games at the end of each of the online courses eight modules, students would create find more relevance in the information being presented and therefore would be motivated to retain the information. Through the assistance of a talented group of subcontractors, this series of mini-games is currently being produced for inclusion.

General Game Concept

PFC Ratner must navigate through a myriad of obstacles and avoid pesky critters in an effort to help his student friends acquire the necessary skills/requirements needed to compete for the upcoming CAS (Cost Accounting Superstars) competition.

BACKGROUND

PFC Ratner is the little secret weapon helping Majors and other students navigate through the acquisition rat race! After years of living within the walls of the Pentagon, PFC Ratner has mastered the art of business financial management by spying, befriending, and sabotaging some of the world's greatest minds in the acquisition process.

Through a freak mishap involving a secret missile guidance system and a block of cheese, PFC Ratner was mistakenly transported to Fort Belvoir, home of the Defense Acquisition University. Now, in a foreign surrounding and no way "home," he scurries through the walls of the DAU, meeting new friends and helping them through the grueling task and competitions connected with the BCF 103 course—helping those who are in need of guidance...and possibly...in return...finding a way home.

Through PFC Ratner's adventures, those taking the BCF 103 course will experience fun, immersive, and entertaining game play that reinforces the lessons learned in Module One.

GAME FEATURES

The point of view is derived from Action Platform/Strategy games, including a diverse mix of fun characters with slightly exaggerated features.

The quest begins with PFC Ratner within a fun rendition of a typical everyday office.

One of the story line's main characters will "pop up" in a two way communications monitor, expressing a need for help finding information pertaining to a number of different topics.

The first person player will guide Ratner across the office, collecting items and post-it-notes that are relevant to the questions posed. Ratner must collect as many correct items as he can, as quickly as he can, to help his friend attain the highest possible score!

There will be items and post-it-notes with incorrect information that will lower the player's score.

Other pesky enemies will be present, trying to slow Ratner down and lower his score! Ratner must jump and duck these nuisances to avoid penalties applied to his game score.

There will be 2 scoring systems. One system will track Ratner's overall progress in achieving 75% or greater response to the learning objectives. The second scoring system, more commonly associated with game play, will be based on time, correct items gathered, and energy level.

After each question/level, the user will be presented with a mini white board that tracks his/her game play progress. It will present to the player: correct items collected; points deducted; current game score; and a summary of the terms/questions asked with reference back to the course for items missed.

After completing all "levels" associated with the learning objectives, a final "grand" white board will present overall score and summary information. Those players that did not achieve a 75% of the cumulative learning objectives throughout the game will be directed to replay the game. Players that achieve a score of 75% or greater will have an option to play again, or move on.

Testing involving both the current course and the enhanced courses outcomes in addition to their retention of the content within the mini-games will be accomplished during the Summer and Fall of 2009.

CONCLUSION

Mini-games are fast becoming an effective and relevant method to deliver game based instruction. They should no longer be thought of as just simple quiz style games embedded in a conventional course to break up the monotony of the information. While they certainly can be included in a web-based course they can also be delivered in the context of a larger game or simulation, or combined with other mini-games to build a training experience with greater depth and breadth than was previously possible. Furthermore, mini-games have become sophisticated enough to stand on their own as a legitimate method of training and education with games.

REFERENCES

Asynchronous Learning Networks Magazine, 1(1). (1997, March).

Baddeley, A. D. (1986). *Working Memory*. Oxford, UK: Clarendon Press.

Baddeley, A. D. (2000). The episodic buffer: A new component of working memory. *Trends in Cognitive Sciences, 4*(11), 417–423. doi:10.1016/S1364-6613(00)01538-2

Bandura, A. (1997). *Self-efficacy: The exercise of control*. New York: Freeman.

Baron, R. A., & Byrne, D. (1987). *Social Psychology: Understanding Human Interaction.* (5th ed.). Newton, MA: Allyn and Bacon.

Bartlett, F. C. (1958). *Thinking: An Experimental and Social Study.* London: Allen and Unwin.

Beck, I., & McKeown, M. (1991). Conditions of vocabulary acquisition. In R. Barr, M. Kamil, P. Mosenthan & P. D. Pearson (Eds.), *Handbook of Reading Research* (Vol. 2, pp. 789-814). New York: Longman.

Bell-Gredler, M. E. (1986). *Learning and Instruction: Theory into Practice.* New York: Macmillan.

Berlyne, D. E. (1960). *Conflict, arousal, curiosity.* New York: McGraw-Hill.

Bower, G. H. (1975). Cognitive Psychology: An introduction. In W. K. Estes (Ed.), *Handbook of learning and cognitive processes, Volume 1, Introduction to concepts and issues,* (pp. 25-80). Hillsdale, NJ: Erlbaum.

Brown, A. L. (1988). Motivation to learn and understand: On taking charge of one's own learning. *Cognition and Instruction, 5*(4), 311–321. doi:10.1207/s1532690xci0504_4

Bruner, J. (1966). *Toward a Theory of Education.* Cambridge, MA: Harvard University Press.

Cameron, J., & Pierce, W. D. (1994). Reinforcement, reward, and intrinsic motivation: A meta-analysis. *Review of Educational Research, 64*(3), 363–423.

Craik, F. I., & Lockhart, R. S. (1972). Levels of processing: A framework for memory research. *Journal of Verbal Learning and Verbal Behavior, 11*(6), 671–684. doi:10.1016/S0022-5371(72)80001-X

Ediger, M. (2001). *Reading: Intrinsic versus Extrinsic Motivation.* An opinion paper supported by ERIC.

Harter, S. (1981). A new self-report scale of intrinsic versus extrinsic orientation in the classroom: Motivational and informational components. *Developmental Psychology, 17*(3), 300–312. doi:10.1037/0012-1649.17.3.300

Lepper, M. R., Corpus, J. H., & Iyengar, S. S. (2005). Intrinsic and extrinsic motivation orientations in the classroom: Age differences and academic correlates. *Journal of Educational Psychology, 91*(2), 184–196. doi:10.1037/0022-0663.97.2.184

Miller, G. A. (1956). The magical number seven, plus or minus two: Some limits on our capacity for processing information. *Psychological Review, 63*, 81–97. doi:10.1037/h0043158

Nash, H., & Snowling, M. (2006). Teaching new words to children with poor existing vocabulary knowledge: A controlled evaluation of the definition and context methods. *International Journal of Language & Communication Disorders, 41*(3), 335–354. doi:10.1080/13682820600602295

Neuman, S. B. (2005). Readiness for reading and writing – What do we mean? *Early Childhood Today 3, 20*(2), 8.

Neuman, S. B. (2006). Building vocabulary to build literacy skills: How to help children build a rich vocabulary day by day. *Early Childhood Today 3, 20*(3), 12-14.

Randel, J. M., Morris, B. A., Wetzel, C. D., & Whitehill, B. (1992). The effectiveness of games for educational purposes: A review of recent research. *Simulation & Gaming, 23*(23), 261–276. doi:10.1177/1046878192233001

Ricci, K. E., Salas, E., & Cannon-Bowers, J. A. (1996). Do computer based games facilitate knowledge acquisition and retention? *Military Psychology, 8*(4), 295–307. doi:10.1207/s15327876mp0804_3

Smith, P., Bowers, C., & Cannon-Bowers, J. (2008). Social Learning Aspects of MMOGs and Virtual Worlds. In R. Ferdig (ed.), *Handbook of Research on Effective Electronic Gaming in Education*. Hershey, PA: Idea Group Inc.

Stahl, S. (1983). *Vocabulary Instruction and the Nature of Word Meanings*. Paper presented at the 27th Annual Meeting of the College Reading Association, Alanta, GA.

Sweet, A. P., & Guthrie, J. T. (1996). How children's motivations relate to literacy development and instruction. *The Reading Teacher*, *49*(8), 660–662.

White, R. W. (1959). Motivation reconsidered: The concept of competence. *Psychological Review*, *66*, 297–333. doi:10.1037/h0040934

Zimmerman, B. J. (2000). Self-efficacy: An essential motive to learn. *Contemporary Educational Psychology*, *25*, 82–91. doi:10.1006/ceps.1999.1016

Chapter 2
Serious Storytelling:
Narrative Considerations for Serious Games Researchers and Developers

Rudy McDaniel
University of Central Florida, USA

Stephen M. Fiore
University of Central Florida, USA

Denise Nicholson
University of Central Florida, USA

ABSTRACT

In this chapter, the authors explore the nature and function of storytelling in serious games. Drawing from the field of narratology, they explore research related to narrative expression and relate those ideas to serious game design and development. They also consider interactive storytelling and apply and adapt traditional ideas about story as a static and predetermined entity into this new setting, a setting which depends in part upon gamer participation to craft dramatic experiences. The authors conceptualize narrative as a combination of plot, character, and environment, and then use that conceptualization to devise a narrative taxonomy that is useful as a heuristic for developing stronger stories in serious games. The chapter concludes with an analysis of the hybrid FPS/RPG game Fallout 3, an analysis included to show that even highly regarded and award-winning games are lacking in the narrative coherence necessary to improve the level of dramatic immersion in virtual worlds.

INTRODUCTION: GRENWIN THE GOBLIN

You are still enjoying your newfound fame as slayer of the great white serpent (and the popularity this earned you with the townfolk of Eleven Isles) when chaos suddenly erupts in the Rusty Hinge tavern.

DOI: 10.4018/978-1-61520-739-8.ch002

Grenwin the Goblin hacks down the door with a rusty axe and crashes into the room. Patrons scatter, shrieking in terror, as the wiry green beast swings the axe about carelessly, smashing flagons of ale and overturning tables, all while cackling maniacally and searching the room for something to steal or devour. If he sees you, all is lost, for he must know it was you who stole his clan's map to navigate through the forbidden mountains. You duck down from your position at the far end of the darkened room, hoping to avoid detection. Alas, it is no use. Grenwin apparently has excellent eyesight. The abomination trains his beady eyes on you. His eyes narrow as he recognizes you and he smiles cruelly. You panic as the goblin reveals a surprising burst of speed and streaks across the pub in your direction. If only you could remember Pythagoras' theorem and calculate the correct trajectories, you would let loose a flurry of virtual arrows and fell this foul creature. As it stands, not having prepared as instructed by Zorak the Bard, you must defend yourself with your untrained peasant's hands and hope for the best...

This paragraph might describe a scene taken from any number of fantasy based role-playing games (RPGs). Though primitive and brief, the example above is also narratively complete. It contains a protagonist: the character controlled by the player. There is a goblin antagonist to provide conflict and there is an environment in which the action is anchored. There is also a plot, albeit a brief one: escape from the pub with your life and wits intact. The central concern of the player is to apply whatever knowledge she has in order to survive the ordeal at hand and then venture out into the fantasy world to continue her adventures. The fact that the paragraph above just happens to be useful for a serious game to teach trigonometry is largely irrelevant. Stories are equally important for serious and non-serious games alike. What is most important is that the game's story offers a chance for the player to project herself into the character of a virtual heroine that is facing an attacking goblin.

Using gaming or simulation parlance, we might call this phenomenon *immersion* or discuss it in terms of *presence* – the replacing of real world cues with virtual cues in successfully crafted fantasy environments. In psychological studies of narrative, it is more specifically known as *narrative transportation* (Green, 2004) when restricted to the influence of the narrative dimension of a system. In this brief and intense moment, the player *becomes* the young heroine, and the story and gameplay merge together as a vehicle for transportation from a real to a virtual identity. The goal of serious games is to create a virtual environment in which this pathway is reversed; by encountering and solving problems in the game world, the player learns skills and builds knowledge useful for problem solving in the real world.

While complete in a narrative sense, the problem with the story of Grenwin the Goblin (which is one of a thousand narrative instances we might extract from an RPG) is that the learning moment used here causes incongruence with the overall story. The learning objectives, which focus on knowledge of the Pythagorean Theorem, do not align with the dramatic objectives, which should showcase the intense, adrenaline-filled process of escaping from a tavern while engaged in battle with a ferocious goblin. Further, the player's actions may serve to further undermine the dramatic quality of the action. She may choose to simply give up, allowing the goblin to tear her avatar limb from limb, or she may decide to take advantage of the pathfinding limitations of the enemy artificial intelligence and find a way to "cheat" the system by standing one step behind the goblin's reach and delaying the conclusion of the scene indefinitely. The inconsistency between the game's story and the game's learning content can be a debilitating problem when trying to motivate players to keep playing and learning. While we cannot always control the actions of the player or the way she plays the game, we *can* adjust our storytelling technique to better align our learning objectives with our dramatic objectives.

As this example illustrates, interactive storytelling is an important craft for serious game developers to understand. When done correctly, storytelling can aid in the game design process in several ways. A strong narrative can improve player motivation by encouraging the player to continue playing in order to resolve the undisclosed elements of the story and explore the nooks and crannies of the game world. Stories can embed learning objectives within the game objectives by positioning the acquisition of learning materials as an active part of quests or missions. Finally, strong stories can tie together various game elements such as artwork, sound, character interactions, gameplay mechanics, and environmental processes into a coherent framework that makes the information presented within the virtual world easier to absorb. As such, storytelling is a very powerful tool that lies within the serious game developer's toolbox.

Despite its usefulness, the narrative form remains elusive to game developers and academics alike. It is both complex and multifaceted, and its study has inspired an entire line of research devoted to better understanding it, the field of inquiry known as narratology, also described as the "science of narrative" (Onega & Landa, 1996, p. 1). Scholars interested in the form and function of stories have considered the nature of narrative for the past several decades; in this chapter, we maintain that there is much important information in this body of work which can be used to improve the narrative aspects of serious games. As game designers and developers, we can use the same critical vocabularies, theories, and taxonomical techniques that have long been established in the field of narratology. While these techniques may require a good deal of revision—due to the interactive, nonlinear nature of video games—they can at least provide a starting point for thinking about game stories using established conventions.

As a first step in this direction, we consider two different ways of conceptualizing serious storytelling: as a theoretical construct and as a subject of critical analysis in existing games. Each of these two perspectives is important for different reasons. On the one hand, the serious games developer may need to consider practical ways for improving storytelling to improve player immersion and motivation. These techniques can be analyzed through the critical analysis of existing games. On the other hand, the serious games researcher might be looking for a way to study narratives in order to gather empirical support for embedding a particular type of story in a particular type of game; a theoretical starting point is important for this type of task. Finally, a recognition that even bestselling and award-winning commercial games fall prey to narrative problems is useful for all audiences as this shows we still have a long way to go to bring interactive story up to the same polished level as other aspects of contemporary games—such as the quality of gameplay mechanics, physics handling, audio and visual fidelity, and so forth.

To address these issues and frame these two perspectives, we first provide an introduction to the field of narratology for non-narratologists. After this initial review, we construct a basic taxonomy for interactive narrative that is useful for thinking about the various ways of creating and studying the narrative experiences found in serious games as theoretical constructs. For our second perspective, we examine the current state of interactive storytelling by performing a brief narrative analysis of a popular commercial title, which we argue is also an example of a compelling framework upon which to build a serious game. Using the video game *Fallout 3* as our subject in this analysis, we study the storytelling techniques used by commercial game development companies in order to illustrate the various elements of narrative in an operational fashion. This analysis suggests that while interactive storytelling in games has made much progress over the last several decades, we are still seeing many types of narrative problems that prevent our players from experiencing fully congruent narrative worlds. We conclude this chapter with thoughts for the future of interactive storytelling as a means for improving serious games. As a brief demonstration of applying the ideas from this chapter to a real game, we suggest several ways in which we might improve our introductory story about escaping from a bloodthirsty goblin while learning basic geometry.

SELECTED NARRATOLOGICAL PRINCIPLES

While narratology itself is a rich field of study characterized by subtlety of analysis and debates concerning the function and nature of narrative in various genres of stories, there are also some major themes and ideas which we can extract and appreciate as being useful for the design and analysis of serious games. In particular, it is useful to have knowledge of some of the basic terminology used in the field and to consider some of the techniques and approaches to narrative structure and narrative taxonomy that have been important in this line of research. This knowledge provides us with some established vocabularies and frames for considering existing serious games and their successes and failures as vehicles for game-story expressions. These ideas can also be useful in constructing preliminary empirical studies to further analyze the effectiveness of story as a scaffold for learning (or as a mechanism for improving immersion through narrative transportation) in serious games.

As much narratological theory emerged from the structuralist perspective, a field of literary study which maintains that stories can be coded, compared, and classified by their structural units, there is a rich history of structuralist work that has value for serious games practitioners and researchers. If we conceptualize serious games as engineered systems that solve problems in particular domains, then, following the structuralist tradition, we can also consider the ways in which stories can serve as modular parts within those systems. Alternately, if we consider serious games from the humanist's perspective, we can appreciate the ability of stories to provide insight into the human condition and to perhaps provide scaffolds for reaching the

"gray areas" of tacit instruction that are not easily taught using learning objectives and engineering design guidelines. Sheldon (2004) expresses this sentiment in terms of affective impact, writing, "if we would like to involve emotions higher than an adrenaline rush, we need to reach the human spirit, not just endocrine glands" (p. 6). Stories can leave a lasting impression of a virtual world long after the gameplay has ceased.

For example, in a serious game designed to teach art history in the Renaissance, some learning objectives might target players' recognition of selected artistic works as recreated in a virtual world. This type of instruction can be embedded into a game without the need for even a minimal story; players match works with titles and are rewarded for successful pairings. Now, consider the same artwork when placed in a narrative gaming environment. This interactive experience is crafted with numerous NPCs and uses a plot involving an up-and-coming artist named Nichola and his quest for legitimacy in 15th century Italy. The player takes control of Nichola in the year 1435, in Florence. The game begins in the church of Santa Maria Novella; after a brief cut scene introduction of Nichola's wife charging him with locating his mentor in order to begin his daily lesson, the player is given control to explore the church and its surrounding artwork. He soon discovers the prominent *Holy Trinity*, a 25 foot tall and 10 foot wide fresco created by the recently deceased painter Masaccio. Later in the game, upon finding Nichola's mentor, the techniques used in the *Holy Trinity* are explained to him through dialog. A minigame then allows him to practice his shading techniques in order to master that targeted skill and improve Nichola's reputation as an artist.

In this type of game, the player sees the artistic material from the context of the story; the fact that Nichola must recognize and master existing styles in order to mature as a painter teaches other, more subtle lessons about Renaissance artistic practices. These are lessons involving traditional training techniques, artistic styles, the integration of architecture and art in fresco works, and even politics (one NPC standing outside the church explains that Masaccio's work was influential to Michelangelo; another suggests the great painter was poisoned by a jealous rival painter, which leads to another quest to collect clues related to his death). With this type of narratological massaging, a serious game can draw a player *into* that world through narrative transportation. He participates in dramatic moments that are carefully chosen to explore the artistic themes of the period and observes the importance of art on the surrounding community of Florence. To be able to recognize important works of art is important, but understanding the cultural and social implications of that art on a Renaissance community serves unstated learning outcomes that may augment the primary learning objectives in unforeseen ways. Even building a modicum of artistic skill through integrated minigames linked to the overall story is possible; these ancillary results can be served through strong storytelling techniques paired with creative gameplay mechanics.

Supporting such grand constructions is not easy. This is in part due to the density of the narrative form. It is the aim of this chapter to partially demystify narrative and to deconstruct it into a set of constituent parts that can then be rearranged for various pedagogical purposes. First, let us consider stories conceptually, using the lens of narratology. Narratology is a term defined by Tzvetan Todorov to refer to the theory of narrative as an academic pursuit. As Prince (2003a) notes, narratologists are concerned with the general study of narrative in terms of its nature, form, and function, and specifically with "what all and only narratives have in common (at the level of story, narrating, and their relations) as well as what enables them to be different from one another" (p. 66). Elsewhere, he notes the theoretical impossibility of defining such a field under a single conceptual model, writing that while "some theorists and researchers believe that everything is narrative; others maintain that everything can be; and still others contend that, in a sense, nothing is (because narrativity is culture-dependent and context-bound)" (2003b, p. 1). Nevertheless, despite the complexity of this expressive mode, there are some ideas we can apply to help better understand the nature of narrative in serious games. First, we can consider the difference between a story and the expression of that story in a particular medium. Narratologists generally distinguish narrative, or the process of telling stories using particular media through an "expression plane," from story, or the "content plane" of narrative (Prince, 2003b, p. 93). Narrative in this sense is perhaps most simply defined by Abbott (2002) as "the representation of an event or a series of events" (p. 12). Important here is the word representation, which further distinguishes the term *narrative*, a specific instance of story, from the term *story*, a generic sequence of events with the potential for narrative expression through media. Collectively, these sequenced events constitute the plot of a story.

Next, we can consider stories structurally. Story plots are composed of a series of events which are related both casually and temporally. Mateas (2001) notes that dramatic stories can be represented along two axes, with a vertical axis used to represent tension and a horizontal axis used to represent time. Moving from left two right across the horizontal axis, one sees a general exposition, a period of rising action characterized by an inciting incident and a crisis, a climax, then a period of falling action culminating in the denouement, or the final unfurling of tension. Aristotle originally provided this treatment of story classification more than 2000 years ago in his *Poetics;* here he provided the most basic distinction between plot types based on the final situation of the hero or heroine. As Booker (2005) explains, in a tragedy, the hero or heroine originally seemed destined for fortune, but there was an eventual disaster at the end of the story leading to catastrophe. In a comedy, complications are introduced early on for the hero or heroine, but by the end of the story, these complications are resolved and the major characters are liberated or redeemed. Booker continues to outline additional canonical plot types such as "overcoming the monster," "rags to riches," "the quest," "voyage and return," and "rebirth" (pp. 17-215). The Hero battling Grenwin the goblin is currently engaged in an *overcoming the monster* plot, while Nichola the struggling artist is on a *quest* plot which may eventually lead to his *rebirth* as an esteemed artist. Nichola's tale may also end as a *rags to riches* plot type, with his newfound reputation as a skilled artist bringing his family fame and esteem. Each of these causal event sequences has various identifying features that enable them to be used in a variety of stories and undergo a variety of transformations and adjustments.

This classical narrative plot structure was further shaped by the German critic and novelist Gustav Freytag in 1863 (Jacobs, 2007). As noted by Meadows (2003), Freytag also charted time along a horizontal axis and plot along a vertical axis, resulting in a pyramid or triangular shape (this structure has come to be known as Freytag's pyramid, or Freytag's triangle). During the first portion of a dramatic story, *desis*, there is a period of rising action, during which tension is introduced and elevated. At the *climax*, or center of the triangle, this tension is at the highest possible level. Then, a reversal of circumstances occurs at the *peripeteia* stage. Finally, during the *denouement*, the tension is released and the story is resolved.

Perhaps more relevant to the structure of serious game stories are the ideas about story that have emerged from the genre of filmmaking. Since many early game stories were told through cut scenes using filmic conventions, proven scriptwriting techniques offer another approach to thinking about the structure of game plots. For example, the work of screenwriter Syd Field offers a formula for building narratives suitable for the screen; in this formula, a three act structure is used to break a story into specific parts (Jacobs, 2007). In the first act, the mood and tone are established and the hero or heroine is introduced. During the second act, confrontation occurs, eventually reaching a climax in which the primary character faces an extreme obstacle. In the third act, the tension is resolved and the hero is either successful or ultimately fails in his journey. Most interesting in Field's model is the fact that minutes of screen time correspond directly to the number of pages in a script, which is useful since Act One and Act Three are roughly thirty minutes each and Act Two is sixty minutes. This suggests an appropriate technique for creating a dramatic primary story in a game might be to use 1/4 of the game for setting up the initial situation, 1/2 for throwing major obstacles at the player, and then a final quarter for wrapping up the story and rewarding the player appropriately. A more detailed example applying Field's three act structure to the film *Star Wars* can be found in Jacobs (2007).

Finally, we can consider stories thematically. In this sense, we organize stories according to their functional characteristics and the relationships formed by various story events. Because of their (relatively) simple structure, fairy tales were often studied as examples from which to devise taxonomies for narrative units and the relations between them. The seminal text *Morphology of the Folktale*, published by Vladimir Propp in 1968, outlines a structuralist approach to the taxonomy of themes and structures present in Russian fairy tales. Morphology refers to the study of forms, and Propp notes in his opening chapter that a traditional way of classifying between narrative forms has separated fairy tales based on content: fantasy stories, stories about everyday life, and stories about animals. But, as he notes, this distinction can be problematic as the content from fairy tales inevitably blurs boundaries (tales with both animals and fantasy elements are common, for example). To counter this problem, Propp suggests a more comprehensive regimen of classification relying on morphology, or using "a description of the tale according to its component parts and the relationship of these components to each other and to the whole" (p. 19). A morphological approach makes it possible to separate story types by the actions that occur rather than by the overall nature of their subject matter, which can be somewhat ambiguous.

Using a morphological strategy, it is possible to classify fairy tales based on their functions, or the actions of characters and the significances of those actions for the overall plot. Propp notes that these functions are the fundamental, stable units from which fairy tales are constructed and affirms that there are a limited number of such functions which can be articulated and studied in relation to one another. Later, he outlines a series of functions of the *dramatis personae*, or the overall cast of characters within a tale. In general, Propp notes that fairy tales follow this structure. First, an initial situation is presented and the members of a family and the hero are introduced (this introduction is not one of Propp's functions, but is important nonetheless). Then, a series of functions define how various tales can develop from the same general stockpile of plot elements. For example, the initial situation is followed by the preparatory section which then leads to a complication often involving villainy or the hero's recognition of a lack or absence of some sort. After the complication, donors will provide aid to the hero and a helper may also provide assistance. At the beginning of the second move, a new villain or difficult task may be introduced into the story. During the continuation of the second move, there is a resolution of the task and the hero will undergo change through recognition, exposure, transfiguration, punishment, or marriage. During each phase of the tale, certain events will make other events more likely or more unlikely to occur. Furthermore, spheres of action may lead particular characters to perform multiple roles within a story (e.g., both a helper and a donor, or both villain and unwitting donor). Overall, Propp devised a series of 31 functions that are used together in different ways within fairy tales.

These ideas regarding narrative structure and form are only a few selected concepts from an important genre of work; there is a rich history of narratological scholarship available from which to study the structure and classification of stories. Narratologists such as Genette (1980) and Bal (1997) have authored comprehensive books looking at narratological principles from a literary perspective, and Onega & Landa (1996) have compiled essays related to the subject in a wide-reaching effort to study narrative from a variety of literary perspectives. The edited collection *What is Narratology* (Kindt & Müller, 2003), collects the results of an international symposium exploring narratology and contains essays from the leading theorists in the field. In terms of other disciplinary contributions, the French anthropologist Claude Lévi-Strauss produced groundbreaking work on identifying universal structures used in myths (Eagleton, 1996), while Campbell (1949) is a must-read volume for understanding mythological types, story structures (most notably the Hero's Journey), and character archetypes. Vogler (1992) takes Campbell's work and distills the ideas into a practical guide suitable for story creation and story analysis, much of which has also been applied specifically to video games in Jacobs (2007).

INTERACTIVE NARRATOLOGY

Narratives take an interesting turn when produced by interactive media. No longer are stories meticulously crafted by a devoted author, polished to an appropriate degree, and released to an audience for consumption. In interactive media environments, the user/player/reader plays dual roles as both recipient and co-author of stories. Immediately, then, we can observe that there are at least two stories that one might talk about when discussing storytelling in the genre of serious games. First, there is the game writer's story, which is typically expressed through cut scenes and other non-playable moments in the game. Second, there is the player's story, which is created by the player's movements, decisions, and actions in virtual space. DeMarle (2007) describes the former story as the "high-level" story and the story co-created by the player as the "immediate-level" story (pp. 77-78). Taken together, the high-level and immediate-level stories express the various narrative possibilities of a given game.

The first dilemma one generally encounters when researching interactive narrative is the rift caused by the supposed "narratology vs. ludology" debate, one that is filled with myths and misconceptions about interactive storytelling. Ludology refers to the study of play (especially in games); ludologists suggest that the narrative framework is inappropriate for the analysis of games since videogames deserve a theoretical framework all their own. Mateas (2002) has offered the term "narrativist" to refer to "a specific, anti-game, interactive narrative position" (p. 34) that suggests that narrative is in some sense privileged over gameplay and game mechanics. This idea spurred contentious debate in some circles about the fundamental nature of interactive narrative; is the medium one in which all action should be analyzed through a narrative lens, or one which is deserving of an entirely new genre of analysis based on gameplay and the unique environment afforded by interactive games? Gonzalo Frasca (2003), who is often (falsely) credited with having invented the term ludology, claims that the debate between the two camps is based on both misunderstanding and erroneous publicity; as he notes, ludologists have never claimed that storytelling is not an important part of games. Similarly, narratologists do not claim that narrative is the only important thing happening within games, but rather that narrative techniques may provide useful perspectives from which to study interactive games.

So-called ludologists argue that game studies need not be framed by narrative constructs and maintain that gameplay is the essential mechanism through which games should be analyzed and studied. Narrativists, on the other hand, are interested in the potential for narrative expression that is offered through interactive gaming technologies and believe that games can be read as texts and studied using narrative conventions. The most reasonable approach in this area seems to be that considered by researchers such as Jenkins (2006) and practitioners such as Sheldon (2004) who follow a middle ground and find value in the balance of both narratological and ludological ideas. What we can take from this discussion as serious games designers is the idea that both gameplay and narrative serve essential functions within learning games. We can also observe, however, that gameplay has made many more advances than interactive narrative, which is still quite immature when compared to the innovations in gameplay mechanics of the last several years. The narratology vs. ludology debate is explored in great detail elsewhere (e.g., Mateas, 2002; Juul, 2001) so we will not spend more time with it here. It is, however, something to be aware of when considering adapting narrative conventions to serious games.

In terms of a more applied overview of interactive narrative for video games, one of the more useful resources on the topic is Chris Bateman's (2007) edited collection *Game Writing: Narrative Skills for Videogames*. Richard Dansky opens the collection by considering the purposes of game narrative: immersion, reward, and identification. He notes that identification serves two different roles, both to give players "context for their actions" that provide "justification for game actions" (p. 6) and to provide "identification in another sense as well, namely the sense of kinship and desire to become the central character" (p. 7). Sheldon (2004) provides guidance for authoring characters in video games, noting the characteristics of good game characters, the various roles available within games, and the different ways in which players can encounter characters in games. DeMarle (2007) explains the process of "gating the story" (p. 74) in which the exponential complexity of nonlinear game stories is somewhat alleviated by reigning the player back in and herding her through narrative gates comprised of major plot points at various points within the game.

In another useful book about interactive storytelling, Crawford (2005) claims that the essential quest in interactive storytelling is to "envision a dramatic *storyworld*, not a *storyline*" (p. 56). In other words, the critical task is to consider the various types of actions a player might take as they encounter a dramatic situation within the virtual world and the various ways in which the overall story can be supported by that world. Recalling the story of Grenwin the Goblin at this chapter's introduction, our task as an interactive designer is to consider how to sustain our fantasy-based environment throughout the duration of gameplay and learning, if indeed these two processes can be separated into distinct tasks. Stories created for games can take a variety of forms depending on game genre; for example, an abstract game like *Pac-Man* or *Tetris* has virtually no story, while complex RPGs generally have complicated and multi-layered stories that unfold over tens or even hundreds of hours of gameplay.

TOWARD A NARRATIVE TAXONOMY: CHARACTER, PLOT, AND ENVIRONMENT

Given that narratology emerged from the structuralist tradition, it seems likely that we can ask and expect an answer to the question of narrative composition for serious games. If we are to break a story down into its various elements, we will find that the stories used in serious games are composed of characters, plots, and environments, which are themselves further composed of times and places. We can decompose each of these subunits down into more primitive types. For example, we can note the differences in primary characters (protagonists), from the contemporary, stoic antihero; to the brave, yet vulnerable fairy tale hero. We can observe the different characteristics of villains, and note how some antagonists present obstacles to the hero in the form of additional characters in the story, while other obstacles come from the environment or even from within the hero as a battle of will or a personal challenge to overcome a character flaw. A major category for the classification of game heroes or heroines is the type of growth that character undergoes over time. For example, Krawczyk and Novak (2006) suggest several binary growth patterns in which less desirable characteristics mature into more desirable characteristics over the span of the game. A few examples of these include a hateful character transforming into a lovable character, a cowardly hero becoming courageous, a dishonorable heroine finding honor, or an immature character learning about maturity. Like in fiction, Krawczyk and Novak explain that "if a character cannot change, the message should be a tragic one. If there's absolutely no growth from any characters within a story, this is usually a sign of bad writing—and a writer who doesn't have a handle on his or her characters" (p. 140).

Plots too are classified according to the particular dimensions of the events they string together. Based on the actions which occur during the game, for example, it is possible to use Booker's (2005) classification of stories to differentiate between, say, an "overcoming the monster" plot and a "rebirth" plot. Furthermore, through our comparisons of game narratives to other genres such as fiction and film, we can use our prior experience with certain plot types to consider the plots most effective for certain types of player reactions. For example, if our goal is to motivate the player to continue immersing herself in the game and eventually enter a state in which narrative learning materials are indistinguishable from non-learning narrative materials, then it might make sense to adapt and use an epic, quest-like plot type like "voyage and return" or "overcoming the monster." Similarly, if our goal is to inspire the appreciation of art and music into a player through allowing her to travel to far off places and see art in its original historical time and place, then a tragic plot might help us serve that goal. Or, our game might be so broad as to warrant the inclusion of multiple plot types within the game. As an example, both the epic fictional quests penned by novelist J.R.R. Tolken in his *Lord of the Rings* books and the various missions scripted by game designer Todd Howard's team in the video game *Fallout 3* offer numerous plot types woven together to challenge the hero or heroine over time.

Finally, in terms of environment, we can classify stories according to the time and place in which they occur. The difference between a science fiction novel and a contemporary thriller may be as simple as location. For example, the science fiction story might share the same essential plot as the thriller, but take place outside the normal geographical boundaries in which we live our day-to-day lives. Similarly, a historical romance and a modern love story may be similar in plot, but dissimilar in time and place. Environment is an important aspect of the game world to consider as much of what motivates players to play games is their desire to explore unknown areas and take risks traveling through mysterious areas that they would not normally traverse in the real world. Environmental changes can be influenced through everything from creative sound design, art, and texturing, to game engine modifications which allow alternate forms of exploration (e.g., swimming underwater in *Tomb Raider* or walking in zero gravity conditions to explore new areas in *Dead Space*). Practically speaking, environments are often constrained by technology. A first-person shooter (FPS) may be more likely to produce certain types of stories due to early graphic limitations of the game engines (this is why many early FPS games were in dark environments, underground, or in space), but we are now beginning to see more immersive, open-ended worlds in both FPSes and RPGs.

Thematic considerations are more difficult, particularly because of the different conventions of game genres. For example, games in which a player builds a simulated city and reacts to emergencies may have less opportunities for creative storytelling than an RPG in which you lead a space team through explorations on Mars. But, we can at least consider some dimensions along which story might be classified in terms of thematic materials. To move towards a morphology of the serious game story, we must acknowledge that such classification is constrained by a number of internal and external factors, including, but not limited to, the overall plot, the subplots, the learning objectives, the environment, the technology and its associated capabilities, and the point of view of both exposition and gameplay. We can, however, begin to construct a preliminary narrative taxonomy keeping this disclaimer in mind.

As we have noted, in general terms, narrative can be defined as the expression of a story through a particular medium. A narrative taxonomy breaks this expression down into a series of fundamental parts. These parts can then be used as independent variables for a variety of research questions in the area of serious games. A sample (and preliminary) narrative taxonomy is shown below, adapted from the work of many of the narratologists discussed above. Below each level on the hierarchy, questions are listed to better specify the important features of each category.

Any of these categories could easily be expanded to fill additional pages with questions and potential options for their answers. For example, the inciting incident category listed under plot could easily be only one of the many narrative functional units identified by Propp (1968) and discussed earlier, and the characters could certainly be described with more nuanced facets (c.f. Dille & Platten, 2007), but the point here is not to exhaustively list every possible unit from which to assemble a serious game story. Rather, it is to suggest an overall approach to the construction of genre-specific serious games along the dimensions of plot, character, and environment. There is no doubt that some important questions have been

Table 1. Narrative game taxonomy

Narrative Development Questions	Some Potential Options
Environment	
Where is the story taking place?	[space, underwater, Florence…]
When does the story take place?	[1435, 1912, 2077, unknown…]
Is the environment fantasy-based?	[high fantasy, mild fantasy, no fantasy…]
What are the sources of environmental conflict in this story?	[biological warfare attack, bird flu, none, volcano eruption…]
How is player action constrained by the environment?	[hot lava limits navigation to a city block, birds will attack outside Central Park…]
How realistic is the environment in depicting real world objects (e.g., graphical fidelity)?	[very realistic, cell shaded cartoons, abstract…]
How does the progression of time affect the environment?	[the environment stays the same, degrades over time…]
Character	
Who is the protagonist of the story?	[Sam or Sarah, Nichola, the unknown pilot, the scientist…]
What is the point of view of this protagonist?	[third person, first person, hybrid…]
What forces (internal or external) is the protagonist facing?	[the player is unwilling to help others, the player struggles with responsibility…]
Does the primary character have a history?	[yes, revealed through dialog with family; no…]
Is the primary character well-defined (e.g., Duke Nukem from Duke Nukem 3d), or relatively "flat" (e.g., Gordon Freeman from Half-Life)?	[the character has a weak personality to allow the player to project, the character has a strong personality…]
Who are the external or supporting characters in the story?	[shopkeeper, bowsmith, mentor, barkeep…]
Can the player control more than one character in the game?	[yes, no]
How does the progression of time affect the environment?	[the character ages over time, the character's youth is shown through flashbacks,…]
Plot	
What type of major overarching plot does the story have?	[quest, overcoming the monster, voyage and return, rebirth, rags to riches, comedy, tragedy…]
What subplots are used?	[quest, overcoming the monster, voyage and return, rebirth, rags to riches, comedy, tragedy…]
How is the high-level plot released during gameplay? At which points in the game are segments of the high-level story unveiled?	[plot point 1: volcano erupts, plot point 2: player sees news video showing extent of damage, plot point 3: player must rescue trapped citizens, …]
How are the plots chronologically connected?	[normal linear time, occasional flashbacks, flashforwards…]
What inciting incident motivates the player to begin her journey?	[a famous painter is murdered, the player's father leaves home without saying goodbye, the princess is kidnapped, the volcano erupts…]

left out and that others may be redundant. We believe, however, that this general taxonomical approach is more useful than considering story as a primitive unit that can be attached to or removed from a given serious game in its entirety.

Also, while they are presented in separate categories, some of these questions overlap. For example, if the environment is fantasy-based, there are more than likely going to be fantastic creatures populating that environment. Similarly, if there is an "overcoming the monster" plot being used in a story, there needs to be a monster character or a metaphorical internal monster that the player is struggling to overcome. A plot that is arranged using extensive flashbacks is also going to require younger versions of characters to be present, and these characters will need to be adjusted psychologically to suit their age in order to appear more realistic. With each seemingly minor decision, there are a number of important considerations to take into account, both in terms of gameplay and production assets, and these issues help to illustrate the complexity of story design for games.

Lastly, in addition to intrinsic elements, there are also a number of extrinsic factors related to the medium (or the exposition of the story) that can be manipulated. For example, some of these items include the type of media in which the designer's story is presented (animated filmic cut scenes, live action cut scenes, textual via dialog, books, and journals left behind by characters in the game, etc.) and the point of view (first person, third person, etc.). Other issues include the degree of interactivity afforded (e.g., can the player still move their character during dialog) and the length of time elapsed between plot points.

A NARRATIVE ANALYSIS OF FALLOUT 3

To make our final point, which is that even highly regarded games are lacking in narrative coherence, we will apply our second technique for considering storytelling in serious games, which is to perform a brief narrative analysis of a video game along the three dimensions of our narrative taxonomy. Commercial games generally exhibit the most polished examples of game story environments as they are produced with multimillion dollar budgets and hundreds of employees. *Fallout 3*, which was released in late 2008, is one such example. The game was developed by the same studio that released *The Elder Scrolls IV: Oblivion*, another award-winning game that was praised for both its open-ended exploration and massive scope. The game was produced by Todd Howard and developed by Bethesda Softworks.

Fallout 3 takes place in the year 2277 in the city and outskirts of Washington D.C. The game-world unfolds in a fictionalized U.S. capitol after the ravages of nuclear war have turned it into a wasteland. The player takes the role of a "vault-dweller" who ventures out of a sealed nuclear fallout shelter and undergoes a journey to locate his or her father in the unforgiving wastes outside the vault. The initial creation of the character is done in a highly creative way, by allowing players to take control of the character from his or her birth and then giving the players influence over a toddler, a teenager, and a young adult as they learn the basic controls and allocate skill points to create a character they will later control throughout the game.

In terms of a high-level plot, *Fallout 3* uses a general "quest" plot combined with several other canonical plot types as embodied by the side quests and even through random encounters in the game. For example, a compressed "overcoming the monster" plot is found in any random battle encounter with a feral ghoul or super mutant, while the "rags to riches" plot materializes through the player's gradual accumulation of wealth and inventory. Similar, the "rebirth" plot may be realized through the player's growth from a neutral karma to a positive or negative karma and the "voyage and return" plot emerges through the player's many questions that lead them to and from big cities in the game such as Megaton, Bigtown, or Rivet City. Comedic moments abound, such as the good-intentioned (but ultimately misinformed) authoring of a Survival Wasteland Guide by a shopkeeper named Moira and the absentminded accumulation of incorrect historical facts by an elderly and eccentric curator, but tragic moments are much more common as the player moves through ravaged fields, abandoned homes, and the burnt out husks of futuristic vehicles. Much like epic fiction, the narrative landscape of *Fallout 3* is both unpredictable and widely varied in terms of both plot structure and character types.

Although not designed as a "serious game," the game possesses many characteristics of serious games, including the use of supplementary learning materials. These are then integrated into gameplay as parts of the main quests. These materials include historical facts, locations, and descriptions, although the game fictionalizes history and employs an alternate historical storyline. Despite this, one can imagine how *Fallout 3* could be repackaged as a serious game. There is ample opportunity for embedding additional legitimate learning materials about U.S. national history into the game. Many of these opportunities are already used in various quests and encounters in post-apocalyptic DC, which require the player to move through locations such as the Museum of History, the Capitol Building, the Museum of Technology, the White House, and the National Archives. While some locations have been renamed and relocated for inclusion in the game, the locations are ripe for embedding information about American history, political science, and even the history and culture of technology. Several of the game's missions direct the player through journeys to retrieve famous historical documents such as the Declaration of Independence and the Bill of Rights, though the catalyst for these missions is an unfortunately misinformed amateur historian by the name of Abraham Washington, curator of the Capitol Preservation Society in Rivet City.

By most accounts, *Fallout 3* is a triumph of a game. Garnering critical success by the mainstream gaming press (Linn, 2008; Tuttle, 2008) as well as high aggregate review rankings (93/100) by MetaCritic.com, the game captured the imaginations of thousands of players and immersed them within the radioactive dystopia of the Capital Wasteland. Much of this immersion was due to the successful narrative transportation enabled by the game's designers (cf. Green, 2004). What is particularly interesting about this fact, though, is that the environment presented in this post-apocalyptic version of Washington D.C. is rather dull. There are not a particularly large number of enemy types to battle and the scenery is mostly dreary: brown, grey, and without color. However, the capacity of the *Fallout 3* environment to immerse the player is well-documented, with players posting to discussion boards after having played the game for hundreds of hours. The game's immersive capacity is due in part to the unique narrative mechanics engaged by the game, but even *Fallout 3*'s stories are problematic for a variety of reasons. We can further deconstruct the successes and failures of *Fallout 3*'s storytelling by considering the narrative functions of the game as they relate to one another at different points during gameplay rather than merely analyzing the simple characteristics of its plots as they exist in isolation. We can also consider the game's narrative problems in terms of environmental, plot-driven, and character-driven storytelling.

To begin, we can consider what works, narratively and holistically speaking, in *Fallout 3*. Perhaps the greatest narrative achievement of the game is its integration of nonlinear storytelling with hybrid game mechanics. While not strictly an RPG and not strictly an FPS, *Fallout 3* combines elements of both genres in order to make innovations in several key areas of the game, one being the cinematic sequences afforded by the V.A.T.S. (the vault assisted targeting system) device used in the game. While it would seem that such a mechanic is more related to gameplay than story, the way in which the device is presented to the player, as a gift given early on to the vault dweller during her initial training, makes for a seamless integration of gameplay and story. There is a narrative reason for the device (to secure the gift as a token of maturity from a donor character) as well as a ludic reason (to learn an important game mechanic early on in the game).

A second impressive aspect of *Fallout 3*'s narrative model is the way in which the world reacts and reshapes itself based on prior decisions of the player. Perhaps unlike any game before it incorporating FPS mechanics, *Fallout 3* offers a nonlinear experience that flexes and adapts in terms of dialog and available quests. If the player chooses to kill a character in the game, future missions that involve that character are no longer available. Sell an important object and you may eliminate future plots which require that object to function. If you are too good or too evil in the game, characters will choose to join you in your campaigns, provide aid, shun you, or attack you. These outcomes depend on your prior actions and decisions. This is quite an accomplishment given the 14 primary quests and more than 70 side quests available in the game.

Despite its critical success and its highly regarded story, *Fallout 3* suffered its share of narrative failures upon launch, and we can learn by considering some of its problems. The first major problem (a fundamental problem of *plot*) presented itself through technical bugs in the path-finding algorithms. It was documented on several forums (and personally experienced by an author of this chapter during gameplay) that important non-playable characters (NPCs) could fall to their early deaths due to incorrect pathfinding in the city of Megaton. Built using scrap metal and old parts, the city presented many locations from which it was easy to fall and cripple a limb or even die. This problem was fixed with an update patch provided several months after the game's launch, but the presence of this glitch caused for many holes to emerge in the various plots of *Fallout 3*. As one simple example, the player is approached by a water treatment repair character, Walter, early in the game. Walter informs the player that she can bring him additional pieces of scrap metal, scattered throughout the Wasteland, in return for a certain number of bottle caps (used as currency in the game). When Walter falls off the ledge to an early death, this particular narrative branch is closed off to the player through no action of her own, and she has no way of knowing whether Walter will ultimately put an end to his offer of scraps for caps or simply continue on indefinitely as long as the player can locate more metal. When Walter suddenly disappeared for no apparent reason early in the game, this was a narrative inconsistency that reduced immersion and contributed to player frustration. Particular plots were no longer able to be experienced by the player.

A second problem, concerning narrative *environment*, is found in the fact that environmental boundaries are not well-integrated into the overall narrative. For example, when reaching the limits of navigable space (to be fair, the world is quite large), a message appears to inform the player that she can no longer proceed in that direction and to please turn back. This could have been addressed through the inclusion of environmental obstacles surrounding the perimeter of walkable wasteland. Another environmental problem is found in Rivet City, where there is a gap in a broken railing on the flight deck. The gap is clearly wide enough for the player to fit through and drop down onto the flight deck, and yet they are unable to. In this case, a gameplay problem contributes in part to a narrative problem, as the environment below looks enticing to explore and yet the player cannot physically go there (at least via that route). When gaming, players expect reasonable actions to succeed (Rouse III, 2005) and such problems with the gameplay also contribute to environmental narrative problems, which can further lead to implausible plots or unlikely developments in the story.

The last category of narrative problems can be found in NPC behaviors. These are problems with *character*. First, NPCs are often unaware of their environment. For example, when questioning a character about the location of another character in Rivet City, an NPC might respond by directing the player to "upstairs on the upper deck" rather than the more appropriate "directly behind where I am standing." This lack of basic environmental awareness reveals limitations in NPC artificial intelligence which undermines the quality of the characters and degrades the overall experience. Further, NPCs sometimes respond inappropriately to environmental cues. For instance, when standing in the player character's home, a fellow party member might proclaim to exercise caution because danger is near, even when it clearly is not. These types of canned responses which do not adapt to the environment are similarly troublesome for recognizing NPCs as realistic characters in the immediate-level stories of the game. Finally, NPCs do not respond appropriately to accompanying party members (characters in the game who have joined with the player as allies). For example, the paramilitary organization known as the *Brotherhood of Steel* professes a supreme desire to rid the Capital Wasteland of all super mutants, and yet when the player arrives with an intelligent super mutant in tow, none of them have anything to say about it and they continue to address the player as though nothing out of the ordinary was occurring. You would expect at least a curious questioning upon arriving with such a companion.

CONCLUSION: IMPLICATIONS FOR THE FIELD

Despite significant improvements in non-linear storytelling in video games, there remains much to be studied and improved upon in regards to the effective use of narrative in this medium. As we have shown with our discussion of *Fallout 3*, even award winning games leave a lot to be desired in terms of presenting coherent stories to an audience. While the willing suspension of disbelief carries audiences along without much complaint, an improved narrative framework would improve the overall quality of serious and non-serious games alike. Rather than there being one particular way of measuring narrative coherence or operationalizing this concept, we must begin to think about serious games storytelling at a more precise level of granularity. How much interactivity is ideal for a particular genre of game that uses a particular type of plot structure? What constitutes an effective use of plot in a particular genre versus a non-effective plot? How can we create more believable NPCs that players can relate to while still leaving them to feel empowered to control their own experiences in the gameworld? Are fantasy-based environments more suitable than non-fantasy-based environments for teaching certain subjects using videogames? Does Propp's morphological study of fairytales translate well to videogame stories, or should we begin analyzing existing game narratives for a new database of dramatic functions suitable for interactive experiences? These are only a few of the numerous questions prompted by a more nuanced discussion of storytelling in serious games.

At a more practical level, we believe that more attention needs to be paid to the role of story in video game environments. For example, a game developer rarely releases a game in which major graphics problems are sustained throughout the entire game, or in which important input buttons on the keyboard or controller cease to work entirely. These things do happen from time to time, but they are relatively rare for highly regarded games. Narrative problems, such as those mentioned previously, are unfortunately all too common, and seem to be taken for granted. In other words, it is okay to have an incomplete or incongruent story because the player is willing to accept it. That may be true to some extent, but if we are to subscribe to the notion of using serious games to better immerse, transport, motivate, and educate the player within a certain educational domain, then we must also consider the potential impact of including a story with incomplete or incongruent information within a specific game environment. To truly design and produce effective serious games for learning, we must also tame the beast that is serious storytelling. While high-level stories are easier to plan and account for, the focus needs to move to considering ways to improve and assess the effectiveness of immediate-level storytelling in terms of plot, character, and environment.

As we suggested several times in this chapter, we must also move away from the habit of considering stories as homogenous entities. Stories are not interchangeable atomic units which can be inserted or removed at will from serious games. Rather, they are sophisticated entities which can be decomposed and reconfigured in a variety of different ways for different dramatic and pedagogical effects in different types of games. It is our hope that this preliminary discussion of a narrative taxonomy for serious games and our conceptualization of the serious game story along both applied and theoretical dimensions will begin a sustained discussion on the subject and lead to additional research on this very important topic. Interactive storytelling, like artificial intelligence, remains a "holy grail" research problem and although some progress has been made in the last decade, the tropes and techniques of this process are far from mature.

So, based on what we have covered in this chapter, how might we improve the story of Grenwin the Goblin we posed in our introduction? The first thing we might do is consider the story in terms of plot, character, and environment. To improve the plot, we need knowledge of the events occurring before and after the player's entry into the local tavern. Based on prior or future events, it might make more sense to move the learning material into a different part of the game, perhaps into a strategic planning session where the player must help townsfolk calculate the right trajectories for catapults to help hold off an invading goblin horde. This would be less intrusive than the current implementation where the player must apply the Pythagorean Theorem as she is working with other complex control decisions such as moving her avatar out of reach of the rusty axe and working with the controls to manage her bow and arrows. In terms of character, it could be interesting to fictionalize and introduce a character named Pythagoras who is a mentor to the heroine and helps her hone her skills using geometry prior to meeting Grenwin in the tavern. Finally, environmental solutions could be explored, such as by envisioning the game in a more contemporary urban landscape and allowing the player to act as a city architect or planner in which knowledge of geometry would be more directly relevant to the type of game. In any case, a variety of approaches could be used to improve the game, but these changes should be carefully considered within the overall context of keeping the player motivated to play and addressing the learning objectives set out by the designer.

In this chapter, we examined the case of serious storytelling along two dimensions: theoretical and practical. Through our theoretical coverage of traditional narratology, we introduced fundamental ideas about how stories are composed, named, arranged, and classified. We next suggested practical ways in which these ideas can be applied to stories used in serious games. By applying narratological principles, we devised a preliminary framework for thinking about storytelling along the three dimensions of plot, character, and environment. We concluded by critically analyzing a contemporary video game in terms of these three dimensions. The rich interplay between these perspectives; and the ability of humankind to use narrative as a technology for communication, entertainment, and education; suggests that this complex form still has much to offer to both researchers and developers of serious games. As we move closer to building truly effective learning games that both motivate players to continue learning and generate transferrable skills and knowledge from the virtual to the real, we would do well to keep in mind the awesome power of interactive storytelling as a means for improving narrative transportation, motivation, and learning. We must also, however, be cautious and realistic. We should acknowledge the work that will be necessary in building our serious storytelling abilities up to the same levels as we currently see in cutting-edge gameplay mechanics such as those used in *Fallout 3*. With a careful attention to the dramatic, ludic, and psychological factors of interactive narrative, we can develop innovative fiction writing and programming techniques that will continue to push the envelope in both interactive entertainment and serious game design.

ACKNOWLEDGMENT

Writing this article was supported in part by the Office of Naval Research (ONR) Grant N0001408C0186. The views and conclusions contained in this document are those of the authors and should not be interpreted as representing the official policies, either expressed or implied, of the ONR or the U.S. Government. The U.S. Government is authorized to reproduce and distribute reprints for Government purposes notwithstanding any copyright notation hereon.

REFERENCES

Abbott, H. P. (2002). *The Cambridge introduction to narrative.* Cambridge, UK: Cambridge University Press.

Bal, M. (1997). *Narratology: Introduction to the theory of narrative* (2nd ed.). Toronto, Canada: University of Toronto Press.

Bateman, C. (Ed.). (2007). *Game writing: narrative skills for videogames.* Boston: Charles River Media.

Booker, C. (2005). *The seven basic plots: Why we tell stories.* London: Continuum International Publishing Group.

Campbell, J. (1949). *The hero with a thousand faces.* Princeton, NJ: Princeton University Press.

Crawford, C. (2005). *On interactive storytelling.* Berkeley, CA: New Riders Games.

DeMarle, M. (2007). Nonlinear game narrative. In C. Bateman (Ed.), *Game writing: Narrative skills for videogames* (pp. 71-84). Boston: Charles River Media.

Dille, F., & Platten, J. Z. (2007). *The ultimate guide to videogame writing and design.* New York: Lone Eagle Publishing.

Eagleton, T. (1996). *Literary theory: An introduction,* (2nd ed.). Minneapolis, MN: The University of Minnesota Press.

Frasca, G. (2003). Ludologists love stories, too: notes from a debate that never took place. In M. Copier & J. Raessens (Eds.) *Proceedings of the Levelup 2003 Conference.* Retrieved 11 Feb 2009 from www.ludology.org/articles/Frasca_LevelUp2003.pdf

Genette, G. (1980). *Narrative discourse* (J. E. Lewin, Trans.). Ithaca, NY: Cornell UP.

Green, M. C. (2004). Transportation into narrative worlds: The role of prior knowledge and perceived realism. *Discourse Processes, 38*(2), 247–266. doi:10.1207/s15326950dp3802_5

Jacobs, S. (2007). The Basics of Narrative. In C. Bateman (Ed.), *Game writing: narrative skills for videogames* (pp. 25-42). Boston: Charles River Media.

Jenkins, H. (2006). Game design as narrative architecture. In K. Salen & E. Zimmerman (Eds.), *The game design reader,* (pp. 670-689). Cambridge, MA: The MIT Press.

Juul, J. (2001). Games telling stories? *Game Studies, 1*(1).

Kindt, T., & Müller, H.-H. (Eds.). (2003). *What is narratology? Questions and answers regarding the status of a theory.* Berlin: Walter de Gruyter.

Krawczyk, M., & Novak, J. (2006). *Game development essentials: game story & character development.* Clifton Park, NY: Thomson.

Linn, D. (2008). Fallout 3 (PC). *1UP.com.* Retrieved February 3, 2009, from http://www.1up.com/do/reviewPage?cId=3170949

Mateas, M. (2002). *Interactive drama, art, and artificial intelligence.* Unpublished PhD Dissertation: Carnegie Mellon University, Pittsburgh, PA.

Meadows, M. S. (2003). *Pause & effect: The art of interactive narrative.* Indianapolis, IN: New Riders.

Metacritic.com. (2009). *Fallout 3*. Retrieved February 3, 2009, from http://www.metacritic.com/games/platforms/xbox360/fallout3

Onega, S., & Landa, J. A. G. (Eds.). (1996). *Narratology: An introduction.* London: Longman.

Prince, G. (2003a). *Dictionary of narratology.* Lincoln, NE: University of Nebraska Press.

Prince, G. (2003b). Surveying narratology. In T. Kindt & H.-H. Müller (Eds.), *What is narratology? Questions and answers regarding the status of a theory,* (pp. 1-16). Berlin: Walter de Gruyter.

Propp, V. I. (1968). *Morphology of the folktale* (2nd ed.). Austin, TX: University of Texas Press.

Rouse, R., III. (2005). *Game design theory and practice* (2nd ed.). Plano, TX: Wordware Publishing.

Sheldon, L. (2004). *Character development and storytelling for games.* Boston: Thomson Course Technology.

Tuttle, W. (2008). Fallout 3 (PC). *Gamespy.com.* Retrieved February 3, 2009, from http://pc.gamespy.com/pc/fallout-3/924348p1.html

Vogler, C. (1992). *The writer's journey: Mythic structure for storytellers and screenwriters.* Studio City, CA: Michael Wiese Productions.

Chapter 3
An Adventure in Usability:
Discovering Usability Where it was not Expected

Holly Blasko-Drabik
University of Central Florida, USA

Tim Smoker
University of Central Florida, USA

Carrie E. Murphy
University of Central Florida, USA

ABSTRACT

Usability testing can be used as an effective tool throughout the design and development of a serious game. User responses to the usability measures can help designers meet the challenges of creating a game that is educational, playable, and entertaining. However, there are a multitude of measures to choose from, some empirically created and tested and some created by developers to test a single game. This chapter will provide a broad overview of usability, two major methods of conducting usability testing and how usability can be applied to testing, and more importantly, improving games. Two possible usability measures are discussed from a game evaluation standpoint and compared highlighting on which aspects of usability heuristics they focus.

INTRODUCTION

Imagine you are playing a simple driving game on your computer. The goal is simple: drive as long as you can without striking any obstacles in the roadway. To do this, you hit the space bar to shift from one lane to another. However, the game itself is drab and simplistic. The car is represented by a colored oval, obstacles are represented with black and white rectangles, and the background is grey. While not exactly exciting or engaging, this game is perfect from the standpoint of usability: all you need to know is how to press the spacebar and you can play.

Now imagine a second, similar game with the same goal of avoiding obstacles in the roadway. However, in this game, the colors are vivid and realistic; the car is futuristic and portrayed in stunning and imaginative detail and large trees and meticulously rendered landscapes fill the screen.

DOI: 10.4018/978-1-61520-739-8.ch003

This game has a more intensive interface as well. Where in the first game a single key press controlled the car's movement, this game requires two buttons to play the game: a move left button and a move right button. Which game would the average user consider more "usable?" Would they judge it on which game is more "real" or more "enjoyable?"

To address these questions, usability analysts turn to the user's behavior as well as their stated preferences. Surprisingly, these are not always the same. For example, the user's behavior in game one would largely be a measure of reaction time. The game, boiled down to its essence, tests how quickly the user can respond. In the second game, though, the user must also make a decision for which button to press to avoid the obstacle. Because of this additional decision component, the user's performance would suffer in the second game. This reduced performance would normally be a symptom of poorer usability. However, when asked, users would rate the second game as being preferable to the first. The truth being that the visual aesthetics of the game would outweigh any perceived loss of performance (if the user even noticed a decrease in their performance level). So what are usability analysts actually testing if not simple performance or preference? If the customer, in this case the user, is *not* always right and better performance is only "good enough", how can usability consider itself a scientific approach and not just marketing with trendy vocabulary?

When it comes to flashy new technology and applications, the gaming industry is usually at the fore. However, while usability has become a buzzword for much technology, gaming has remained more cautious in its adoption of explicit standards. In fact, in the gaming industry it is an oft-debated topic, usually concerning if it should be considered at all. Usability is often seen by serious researchers and developers as the stepbrother who only shows up for money, the metaphorical black-sheep of the development cycle that nobody really wants around but it looks good for the family pictures. This arises from a single problem; there is no universal standard for usability. Usability is itself a ubiquitous term but it is usually coupled with multiple standards, variations, and definitions that are often not only industry-specific, but company-specific as well. One company may use validated measures like the *Systems Usability Scale* (*SUS*) or the *Questionnaire of User Interface Satisfaction* (*QUIS*) but many companies choose to develop their own standard for usability measurement, and sometimes their own definition of just what a "usable" product entails.

So, why is there so much disagreement? One reason is because there are no single, simple answers to just what the best measure of usability is. Two of the more popular measures, the *SUS* and the *QUIS*, have several advantages and disadvantages. Because these measures have been validated they give an accurate scale, but they may not have the ability to test what the developers care about. More specific tests that have been individually developed preserve the developer's rights and may test specific aspects of interest but the accuracy of these measures is often dubious at best. When attempting to apply a usability test to a new product, these costs and benefits must be weighed against further constraints of resources and the time available to conduct the analysis. In the 24-hour development rush to create games, this is especially the case. Usability is often a last minute, rubber stamp process, more concerned with accessibility than usability professionals would prefer. Ideally, usability would be based off of an industry standard that can be applied universally to software development at all levels. For example, tests that that can measure the effectiveness of a webpage while still having the capabilities of measuring user satisfaction in gaming could increase the effectiveness of usability immensely by allowing comparisons currently impossible.

As a result of this, developers will spend thousands of hours developing a new game that may be visually pleasing, then turn around and ignore or forget user reactions to the game beyond its aesthetics. This "bloat" arises from the attempt to add as many applications as possible without looking at how those applications interact or behave during normal play. For example, in *World of Warcraft* (*WoW*) the designers wanted to have a socially interactive game, a goal they achieved admirably through the use of in-game chat. To include this feature, developers created several automatic chat channels or the option for more advanced players to create their own. On paper, this is a great feature and adds a unique and appealing aspect to what would otherwise be a traditional Role Playing Game (RPG). In practice, however, the attempt is hamstrung by the lack of a graphical interface or abbreviated list of commands to allow the player to switch from one chat channel to another easily. Instead, in order to switch channels the player must remember the correct command, format, and channel name that they want and type the command into the chat window. This must be achieved without any support from the game in terms of auto-fill, predictive text, or menus. The end result is a game that has many features that may increase the capabilities of the game itself, but only if the user has the knowledge and skill to use them (Krzywinska, 2006). In other words, the implementation of how the user applies this function or others is an afterthought to a new feature. Because of this, as the applications of a game increase so do the inputs and commands required to play the game, resulting in an upward spiral that new players must climb before they can play. And clearly this is a concern for gamers. For example, players will often create or use work-arounds to address problems with the game. In the chat example above, many gamers choose to download a special add-on program to make chatting in game easier to play. These add-ons display all the channels and have a user interface where players can click to join a chat channel. Other more expert computer users that play the game create their own specialized mini-programs, or macros, to make the game more usable in more complex ways. These extra steps may not be an issue for power users or expert gamers because there is often an overlap between gaming commands for *WoW* and other games or programs and these high level users generally know what will be required. Because the developers themselves are often in the category of power user it is easy for them to forget that the needs of the novice user during the development and design cycle and that these needs will be different from their own (Cornett, 2004; Desurvire, 2008). This becomes especially important when a game is developed to instruct or educate. The necessity of a broad appeal means that all levels of users must be taken into account.

With the increased accessibility of game systems and technology, and the broadening of diversity in the types and styles of games being created, the gaming culture is growing. The gaming industry has grown exponentially in the United States since 1996 when there was an estimated $2.6 billion in sales. In 2007, industry profits had leaped to an estimated $9.8 billion in the United States (Invetstment-U, 2008), with similar gains around the world. For example, in the United Kingdom, growth in the industry showed a 100% gain over a six-year period reaching an estimated 4 billion pounds in sales (ISFE, 2005). In 2007, Australia broke the billion dollar mark for the first time in game sales (IEAA, 2008). With this many people reaching for games as a form of entertainment, and markets of this size and growing, it is especially important to develop video games wisely. When competition is this fierce, the playability of a game becomes a key component. And playability in this context often rests on the shoulders of the game's usability. In this, the importance of usability can decide the future of a game. Game concepts can only take you so far and after that it is the nuts and bolt of whether the game is as good in practice as it is in theory.

As games are becoming an omnipresent aspect of our society, they are being to be applied in new and creative ways. Increasingly, games are being seen in areas not traditionally known for being fun, in other words, work. Traditionally just the province of entertainment, a new genre of games, called serious games, use a gaming format or platform to convey information or training that might have otherwise remained in a textbook or presentation. Serious games are developed to educate, train, and inform the user on topics important beyond the environment of the game itself. In other words, serious games reach beyond just teaching the user how to play the game to become a vehicle for more effective learning. This creates a unique situation as the game must be both work and play. The serious content must be presented in an entertaining fashion but without becoming so caught up in the gaming aspect as to lose the purpose or message of the games intention. Serious games also create a unique set of headaches for usability. Because serious games are not developed solely for entertainment value, and therefore need to reach a much larger user base, the game cannot be designed with an expert gamer in mind.

Obviously, the application of this to the education of children is a target of many research paradigms in both education and usability. One goal of serious games is to supplement current curriculums in school and increase student focus in learning and retention of material. Few parents would disagree with the observation that children will be more motivated to reach the next level in points or character development than the next grade level. Children will also spend hours glued to the screen repeating the same level over-and-over with a minute attention to detail that is rarely, if ever, given to their math homework. While the result of this is children with a fair amount of expertise in a specific game, this expertise is not necessarily transferable to other games or computer applications. During the development cycle, this becomes an excellent example of needing to design a game for multiple user-skill levels. If a game is developed for elementary education then chances are this level of user is going be below expert level, if not in game playing then in other supporting abilities. Therefore, if the usability of the game surpasses the computing knowledge of the user, they will spend most of their time learning the game process at the expense of the educational purpose of the game. This can be a double-edge sword, however, if the game is not engaging enough then boredom will set in and the user will not be motivated to continue playing. Either way the educational value is lost. The game has to be aesthetically pleasing, easy to learn, difficult enough to keep attention and functional enough to maintain that attention without losing it to frustration or boredom. The true difficulty occurs when a usability expert attempts to walk the fine line between aesthetic value and usable design.

In an attempt to walk this knife's edge, developers have been working with the academic community to use or adapt current and popular games to serious gaming purposes. Take *Sid Meier's Civilization IV*, for example. This is a popular game with a large and devoted following where players choose a civilization based on each civilization's different historical, cultural, research and social advantages and disadvantages. Players then compete to raise their civilization from nomadic tribes in 5000 BCE through several epochs of social and technical development making decisions along the way regarding resource allocation, military action, and society organization. Game play teaches the player history, critical thinking or problem solving, as well as societal evolution (Papaloukas & Xenos, 2008) under the guise of competition. The player interacts with computer controlled characters and nations or other players to determine treaties, building alliances, fighting wars, and creating trading routes. The amount of control the player exerts can vary based on the user's experience playing similar games, their age level or to highlight a particular lesson. For example, one feature of *Civilization IV* that might make it useful to researchers is that many adjustments can be made to the game from the defaults settings at the beginning of the game. For instance, researchers could allow the players to win the game in up to five different ways, each with their own lesson. In one game, players could vie for market or economical dominance tied into more traditional lessons on free trade and exchange, lessons on diplomacy and balancing the demands of neighboring countries could be successfully rewarded with a diplomatic win that is achieved by being elected planetary governor. Technological development and the relationship between technology and societal production are demonstrated through a technology-based win with a space race beginning with the wheel and ending with the space shuttle. The game seems well designed to allow players to learn as much as they want. Those interested in learning can click links to read about important people, places, history, often learning tips or special methods that might be used to give their nation an edge. For those more casual gamers, there are options that allow them to see flow charts so they can plan for the future of their society and manage resources to their advantage. The competitive gamers, who seek to win as quickly as possible, may do so without being forced to scroll or click through pages of "educational" text and materials or scrolling through charts, in fact, they can elect to use automated features to manage their cities, research, or economies.

Can something as innocuous as increased usability increase the potential these games have for educational purposes? Adapting usability methods to popular games like these can be difficult. This chapter will attempt layout a history explaining the importance of usability and the correct methods, measures, and heuristics needed for a usable and playable game.

BACKGROUND

Usability as a field and profession began in the late 1980s when John Whiteside and John Bennett published a series of articles under the topic of "usability engineering" (Whiteside, Bennett, & Holtzblatt, 1988). As they defined it, usability engineering focused on using a quantitative and practical approach to product design as opposed to the more traditional methods that relied on an expert's "gut instinct." Whiteside and Bennett recommended clear methods and early goal setting, prototyping, the iterative evaluation processes that focused on more than just what *could* be done, but also what *should* be done. Their method takes into account both the user and the context the product would be used in. By the early 1990s, trends were being observed and the old and new methods of product evaluation and design models were being summarized and grouped into heuristic guidelines that developers could follow when designing products (Neilson, 1992).

Since the beginning, the goal of usability has been to improve the quality of interactive systems and the challenge has been defining what these improvements may be. Based on the product or system, the definition and constraints of a possible improvement may change. For example, when a developer is designing a webpage they are going to focus on ease of use. This type of software has a wide population and is limited only by the availability of a computer. In gaming however, there are other matters that must be taken into account. The gaming population is a restricted and niche-based culture. This means that the usability of a game is going to be defined by the gamer or the type of game being developed. When expanding our definition of games to include console games, this selectivity extends beyond just user traits to encompass technological constraints as well. While developers of a driving game may want to focus on the tracking capabilities and how best to represent the motion of the character over time, a game teaching math would focus on the best way to display numerals to be legible and easy to input to the device. There are several ways to approach this task, the first of which is getting to know the user.

One goal with usability is to define the target market for the usability test. There are methods for targeting this information beyond posting a memo to the sales department. Currently, experts in usability have been developing the concept of personas. This buzz word represents an attempt to define the user of a specific type or application of software. Demographic variables are of particular interest, such as age, gender, family unit, and social economic status. These variations are collected for the specific population of users of the software and correlated with user behavior trends. The data is then assembled into the perfect user and are applied by developers or engineers to create an ideal product for this population. For example, elderly users will often display hesitance at learning new technologies. By understanding this behavior, developers can tailor a system to have a more traditional appearance or interface if it is intended to be used primarily by an elderly population (Holzinger & Searle, 2008). This is how personas have the capability to assist in usability. Current usability measures can benefit from a defined user such as the "avid gamer", "young novice," or "occasional player."

Currently, software is being developed to define usability mapping. The *Quality in Use Integrated Map* (QUIM) was developed by Concordia University in 2006 (Donyaee, Seffah, & Rilling, 2007). Researchers also created the *QUIM Editor*, an interface that allows the analyst to select pertinent variables for measuring usability then tests for the variables defined in the system. In this application, the use of personas could be highly effective by creating for the analyst a template for the necessary variables needed for an accurate usability measure. This concept is especially important in the area of gaming because of the drive within the domain to develop games for specific markets. Usability researchers can use personas to execute the QUIM effectively to create an example of the gamer for a specific market thus allowing researchers and industry developers to select the most important aspects of a game to test.

The adaptation of popular games to educational applications can be refined with usability. For example, research using first person shooters has demonstrated a significant increase visual attention. Visual attention is a perceptual process that is important for many everyday activities and work environments. In particular, driving a car requires extensive visual attention to safely control the vehicle, avoid hazards, and operate under adverse conditions. The ability to track other cars, road signs, and other potential hazards can increase driver awareness and reduce accidents. Properly adapted usability measures and methods can be used to refine these ideas further and to test new technologies. Usability allows researchers to identify what aspects of the game specifically increase visual attention and what aspects of the game may be potentially detrimental to these abilities without ever needing to create a physical prototype of the technology. If, for example, the use of maps in *WoW* increases the gamers' visual attention through their ability to divide attention between the informational aspects of the map and the action of the game usability measures could be used to identify the most efficient design of these maps to maximize their training abilities in visual attention. The capacity of usability to tweak these popular games without removing their playability and recreational appeal may cause improvement in these abilities or in the development of technology to support them.

Usability testing and traditional empirical research use many of the same methodologies, but with several, very important, differences. For example, usability testing and research are similar in the wide array of different available approaches and similar qualities. However, how these methodologies differ is the basis of their hallmark differences. Research focuses on well-crafted experimental studies to test or evaluate an idea through previously validated and reliable measures. To achieve this researchers attempt to control for extraneous effects and variables that may have an effect on what is being researched.

To this end, research is often conducted in a very simplistic environment, and often in a laboratory. Usability testing, however, takes a more ecologically valid approach and conducts testing in a more realistic environment on a broader range of tasks. The end result of this is that usability tests rarely look like an actual study. Instead, usability testing often looks like structured observations of the user interacting with the product. The use of user observation is not without reason. While empirical research seeks to determine the average behavior of a user under ideal conditions, usability testing attempts to quantify the extremes. For example, most usability studies are used to identify flaws in the interface along with other factors that my decrease the performance of the user.

In gaming this relates more the concept of the playability, instead of usability, though the concepts overlap to a significant degree (Desurvire & Wiberg, 2008). What is considered playable is frequently also considered usable, though times when this is not this case are especially notable. In a manner similar to the *WoW* chat interface, or lack thereof, mentioned previously, you would have a difficult time classifying many of the popular games currently on the market as "usable" from a testing standpoint. Also in *WoW*, for example, the player basic interface easily overwhelms the player with information including chat windows, action bars (i.e., clickable buttons arranged on the bottom of the screen), menu buttons, a small map, and character details. This information floods the player with choices and options while they juggle coordination with other players, their class role or function (e.g., healing, attacking, etc.), terrain features, and movement or maneuverability (Lowood, 2006). In order to master their classes and abilities, the gamer has so many things to keep track of, plan, and coordinate that they often change the interface to help manage incoming information. While technically less usable, the trade off in playability outweighs the poor levels of usability. For instance, many gamers often add action bars, custom warnings, and trackers. Others attempt to minimize the number of actions they must take by running a separate voice-over-IP (VoIP) programs that lets players talk to each other, thereby eliminating the need to read or type text chat in the middle of the action.

USABILITY METHODOLOGIES

Two methods are most often used to calculate usability: the general or traditional method and the Rapid Iterative Testing Evaluation (RITE) method. The general or traditional method relies heavily on the experimental process. To use this method, the analyst conducts usability testing to collect a large data set based on a specific game. They then analyze the data and make changes to the game according to the information collected. This method focuses on making changes to the game only once. At the other end of the spectrum, the RITE method will make many small changes over time based on the results of short interval testing. Short interval testing collects data from individual participants and then adjusts the game after each participant before moving to the next participant. In the end, the strengths and weaknesses of both approaches come down to practicality and the situation in which the analyst finds themselves. While the traditional approach results in highly accurate results that can be applied more flexibly, the RITE method requires much less time and money. Developers must decide on the goals of their testing and what they want to learn so that they can choose the correct method to fully benefit from either method.

Defining the correct method for usability purposes requires researchers to focus on five general aspects:

- Target population
- Available participants
- Time available
- Expenses
- Measures

Figure 1. Representation of the traditional usability testing method

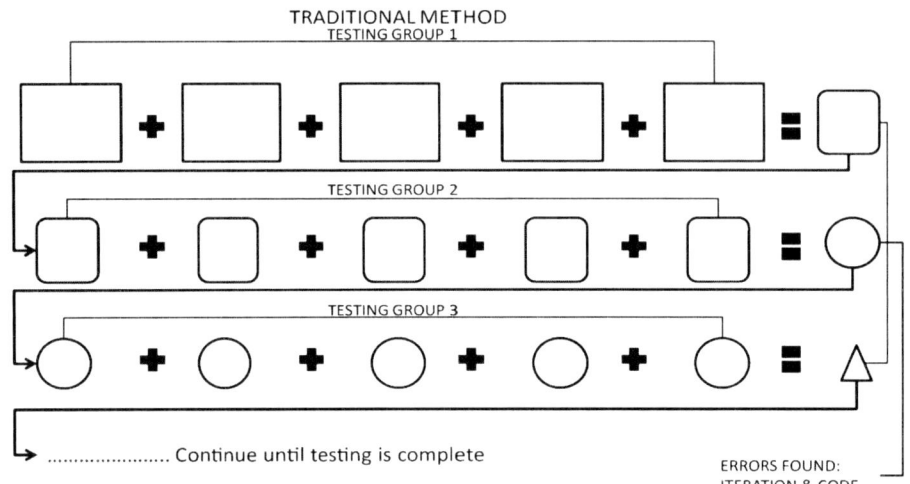

All of these characteristics are going to be defined by the level of game being tested. High level games such as *Civilization IV* or *WoW* are going to have a target population related to more advanced gamers. Whereas a driving game like *Need for Speed* is going to have general target population. The knowledge needed to play the game is minimal and there are very few commands needed to control the game, but there are more gaming platforms and therefore may require a different method of analyzing the game's usability.

As mentioned previously, the traditional method requires large subject pools, more time, and costs more on the whole. This is because the traditional method uses groups for testing purposes instead of the individual and would be of more use if, for example, a company is testing a game for all-around usability. To use the traditional method the researchers plan and make decisions on what type of surveys, materials, and measures are going to be used to collect the data. After the planning is completed, the researcher must run a Phase 1 test. During this phase, the software will be tested by several different participants, then the data is analyzed and the errors in the software interface are identified. The developer must then re-code the data to correct for the errors identified. Then testing repeats with the corrected software, this process will continue to repeat and refine the software until all problems have been identified (see Figure 1). Generally this method only takes two iterations, however, if the company has the resources to continue testing the software it will only improve the usability of the software. The use of continuous iterations of testing is most common in software development were there are likely to be future versions of the program. However, generally a fewer number of iterations are required for each version because there is a larger number of participants being used.

The RITE method takes a slightly different approach. This method was developed by Microsoft in 2002 (Medlock, Wixon, Terrano, Romero, & Fulton, 2002). RITE was originally developed to allow for more efficiency in usability testing. Though it requires less time, on the whole, as well as fewer participants it is still very similar to traditional usability testing. To conduct a usability test using the RITE method, the analyst defines their target population and gathers materials (e.g., surveys, scripts, and measures) for data collection. The real differences occur in the process of test-

Figure 2. Representation of the RITE usability testing method

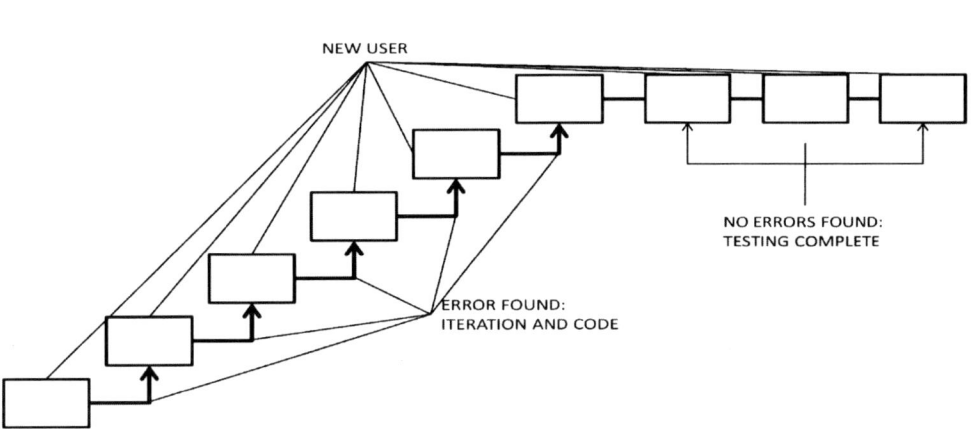

ing itself. Instead of testing groups of participant before each update of the software, RITE updates the software after each participant (see Figure 2). This method of usability is ideal for testing small games, programs, or specific aspects of a program or system but is rarely used for a complete game. This is because the RITE method can be tedious if the researcher is attempting to recode a video game that took thousands of hours to code in the first place. For example, the RITE method is ideal for the purpose of testing the usability of steering wheel interface or the legibility of a gauge in the vehicle, but not for the entire look or feel of the game or in-game story.

Gaming usability can use both of these methods efficiently. The objective is to select the correct method of testing needed for the game being tested. To test overall usability and for a large scale RPG like *Civilization IV* it would be better to use the traditional method because making adjustments to a game that large could be a hefty undertaking. The company who developed the game *Civilization IV* would also more than likely have the resources to assume the financial requirements needed for traditional usability studies. However, if the researcher wanted to test a flash game attempting to teach physics, the RITE method may be a better method. The written code for flash is generally much more manageable than engines that would drive a game like *Civilization IV*. The ability to make minor or even extreme changes in a flash game could take mere hours instead of days. The RITE method may still be used for large scaling gaming, but it is better to focus the testing on certain aspects of the game. *Civilization IV* uses a mapping system known as a God Map, which means the user can travel to any location in the game by opening the map. The RITE method would be perfect to define the correct layout of this map in the game.

USABILITY MEASURES

Several years ago, a *WoW* player became an internet sensation. Famous not for his abilities, skills, or achievements, this player achieved infamy and prompted a gamer catch phrase for charging into battle unprepared without considering the consequences. After becoming impatient with a group planning session, the player shouted his name and rushed into battle. Since then, Leeroy Jenkins has become a household name among gamers and a classic example of what not to do.

The chaos resulting from his actions as the group tried to follow their original battle plan, then tried to save Leeroy, doomed them to failure. Finally, strategy broken and with the majority of team members dead or seriously wounded, the team was overwhelmed by their enemies and Leeroy had gone down in history as an example of what occurs when planning and tactics are thrown away in a rush to achieve a goal (Lowood, 2006). This parable is especially applicable to usability testing. Often, developers will rush into testing their products without a plan of action or careful analysis of their methodology. The stress of trying to meet production deadlines on something as complicated as a game, makes shortcuts especially enticing. When shortcuts are taken, it is frequently in the area of usability evaluation. However, taking the time to plan out an analysis conscientiously ultimately saves time and resources.

Usability testing is able to save developers time and effort by allowing them to tactically plan and test game features and goals (Pinelle, Wong, & Stach, 2008). This is even more important when designing serious games. Developers of serious games must find a balance between the learning objectives they are trying to convey and keeping the players interested and enjoying themselves. Because usability measures can test both the subjective ("did adding a new feature really make the game more fun?") and objective ("did the feature improve performance or learning?") aspects of the game it can quickly show if is worthwhile to implement game-wide changes (Hornbaek & Lai-Chong Law, 2007).

Being able to select the correct usability measure by evaluating if it is appropriate to use and that the measure will be sensitive enough to detect changes can be a challenging task. Developers must select from measures that cover a wide range of options based on the type of games they are designing. These are often designed specifically for a single system rather than for a particular game. This means that the measures rarely "fit" the game tested, but instead tap into an underlying component (e.g., realism of a vibration feature) that is present in the game but that may have been significantly modified from the original implementation of the component. Many measures that are available to use were created or modified by people without research or psychometric training in order to tailor them to the system they want to test. At first glance, it may seem that these new measures are perfect for testing the game, however items on these measures may have low validity and reliability, seem ambiguous or confusing to the user, or not be designed to target important features in the game.

One way to try and avoid these problems is to use an empirically created and tested measure; unfortunately, there are no standard usability measures that will work across all games, or even genres of games (Hornbaek & Lai-Chong Law, 2007). A thorough review of available usability measures demonstrates that even the experts have different definitions of usability and there is an active debate over which variables are most important for each implementation. This overabundance of options, and limited guidance as to what to look for, makes choosing the correct type of usability measure a search for a needle in a haystack (Winter, Wagner, & Deissenboeck, 2008). However, there are ways of making this haystack a little smaller. For example, it is best to use a general, reliable, and easily implemented measure for initial testing. A good example of this style of test is the *Systems Usability Scale* (*SUS*). The *SUS* is notable for being a particularly fast and easy measure of usability with a long track record of strong usability prediction. Initially created in 1986 by John Brooke, the *SUS* was designed to be a *"broad general measure which can be used to compare usability across a range of contexts... a scale that can be used for global assessments of systems usability"* (Brooke, 1986). Brooke's goal was to design a measure that was quick and reliable, but sensitive enough to show contrasts between versions of a product. Since its creation, the *SUS* has been used mainly to evaluate electronic office systems, specifically computer hardware and software. Recent new applications have included evaluations of complex user interfaces such as comparisons of mobile phone interfaces (Jokela et al., 2005), the e-government *Advanced Travel Information Systems* (Horan, Abhichandani, & Rayalu, 2006), and a multimedia training system for deception detection (Jinwei, Crews, Nunamaker, Burgoon & Ming, 2004).

We can learn a lot from looking at previous applications of a usability test. Even though the cases mentioned above did not involve a game, because the *SUS* was used to look at a different facet of the user's interaction with a product they provide a wealth of lessons for how to correctly implement it in other situations. One of these lessons learned is that it is necessary to identify exactly what the measure was initially designed to test and compare that with what the measure will be testing in the future. For example, in the comparisons and product developments mentioned above, the *SUS* was geared towards development of a "system" rather than a game. If, however, we wish to modify the measure to test a serious game, we will need to modify the terminology and phrasing used. When modifying the measure, be aware that the changes do not change the meaning of the question. For instance, the first question on the *SUS* is "I think that I would like to use this system frequently." Changing "system" to "game" would be a common way of clarifying the question though the phrasing may seem unnatural. The important point to note here is that the change in the wording does not change the underlying meaning. This is most easily achieved by consistently making the wording change for each question and making as few changes to the questions as possible. It may be tempting to change the item to something that sounds more natural (e.g., "I would enjoy playing this game), but even these few seemingly minor changes may have major impacts on how the user responds. This question may not mean the same thing to the player and depending on how the items are grouped the question's response may no longer be grouped in the same category and might be scored slightly differently.

Another consideration when adapting the *SUS* to a serious game is the creator's goals when developing the items. Originally, the *SUS* was tested to ensure that each item was easily understood by respondents with a wide range of education, age, and intelligence. Also, the *SUS* should typically be given to the end-users of the product in a consistent manner. It is designed to be given to the testers immediately after they have viewed or used each system. This testing should be easily completed in less than 5 minutes. It is precisely because of how quickly this measure can be used, and that it returns a single score, that the *SUS* has become so popular with many companies. This straightforward utilization successfully provides a quantitative measure of which version of the system was preferred by the user. Consequently, it provides an excellent global assessment of the product. However, because the *SUS* cannot be broken down by question it is limited in the amount of fine-grained feedback that can be interpreted from the results. Many times, users will tell usability testers that the game or system is confusing or hard to learn, and while these are serious issues, they provide nothing specific that the game developers can fix.

The *SUS* is best applied to detecting the result of significant changes to the game rather than several changes because of the single score result. For this reason, the *SUS* is the recommended measurement method for simple games (e.g., first person shooter games) where there player is attempting to achieve a single goal at a time (e.g., one major quest line or objective) through a relatively standard method of completing the objective or winning the game (e.g., total number of points or "kills"). The *SUS* is ideal for this style of linear game narrative because the more rigid gaming environment makes testing the effects of one change easier interpret than possible in "branching" styles of game narrative. This does not preclude the use of the *SUS* for more complex games like *Civilization IV* or

WoW. It is possible to extract meaningful data from the *SUS* for multi-faceted games, it is just necessary to do so cautiously and with rigorous pre-planning in order to control the possibility of unintended effects. In this way, the *SUS* can be applied to a complex RPG such as *Civilization IV*. Usability analysts or developers can evaluate if *a* limited number of game options or features liked equally by the gamer. For example, the *SUS* would be very useful to determine which of the five victory conditions were preferred and if players felt one was more achievable or significantly harder than the others. In *Civilization IV,* for example, a research-based victory involves little to no interaction with other nations. Instead, it requires a considerable number of turns to develop the essential technology. This victory condition is, in essence, an option of "running out the clock" so that games do not continue interminably. A potential application of the *SUS* for this condition would be to determine if modifying the number of technologies required or adding other special features (e.g., research bonuses or adjusted timing for different chronological epochs) would significantly increase the gamers' enjoyment and the game's difficulty. This is especially the case if we wanted to develop *Civilization IV* as a serious game, because the ideal design would ensure that each victory conditions is as enjoyable, challenging, and interesting as the next. By keeping the various conditions equally challenging and rewarding, the gamer will be equally willing to choose each of the different methods to win, thus insuring a balanced lesson. In this way, the gamer will be unconsciously learning important historical facts, the basics of managing an economy, political maneuvering, the ecological impact of technological development, or military costs without an undue preference being given for one lesson or another.

It is not always feasible for such simplistic research designs to be employed. Instead, because they want to be able to target several problems or determine if a new fix is effective, game developers and researchers are often interested in more complex issues that cannot be addressed by the *SUS*. In these cases, a second measure can be administered in order to get more detailed information. These measures typically include many more questions and require more complex scoring and analysis to use and interpret properly. Of these measures, the *Questionnaire for User Interface Satisfaction* (*QUIS*) was specifically designed to determine the user's satisfaction with computer software and interfaces. Since its original conception, the *QUIS* has been expanded with additional questions and sections and further refined with new validation in order to capture the new technology and features being utilized by game developers. Currently the *QUIS* is comprised of 12 parts containing 125 Likert scale questions, 12 open-ended response questions, and several questions pertaining to the users' past experience with the current or similar systems. Giving the users the entire *QUIS* would be time consuming, possibly taking thirty to forty minutes to complete. This time requirement is unlikely to be feasible in most cases, especially if users are testing several games or versions. Instead of adding several hours of testing time per session, inapplicable or redundant sections can be removed depending on the types of features being tested in the game because the *QUIS* is broken down into sections and subsections. It is easy significantly shorten the testing time by removing large topics or subsections that are not of interest. This gives usability testers a greater flexibility to target specific things that might impact the gamers play. This recommendation does not come without caveat, however, as removing sections, or changing the order of sections, may have an effect on the results. Therefore, sections should only be removed after careful consideration of the effects removal may have on the results of the remaining items. Though practically, these effects may be negligible, any changes such as removing sections or, as previously discussed, rephrasing items must be noted and any changes made deliberately. In summary, when analyzing more simplistic games, usability

Table 1. Comparison of two usability measures for 10 major usability heuristics

		Measure	
Usability Heuristic	Definition	SUS	QUIS
Visibility of system status	Keep users informed about what is going on, through appropriate feedback within reasonable time.		x
Match between system and the real world	Words, phrases, and concepts are familiar to the user.	x	
User control and freedom	System navigation is clear and allows users to undo and redo.		x
Consistency and standards	Words, situations, or actions are consistent and mean the same thing.	x	x
Error prevention	Attempt to eliminate error-prone conditions or present users with a confirmation option before they continue.	x	
Recognition rather than recall	Minimize the user's memory load by making objects, actions, and options visible. Instructions & dialogue should be visible or easily accessible.		
Flexibility and efficiency of use	System can cater to both inexperienced and experienced users. Allow users to tailor frequent actions.	x	
Aesthetic and minimalist design	Dialogues should clear and not contain information which is irrelevant or rarely needed.	x	
Recognize, diagnose, & recover from errors	Error messages should be expressed in plain language, precisely indicates the problem, and constructively suggests a solution.		x
Help and documentation	Provide help and documentation as needed. Any such information should be easy to search, focused on the user's task, list concrete steps to be carried out.		x

Note: Usability heuristics & definitions were adapted from Neilson (2005)

analysts should use less complex questionnaires like the *SUS* because when the game is too simple very few questions and subsections categorized by the *QUIS* will apply. In these cases a single overall usability score is generally more beneficial. However, when testing higher level games more complex results and scores are needed to identify important outcomes and effects.

If any general principle could be advanced regarding usability testing, it would be that it is best to make decisions purposefully, methodically and consistently. Though usability testing does not involve test-tubes or beakers, applying the same scientific reasoning and systematic approaches will only strengthen the validity of the analysis. It is important to select the correct measure in the development of a usability test and carefully consider the strengths and weaknesses before implementation of the testing agenda. Different measures have the capability of defining separate aspects of the developed game and target distinct characteristics. Many of the important features of gaming can be defined through heuristics created by Neilson (2005). Neilson's top ten usability heuristics are often thought of as basic guidelines that should be followed when designing a system and are particularly appropriate when applied to software interfaces. Many usability measures were originally created to test these heuristics and identify areas that needed attention or revisions in the system being tested. For example, the list below is Neilson's ten heuristics followed by a simple definition of each (see Table 1). How each of these heuristics is represented by the *SUS* and *QUIS* is also noted.

A few things need to be highlighted regarding the *SUS* and *QUIS*, and how they interact with Neilson's heuristics. First, the *SUS* and *QUIS* are only examples of the many options currently available to test usability in games. Though these measures are examples the most validated and frequently used usability measures currently available. Second, despite their popularly and history, neither of these measures have the ability to accurately test everything a usability analyst may be interested in, though they can be combined to cover many different dimensions of usability if implemented through a well structured test methodology. Finally, it is important to remember that often the described pitfalls of usability are not due to its concepts but to a lack of strong testing methods and measures. And these hazards, once aware of them, are particularly responsive to conscientious testing, design, and procedure.

DISCUSSION

Once again, you are playing a simple driving game. However, since the last time, the game has evolved into a serious game to practice simple mental arithmetic. The car you have been operating has been labeled with a number, and your goal is to drive through walls that are labeled with matching mathematical equations. On your first trip out, you are given car number 4 and successfully drive through the matching equations (e.g., "2 + 2" and "9 – 7", etc.) while dodging the other equations (e.g., "3 × 5" and "1 + 7, etc.). Unfortunately, just as the game is starting to heat up, you make an error in your order of operations and crash through the barrier labeled "6 + 2 ÷ 2". Game over! This version of the game has advanced from the previous versions in a number of ways. Not only is more presented to the gamer on the screen, success has been redefined to include problem solving and more complex forms of decision making.

But, does it work? By looking at the previous versions of our driving game, we were able to deduce reaction time (in the first version) and basic decision making (in the second version). Now, we are attempting to determine if the newest version achieves our goal of providing effective practice of mental math. Once again, a usability analyst can answer this question through testing. By examining user performance and preference, and how it changes between the separate versions, analysts can determine the effect of more elaborate aesthetics, different interfaces, and more difficult tasks. For example, if the gamer is focusing on the vivid colors and the foliage lining the road the pertinent information required for learning may become obscured by aesthetic value. On the other hand, slower reaction times may be due to the addition of more complicated interfaces, requiring the user to make more decisions to control the car, for example, which button is required to move from one side of the road to the other. The additional components that the developer has added to the game may detract from the game's educational value. Careful analysis can discover which aspects are important to the game now and if performance is accurately reflecting the player's aptitude towards math. By using the methods previously discussed, usability analysts can extract what information is extraneous while still leaving the entertainment value needed to keep gamers coming back for more.

This example demonstrates how much of a place in the development cycle of serious games usability holds. As games become more complex and are implemented in more diverse situations, the applicability of usability testing for new technologies, interfaces, and purposes becomes more apparent. This is especially the case for serious games that attempt to embed an educational purpose in an engaging framework. Usability analysis that targets these aspects provides a unique opportunity to ensure that games are meeting their supposed purpose in a way users enjoy and prefer while minimizing the amount of resources and time developers must devote to near-endless iterations of the game. Meanwhile, not only can usability enhance the development of current versions, lessons we learn from usability analysis can be applied to future iterations, making the development process increasingly efficient and effective.

Games like *Sid Meirer's Civilization IV, WoW,* and *Need for Speed* do have value in the world of serious games, but defining their educational potential will be difficult. We are not engaged in a Sisyphean challenge; through the correct use of usability methods and measures gaming's value to education can eventually be defined. As we see with *Civilization IV*, and it's potential to demonstrate important historical lessons or *Need for Speed's* application to hand eye coordination and visual attention, these games and others can be quantitatively linked to important aspects of education. Further development of more complex usability measurements and systematic application of them to vivid game environments, such as *WoW* and *Civilization IV*, will lead to the expansion of new test subsections. These improved testing methods can then, in turn, be used to obtain information critical to the definition of important educational game models allowing refinement and prediction. More research and progress in this area may not see an immediate benefit, but as more evidence is generated, synthesized and shared, our momentum will result in clear benefits to serious games and assist their integration into accepted educational techniques.

REFERENCES

Brooke, J. (1986). *SUS- A quick and dirty usability scale*. Earley, UK: Readhatch.

Cornett, S. (2004). The usability of massively multiplayer online roleplaying games: Designing of new users. *Computer Human Interaction*, 703-710.

Desurvire, H., & Wiberg, C. (2008). Evaluating User Experience and Other Lies in Evaluating. *Computer Human Interaction Conference*, Florence, Italy.

Desurvire, H. W. (2008). Evaluating User Experience and Other Lies in Evaluating Games. *Computer Human Interaction Conference*, (p. 6). Florence, Italy.

Donyaee, M., Seffah, A., & Rilling, J. (2007). Benchmarking Usability of Early Designs Using Predictive Metrics. *IEEE*, (pp. 2514-2519).

Holzinger, A., & Searle, G. (2008). Investigating Usability Metrics for the Design and Development of Applications for the Elderly. In *11th International Conference on Computers Helping People with Special Needs* (pp. 98-105). Linz, Austria: Springer.

Hornbaek, K., & Lai-Chong Law, E. (2008). Meta-Analysis of Correlations Among Usability Measures. *Computer Human Interaction* (pp. 617-626). San Jose, CA: ACM.

IEAA. (2008). Retrieved Feb 12, 2008, from The Interavtive Entertainmnt Association of Australia http://www.ieaa.com.au/

Invetstment-U. (2008). Retrieved April 25, 2008, from Investment U http://www.investmentu.com/

ISFE. (2005, June). Retrieved April 20, 2008, from Interactive Software Federation of Europe http://www.isfe-eu.org/

Krzywinska, T. (2006). Blood Scythes, Festivals, Quests, and Backstories: World Creation and Rhetorics of Myth in World of Warcraft. *Games and Culture*, 383–396. doi:10.1177/1555412006292618

Lowood, H. (2006). Storyline, Dance/Music, or PvP?: Game movies and community players in World of Warcraft. *Games and Culture*, 362–384. doi:10.1177/1555412006292617

Papaloukas, S., & Xenos, M. (2008). Usability and education of games through combined assessment methods. *PETRA*.

Pinelle, D., Wong, N., & Stach, T. (2008). Heuristic Evaluation for Games: Usability Principles for Video Game Design. *Computer Human Interaction Proceedings* (pp. 1453-1462). Florence, Italy: ACM.

Winter, S., Wagner, S., & Deissenboeck, F. (2008). A Comprehensive Model of Usability. *IFIP International Federation for Information Processing Proceedings* (pp. 106-122). EIS.

Chapter 4
Development of Game-Based Training Systems:
Lessons Learned in an Inter-Disciplinary Field in the Making

Talib Hussain
BBN Technologies, USA

Wallace Feurzeig
BBN Technologies, USA

Jan Cannon-Bowers
University of Central Florida, USA

Susan Coleman
Intelligent Decision Systems, Inc., USA

Alan Koenig
National Center for Research on Evaluation, Standards and Student Testing (CRESST), USA

John Lee
National Center for Research on Evaluation, Standards and Student Testing (CRESST), USA

Ellen Menaker
Intelligent Decision Systems, Inc., USA

Kerry Moffitt
BBN Technologies, USA

Curtiss Murphy
Alion Science and Technology, AMSTO Operation, USA

Kelly Pounds
i.d.e.a.s. Learning, USA

Bruce Roberts
BBN Technologies, USA

Jason Seip
Firewater Games LLC, USA

Vance Souders
Firewater Games LLC, USA

Richard Wainess
National Center for Research on Evaluation, Standards and Student Testing (CRESST), USA

ABSTRACT

Modern computer gaming technology offers a rich potential as a platform for the creation of compelling immersive training systems, and there have been a number of game-based training systems developed in recent years. However, the field is still in its infancy. Improved understanding is needed on how to best embed instruction in a game and how to best use gaming features to support different types of instruc-

DOI: 10.4018/978-1-61520-739-8.ch004

tion. Further, the field is inherently inter-disciplinary, requiring instructional system designers, software developers, game designers and more, yet there are no established development methodologies to ensure effective coordination and integration across these disciplines. The authors introduce a collaborative effort that is investigating how to improve the craft and science of game-based training. They present their experiences in creating a flooding control training system for the U.S. Navy Recruit Training Command, and discuss the inter-disciplinary development issues that they encountered. They present the lessons they learned and their views on how to advance current methods to support the consistent production of effective game-based training.

INTRODUCTION

Computer games of various kinds have been used for education and training purposes for over two decades with varying degrees of success (O'Neil et al., 2005; O'Neil & Perez, 2008). As computer gaming technology has matured and increased in capability, the opportunities available for delivering immersive learning experiences have increased (Bonk & Dennen, 2005; Hill et al., 2006; Hussain et al., 2008; Johnson et al., 2007; Roberts et al., 2006), and so has the challenge of creating experiences that are pedagogically effective (Diller et al., 2004; Hussain & Ferguson, 2005). A training game is imbued with a purpose - to provide experiences which lead to specifics gains in the student's knowledge and/or skills. A good training game will consistently balance the instructional goals with the goal of motivating the player. However, a poorly designed training game will sacrifice one or more fundamental elements of the gaming experience in order to attempt to satisfy the training goals, or will sacrifice effective pedagogy in order to attempt to keep the game compelling. The former may be a great training system, and even a great simulation-based training system, but doesn't pass muster as a game-based training system since the players don't enjoy it. The latter may be a great game, but doesn't pass muster as a training system since it does not produce the desired learning outcomes. Developers of game-based training systems know this, but achieving this synergy between instruction and engagement is a poorly understood art.

The challenges facing us as a discipline are:

1. An enhanced understanding of the elements of game design and pedagogical design that are crucial to game-based training and how to balance those elements effectively,
2. An enhanced understanding of how to assess the success of a game-based training application, and
3. The creation of development methodologies that lead to repeatable successes, especially for non-commercial training programs that are limited in the scale of effort that can be supported.

We introduce the initial results of a multi-disciplinary effort sponsored by the Office of Naval Research to directly address the issue of how to best create effective educational immersive computer games. The team for our project included researchers and content developers from the fields of instructional design, story-based training and entertainment, movie production, human performance assessment, game engines, commercial games, game-based training systems, simulation and modeling, intelligent tutoring systems, graphic design and modeling, system integration and educational science. As professionals in our respective fields, we each brought different perspectives on the interactions of different aspects of gaming and pedagogy to the table.

Our effort had two mandates. The first was to conduct applied and empirical research on tools and methods for enhancing the art and science of educational games. In particular, our initial focus was on identifying extensible design and development methods that support efficient creation of training games with strong pedagogy and embedded assessment. The second mandate was to create prototype training systems for the Navy that met real training needs and embodied sound pedagogy. The customer for our first training prototype was the U.S. Naval Service Training Command. The goal of this prototype effort was to provide training on how to perform effective flooding control aboard a simulated naval ship. The Flooding Control Trainer that we developed is intended for use at the Navy's boot camp to augment the classroom-based and hands-on instruction that is currently provided to over 40,000 recruits per year.

In a seven month period, our team members worked closely together, with subject matter experts and with our Navy customer to produce a prototype that met our customer's needs and was effective. We present here a description of the process that we followed, the tensions we encountered during the effort, and the lessons we learned on how to work in an interdisciplinary manner to achieve an instructionally strong and enjoyable outcome. We discuss our next steps in the project to refine and formalize our process, as well as our thoughts on where the field needs to focus going forward to ensure longevity and success as a discipline.

BACKGROUND

From a theoretical perspective, games hold promise as effective teaching devices because they can provide instructional environments that embody key principles derived from learning science. For instance:

- Interactions that facilitate player engagement in a *compelling task environment* should facilitate learning. Practicing in this type of environment is consistent with notions about the development of expertise (Bransford et al.,1999; Chi et al., 1988; Glaser, 1989) and anchored instruction (e.g., Bransford et al., 1990; CGTV, 1992; CGTV, 1997; CGTV, 2000). The game world provides a context in which appropriate mental models are developed and refined through repeated exposure to important cue patterns.
- Games provide an excellent *model-based world* to foster reasoning. Students are able to manipulate variables, view phenomena from multiple perspectives, observe system behavior over time, draw and test hypotheses and compare their mental models with representations in the external word. These features are consistent with the model-based reasoning concepts advocated by learning researchers (Cartier & Stewart, 2000; Gentner, 1983; Leher & Schauble, 2000; Raghavan et al., 1997; Raghavan et al., 1998; Stewart et al., 2005; Zimmerman et al., 2003).
- Game-based tasks are expressly designed to help players *progress toward goals*. Moreover, the goals are concrete, specific, and timely. The vast literature on goal setting in instruction suggests that this characteristic property of games should enhance learning (Locke et al., 1981; Locke & Latham, 1990; Schunk & Ertmer, 1999).
- Interaction frequency is very high in games. These often require decisions and inputs by players several times a minute. Thus, games provide a *highly active learning* environment, the kind of environment associated with effective instructional system

design (Chi, 2000; Mayer, 2001; Rothman, 1999; Vogel et al., in press).
- Well-designed games provide the player with constant successes. Many small tasks are embodied along the way in the pursuit of a greater goal. The result is that the game generates a feeling of constant accomplishment, a feature likely to *enhance self-efficacy*, which has been consistently shown to improve learning and motivation (Bandura, 1977; 1986; Gist et al., 1989; Gist et al., 1991).
- Games provide a *continuous source of feedback* so that players know where they stand with respect to their goal accomplishment. This is crucial since feedback improves learning through both its informational and motivational qualities (Bransford et al., 1999; Salas & Cannon-Bowers, 2000).
- Game play tends to be *self-regulating*, an important feature of effective learning. Players are motivated to accomplish the next challenge and they know where they stand relative to their goals (Kanfer & Ackerman, 1989; Kanfer & McCombs, 2000; Pintrich & Zusho, 2000; Schunk & Ertmer, 1999; Schunk & Zimmerman, 2003).
- *Engagement and immersion* are high in well-designed games. Literature is beginning to investigate these concepts as psychological states associated with effective performance and learning (Csikszentmihalyi, 1990; Gerhard et al., 2004) and to examine what contributes to them (Baylor, 2001; Gerhard et al., 2004; Moreno & Mayer, 2004).
- *Challenge and competition* are hallmarks of good games. These features have been found to be motivational under certain conditions (Epstein & Harackiewicz, 1992; Reeve & Deci, 1996), and may be useful to motivate learning.
- Perhaps due to factors listed above, *time on task is very high* for games. It is not uncommon for players to spend hours a day on engaging games, and to continue interacting with a single game for years. From a strictly time-on-task perspective, we would expect that learning would be enhanced when trainees engage quality learning content for longer periods of time.

Despite this promise, however, many game-based training applications suffer key deficiencies leading to poor learning outcomes, such a poor instructional design, poor (or no) performance assessment, limited training scope, low levels of reusability, and lack of appeal to students (Hussain et al., 2008; O'Neil & Perez, 2008). These problems are due in part to the limitations of current gaming technology - for example, commercial games typically do not provide the ability to capture the relevant performance data at the right level of detail for tracking and assessment of trainee performance. However, they are also due in large part to the fact that there are no clearly proven methods for the development of effective training games. Further, most organizations and companies developing game-based training systems have only a few years of experience doing so and their experience is usually limited to work on one or two systems. Generally, designers and developers apply methods that have been appropriated from other fields, such as general software development, simulation-based training, computer-based training, commercial game development, and intelligent tutoring systems. The lessons learned from these ad-hoc approaches to game-based training development are rarely shared. Hence, insights are lost, pitfalls are re-encountered, and useful community-wide guidelines are not formed. Further, different development teams contain expertise in different fields, and hence some elements of game-based training design tend to be emphasized at the cost of others.

Our project team was a highly multi-disciplinary set of experts, each with varying degrees of expertise with game-based training, but all with deep knowledge within their core disciplines. The development team included members from seven different organizations that together provided expertise in game-based training, educational science, commercial game development, instructional design, human performance assessment, story-based training, simulation-based training, computer-based training, military training systems, graphic design and modeling, animated movie development, entertainment and media production, intelligent tutoring systems, game engine development, agile software development methods, and systems integration.

Collectively, these experts brought to bear a rich subset of the practices currently used for developing game-based training. The key practices that we chose to start with based on preliminary discussions included:

- The manner in which the system will be used impacts instructional design choices, so pay close attention to requirements gathering.
- Remain focused upon learning objectives throughout system design and development.
- Develop the story outline early on, base it on the learning objectives, and iterate as system design proceeds. Taken to an extreme, one of the teammates believed that "It all starts with story." According to Jerome Bruner (1990), plots are a creation of "transactional relationships" between reality, memory, and imaginary/narrative worlds. Transactional connections help learners use what they know in order to contextualize what is unknown, meaning that since the human brain needs story to provide context. An effective story provides a basis for addressing requirements and framing all content development in order to produce a consistent and compelling product.
- Incorporate assessment needs as early as possible during system design and development. Post-development efforts to graft assessment on a system not designed for it lead to significant problems in capturing the type, quality, and quantity of data required for effective assessment.
- Agile development methods work very effectively in fast-paced game-based training development efforts.
- Early involvement of all stakeholders in design and ongoing involvement during design and development iterations leads to a more robust product and helps avoid significant pitfalls.
- Keep the customer involved throughout to ensure the product meets stated and unstated requirements (and adapts to meet changing requirements).

DEVELOPMENT PROCESS

In March 2008, our team began the development of a game-based training prototype. Over the course of seven months, we followed an agile development methodology to iteratively create and refine our instructional design, and develop and refine our initial product prototype, which we delivered successfully at the end of September. During our effort, we encountered a number of issues, many of them due to tensions caused by differing perspectives of the project stakeholders based on their backgrounds and relative priorities. All key issues were resolved, practically-speaking, to result in a product that met our customers' needs and satisfied the immediate goals of the project. However, a number of the tensions remained as background issues and recurring discussion themes. In this section, we summarize the key tasks of the project in a roughly chronological manner, and identify the key tensions that occurred along the way.

Figure 1. Timeline of project milestones

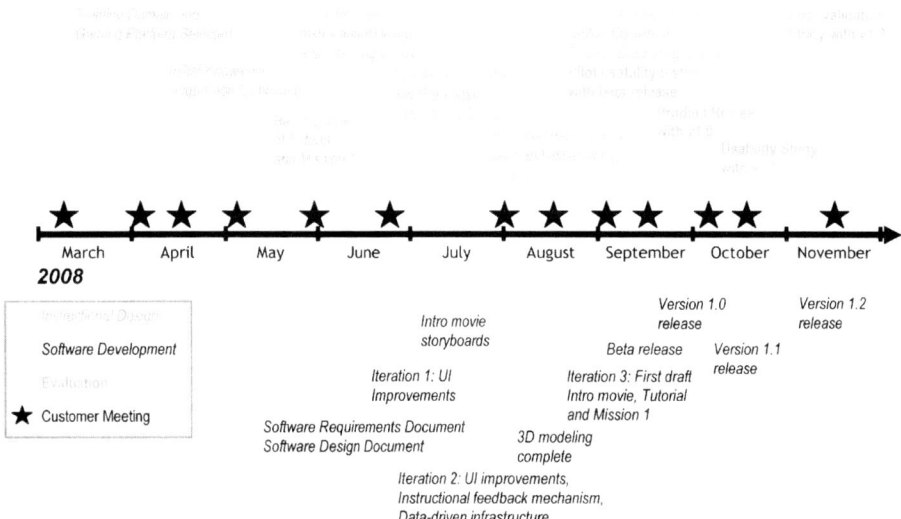

Figure 1 illustrates the project milestones that are discussed in more detail in this section. The following section expands upon these tensions and identifies the methods used to address them and/or the lessons learned from our experience.

Our project, a three-year effort that started in February 2008, had two mandates - to conduct applied and empirical research into tools and methods for enhancing the state of the art and science of educational games, and to ensure that these developments would have long-term value to the Navy. In order to achieve the latter mandate, we created a prototype training system for the Naval Service Training Command (NSTC) to enhance the Navy's Recruit Training Command (RTC) curriculum. Our NSTC customer has a background in educational science and deep knowledge regarding the training needs and culture at RTC. From RTC, we drew upon the training staff as subject matter experts (SMEs). Thus, the stakeholders of our effort included our transition customer (NSTC), our program manager (ONR), the subject matter experts and the members of our team.

To support the first mandate, we identified a need for a game-based training test bed allowing the explicit control of diverse variables in order to empirically study the impact of gaming and instructional features upon learning outcomes. This test bed capability formed an additional requirement for the initial training system we created. The focus of the paper is the process we followed in developing the training system product and our lessons learned from that effort. We describe certain interim and final elements of the product to support our discussion. However, a complete description of the final product is not given.

Selection of Training Requirements, Domain, and Gaming Platform

In developing our training product for RTC, our initial step was to identify the specific training needs. At RTC, recruits currently undergo six weeks of intensive training on fundamental navy skills and end their training with an intense exercise called Battle Stations 21 (BS21). BS21 is a real-life simulation environment in which recruits are exposed to seventeen simulation scenarios during a single night. The facility is a building designed to give recruits the experience of being aboard a ship, and contains a simulated dock, a simulated exterior of a destroyer, and several floors that simulate different decks of a ship.

The goal of our effort was to provide supplementary training to augment the current classroom-based and hands-on instruction in order to produce better prepared sailors. While overall performance of recruits in BS21 is excellent, students frequently exhibit key errors in several of the BS21 scenarios due to training gaps on those skills. Our requirement was to provide a compelling virtual training system to address some of those gaps.

At this earliest stage of product development, past experience has shown that it is critical to fully involve the customer. With any training system, the focus of the training must be driven by customer requirements, and the requirements must have a direct relationship to the learning outcomes desired. With any training application, it is important to choose a training domain that is suitable for the type of training possible with the technology used, or, alternatively, to choose the right type of training technology for the type of training desired. In our case, the customer desired an initial product delivery as soon as possible and wanted the product to provide immersive training geared toward single players. The customer desired that the training address a domain that would have high payoff in terms of improving recruit performance in BS21, but also wanted the application to provide familiarization with operating within a (virtual) ship environment. Within those constraints, we had a fair amount of leeway.

In initial discussions with our customer and with SMEs, four key training domains were identified as high benefit: controlling a flooding situation, standing a bridge watch, handling rope and navigating within a ship. Of these, it was determined by all stakeholders that handling rope was the least appropriate for an immersive gaming environment. Navigating within a ship was determined to be of secondary importance in that it could be embedded within training focused on either of the other two domains.

In these early discussions with the customer, it quickly became apparent that the high-level goals of the customer and the project could be best attained by leveraging an existing, open-source game-based simulation prototype developed for the same customer. That application was based on the game engine Delta3D and contained a high fidelity representation of the ship interior of BS21 with a first-person perspective (see Figure 2). The drawback to the application was that it had minimal pedagogical infrastructure and minimal gaming elements. The advantage of the application was that it would avoid the need for an intensive graphics development process, and, since it was open-source, would allow us to add the pedagogical elements we needed. One of our team members had been the software developer for that application, which also provided us with immediate expertise. From the customer's perspective, reusing the application would justify earlier investments. Further, the customer already knew its strengths and weaknesses and was able to give us specific feedback on what he wanted improved for our product. The earlier application did not, however, contain virtual characters. Preliminary task analysis determined that it would be pedagogically appropriate to provide flooding control training without virtual characters since those skills can be performed by a single individual alone in real-life situations. Bridge watch, in contrast, inherently involves a team of people and thus the application would require augmentation with simulated characters to provide effective training. The bridge watch domain was deemed too risky to meet our aggressive development schedule.

The final choice of training domain for our initial product - flooding control - was made in early April for practical reasons based on the nominal choice of training platform and the customer's view of which domain would have higher product acceptance at RTC. The process of making this decision revealed one of the first key tensions in the project:

Figure 2. Available 3D simulation environment of BS-21 interior

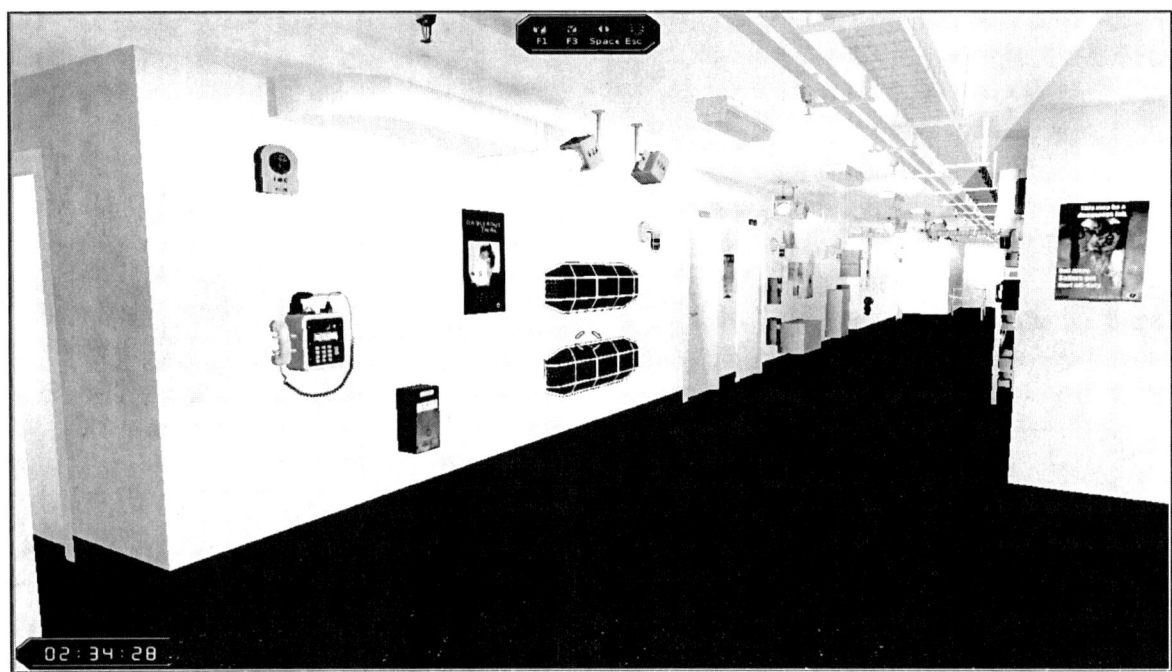

- The basis for technology decisions – pedagogical or technological

Knowledge Acquisition

Once the choice of the flooding control domain was made, we undertook the task of knowledge acquisition for the flooding control domain. The knowledge acquisition methods we employed included a traditional, formal cognitive task analysis (CTA) as well as a subject matter analysis focused upon the potential elements for flooding control game scenarios, termed here a scenario element analysis. The analyses were based upon reviews of training materials in use at RTC, observations of recruits performing the flooding control scenario (and others) in BS21, and discussions with SMEs. In mid-April, a SME session was conducted face-to-face with six trainers from RTC. During this session, the questioning was led by our instructional designers, but representatives of most team members were also present and contributed to the discussions. This session, as well as a couple of follow-up face-to-face and teleconference sessions, provided two initial products.

The first was a traditional breakdown of learning objectives and the identification of expected behaviors, common errors, and consequences. The CTA determined that there were four key categories of learning objectives—those related to communication, situation awareness, appropriate decision-making, and leadership. The communication learning objectives, for example, included reporting a situation immediately, communicating relevant information, and reporting when an action is completed. Decision-making objectives included maintaining watertight conditions, requesting permission before taking action as appropriate, and performing proper containment procedures. In addition, the CTA determined that there were certain skills that the recruits were expected to know based on the current curriculum, certain skills that it would be desirable for them to know if training could be enhanced, and certain skills that were very important in real-life, but that were beyond what could be expected of a normal recruit.

Figure 3. Learning objectives and common errors in context of flooding mission phases and roles

Learning Objective	Expected behaviors (enabling objectives)	Discovery Phase	Preparation Phase	Transit Phase	Casualty Combat Phase	Completion Phase
Communication - Demonstrate they know (a) whom to contact and (b) what to say and (c) how to listen and repeat back							
Report situation immediately	Use best available means to communicate Report hazards upon discovery Report new factors in situation immediately	**Role**: First on scene **Specific objective:** Notify someone of the flooding situation immediately **Expected behavior:** Report flooding immediately upon discovery. Use best available means to communicate (phone, yell, send messenger) **Common errors:** Do not communicate. Ignore problem, isolate location without permission, combat casualty on own. **Consequences:** No one else knows. Leak/flood progresses quickly and DC has less time to get there. **Scenario variations:** Individual injured or dies while combating. Individual electrocuted due to unperceived electrical hazard. Ship capability reduced. Ship sinks.		**Role**: DC team member **Specific objective:** Report any hazards en route to flooding area. **Expected behavior:** Report hazards upon discovery **Scenario variations:** Additional leaks, additional flooding locations, unsecured items in hallway, injured personnel, watertight doors not closed appropriately	**Role**: DC team member, DC team leader **Specific objective:** Keep DCC informed of any changes in situation. **Expected behavior:** Report new factors in situation immediately **Scenario variations:** Additional leaks, additional flooding locations	

The second product was a breakdown of the elements of the flooding control mission that suggested the potential structure, actions, and variations for a simulated flooding control scenario. The scenario element analysis determined the typical timeline involved in a flooding control mission and the specific elements of the different phases of the mission that suggested the potential actions and variations to be included in a simulated scenario. These phases were broadly captured as: the discovery phase (the actions surrounding the event that identifies the need for flooding control), the preparation phase (actions in getting assigned and ready to combat the flood), the transit phase (actions taken en-route to performing flooding control), the casualty combat phase (the procedures and options available while trying to control the flood), and the completion phase (what actions are performed at the end of the situation). These phases summarized the SMEs' views of the typical way in which a flooding mission occurs and what they felt needed to be reinforced with the recruits. Within each phase, a number of activities, possible actions and scene variations were identified. For instance, in the transit phase, additional hazards could be encountered, such as additional leaks, additional flooding locations, unsecured items in hallway, injured personnel and watertight doors not closed appropriately. During different phases, inappropriate setting of a watertight boundary could lead to various complications, including allowing a flood to spread or trapping a shipmate. In addition to multiple phases of a mission, the scenario element analysis identified multiple roles involved in different phases of a flooding control mission, including a first-on-scene sailor, a damage control (DC) team leader or member, and damage control central (DCC). The SMEs indicated that the recruits should be taught certain skills associated with certain roles, even if they would typically not be in those roles for some time.

The products of the CTA and scenario element analysis were then merged to form a third product that mapped the learning objectives in context across the phases and roles (e.g., Figure 3 shows one of 26 learning objectives mapped). This product was prepared by May 1 and provided inputs to our story development process.

During the knowledge acquisition process, several tensions arose, including:

- Identifying all the learning objectives versus selecting the objectives to be addressed in the first game prototype
- Balancing cognitive elements with experiential elements

Story Development

Great stories share at least six elements that set them apart from "scenarios" or "case studies". They are: setting, characters, conflict/resolution, plot, voice, and emotion. Compelling stories, and those experienced in game environments in particular, are often called "immersive." However, immersion doesn't just happen; it takes purposeful effort on the part of the storymakers or designers. To achieve a great story, one must be intentional about combining the six elements in a way that will draw the audience (in the case of a game, the players) into the story so that they can see themselves as part of it. To do this, the story must be easy for them to relate to. To be successful as a learning environment, it must also create an "envelope" that will carry the learning tasks so that the learner literally "embodies" them.

Generally, fodder for narrative development in a learning product can be gathered in early conversations with the customer. In our effort, during the knowledge acquisition process, we determined through discussions that the basic objective of our initial product would be to teach recruits the steps required to stop (or mitigate the flow of) a leak on board a ship. We were thus able to begin to zero in on some of the key story elements from the beginning. For instance, from these early discussions, we knew the following about the eventual story:

- The story *setting* would be on (some type of) a ship.
- At least one of the *characters* would need to be a sailor on board that ship.
- The *conflict* to *resolve* would be a leaky pipe or breach in the ship's hull that the sailor would need to fix or at least mitigate the effects of.
- The *plot* would need to include some sort of potential flooding scenario.
- We believed that the *emotion* of fear could be used to engage the learner as a crisis ensued. We also believed that the Navy values of "Honor, Courage, and Commitment" that were instilled in recruits during basic training could be used to provide motivation to overcome the fear and lack of initial skills, competence, and confidence to perform the mission.
- Voice, or who the "teller" would be, was not established until later.

On May 6-7, we held a two day story conference. We came together as a team of instructional designers, programmers, writers, and artists to begin to create the metastory for the game. Our customer participated and provided subject matter expertise as well as general guidance. The instructional designers kicked off day one by reviewing the learning objectives with the group so that we would stay focused on those throughout the day. The group was led by experienced story development facilitators through activities chosen to help us experiment and "play" with character development, setting, and plot points that would drive the learning objectives involved with a flooding scenario. It was important to develop a metastory that would be believable and that a new recruit would identify with as well as one that would set us up for the kind of conflict or crisis that could incorporate other possible disaster prevention/ recovery scenarios (e.g., fire fighting) over time.

During the discussions, our customer emphasized the need to reinforce the pattern of "look-report-do" when training recruits. This led us to focus upon the situation awareness, communication, and decision-making learning objectives, and to de-emphasize leadership. To further focus the story, our customer in particular and the team in general prioritized all the enabling learning objectives based on training requirements and perceived utility.

After the story conference, the story developers used the notes from the two-day event and developed a draft of the metastory which included several possible game levels. During the first pass at the story, we decided that the background story ("backstory") would be told in the third person by a narrator that was not a part of the story itself. This backstory eventually became the introductory movie or "intro cut scene" of the game.

During story development, a key tension was:

- Story first versus design first

Game Design

On May 29-30, we held a two-day team meeting to flesh out the specific training scenarios for our product. A representative from the customer attended, but no SMEs did. Using the learning objectives and story construct as a basis, the team brainstormed on how we wanted to provide training within the game. We made the decision to couple a guided discovery instructional strategy with an adventure-style gaming strategy. Following a common gaming approach, we decided to create a game with multiple levels of increasing difficulty. Based on the scenario element analysis, we decided to structure each level as a single mission, possibly with some "twists" to add to the sense of adventure. To support our instructional and gaming objectives, we identified a general mechanism for using dialog and messages from game characters to provide mission objectives, story and adventure elements, and instructional guidance and feedback. To avoid the need for animating the non-player characters, we designed the game in such a way that these characters would be elsewhere in the ship, out of view. Further, we incorporated the capability to provide students with opportunities to fail, possibly catastrophically, depending upon their errors.

Early on, we realized that we would need a tutorial level to train students on how to play the game. This is a standard mechanism used in commercial games to bring players up to speed. The instructional designers deemed it important that this tutorial level would introduce the basic instructional mechanisms to lay the foundation for how the game would deliver its training. The game designers emphasized that the tutorial level should reinforce the mission metaphor to introduce the gaming strategy and to set student expectations. The story developers encouraged the incremental introduction of story elements throughout all levels, including the tutorial level. We converged on a tutorial level design that met all these goals.

Using the multiple levels of expertise identified in the CTA and our earlier rankings of the relative priorities of different learning objectives, we fleshed out the basic enabling learning objectives for the initial game levels. By the second day, we had designed the outline of tutorial level and three training levels of increasing difficulty. For each level, we identified which behaviors we wanted to assess, the key story developments, potential instructional scaffolding, desired im-

Figure 4. Potential game levels produced during game design session

```
Story Development: Approaching area of operation, prepare for underway replenishment (UNREP)
Tutorial Level
    Inspection: Inspect and secure repair locker
    Objectives: Game mechanics, Communicate using phone.
    Assess: Time taken. No failure and no negative feedback. High guidance.

Story Development: Ship collides with oiler during UNREP
Mission 1 Level
    Investigate and Repair: Assess and repair leaking pipe
    Objectives: Communicate with DCC, Maintain watertight integrity
    Assess: Intent and accuracy of actions
    Feedback: Guidance primarily in response to errors. Negative feedback and potential catastrophic failure
Mission 2 Level
    Repair Complex Flood: Assess and repair multiple leaks
    Objectives: Establish boundary, Apply appropriate repair, monitor and report situation, prioritize
        decisions, dewatering
    Assess: Intent, accuracy of actions and effectiveness of decisions
    Feedback: Low guidance, multiple ways to fail, complete success possible but order of decisions matters
Mission 3 Level
    Rescue: Combat flood and ensure safety of equipment or personnel at risk
    Objectives: Maintain safety of ship. Assess and report changes
    Assess: Appropriateness of decisions
    Feedback: Low guidance, complete success not possible – hard choices necessary, multiple "passing"
        mission completion states
```

mersive elements (such as sounds, lighting and game objects and effects), and the general flow of the gameplay.

At this point we captured and prioritized all the desired instructional and gaming capabilities. We determined that the initial prototype would focus on the tutorial mission and the first training mission ("Mission 1"). This provided the basis for the software requirements and the launching point for further instructional design.

The tension revealed here was:

- Gaming strategy versus instructional strategy

Initial Instructional Design

Following our May 29-30 meeting, the instructional design process began in earnest. Our instructional designers, assessment experts, and game developers worked iteratively to identify the specific instructional elements of the proposed "Mission 1" game level based on our cognitive task analysis and our guided discovery instructional strategy.

An initial instructional design was developed by mid-June that elaborated on the learning objectives associated with each level and suggested the general instructional approaches to be used in the product. The guiding principles of this initial design were to:

- Address objectives in story, rules, consequences
- Design consequences to dispel misconceptions
- Provide feedback to promote learning
- Use scaffolding to guide players and allow them to experience actions beyond their level to reinforce learning objectives
- Use gaming technologies to facilitate the development of mental models

The instructional approach was a holistic design for providing part-task training within the context of the whole task. It specified multiple forms of feedback for alerting players to deficiencies and providing opportunities for players to demonstrate skills again in similar but not necessarily identical situations. It attempted

Table 1. Mission 1 level scenario constraints and learning objectives

Event	Environmental Requirements	Learning Objectives	Assessment Requirements
Readiness Set Readiness condition is set as a result of ship-wide "event" –General Quarters	• Alarm sound • Lighting effect • Announcement over the sound system	**Decision-Making** • Respond to readiness condition **Situation Awareness** • Report to proper location	Learner immediately leaves for general quarters assignment
Navigate Ship (recurs throughout game) Learner navigates within the ship	• Circle X, Y, & Z doors to open/close • Limit space to task area – Control where they can go. • Phone on wall in destination compartment or passageway	**Decision-Making** • Comply with restrictions on doors **Situation Awareness** • Identify location **Communication** • Ask permission before opening door	• Used phone to ask permission for opening doors • Opened/closed correctly • Path taken • Used phonetic alpha/num • Repeated back instructions • Located the proper compartment, as ordered • Reported location
Detect Flooding Situation Learner encounters flooding situation in a compartment	• Pipe w/ slow, obvious water leak • Other pipes not leaking • Identifier for compartment location	**Situation Awareness** • Conduct assessment • Recognize flooding cues • Identify location • Identify type of pipe leaking **Communication** • Ask permission before opening door, entering compartment;	• Asked permission to open door • Did not enter the compartment without permission.
Report to DCC (recurs throughout game) Learner reports to DCC	• Phone to use • Way to conduct two-way communication. Phonetic alpha/num • CCOL that can be read.	**Communication** • Communicate relevant information • Listen and repeat back instructions • Listen and acknowledge or correct repeat back **Situation Awareness** • Locate CCOL • Recognize equipment and material settings in CCOL	• Reported correct pipe/liquid type • Reported correct location • Corrected errors from DCC • Repeated back • Look at CCOL and used CCOL information correctly • Reported contents of the room
Minimize/Contain Leak Learner is ordered to repair/contain the leak	• Repair locker (to select & open) • Containment equipment (jubilee, shoring equip) to select from	**Situation Awareness** • Find repair locker **Decision-Making** • Perform proper containment procedures • Select appropriate equipment • Use proper technique for attaching patch	• Located repair locker (path taken) • Opened locker • Selected correct equipment (# of attempts, order of selection) • Placed patch and & tightened with wrench

to carefully control variations to ensure that the player did not just learn the game mechanics, but also understood the underlying facts, concepts, or principles.

The initial design also identified a general approach for assessment within the game. The types of assessment envisioned included completion of tasks, accuracy, time taken and steps taken. The intent was to assess learning objectives throughout the game by the actions the player would take. These assessments were also expected to affect play. The player's actions could be limited until he or she demonstrated a predetermined level of mastery. A final performance assessment would be given at the end of the game that would detail strengths and weaknesses. In this final assessment, the player would be informed of how well he or she did on each of the game performance criteria and how to improve performance in the future. Drawing upon the CTA, the scenario event analysis, and the basic game level design, the initial instructional design identified a series of key scenario events, the associated learning objectives and the associated assessment requirements (see Table 1).

The design specified several different types of feedback to be used in the game, such as:

- Natural consequences would be used, when possible, to indicate the effects of the player's action. The player would be made aware of the relationship between their action and the consequence (e.g., the player must know that turning the valve cuts off the water supply for fighting a fire).
- Hints would be given to reinforce or correct declarative knowledge, for example by telling the player the general rule or protocol.
- Catastrophic failures that immediately stop the game would be used to emphasize critical errors made by the player.
- Game scores would be used to provide feedback on the key behaviors to be developed and to reward the attainment of desired learning outcomes

Finally, the design specified several possible types of scaffolding to consider, including:

- Limiting the play space and/or choices offered to avoid cognitive overload and focus attention on key cues in the environment
- Providing hints that focus on the facts, rules, and or steps in a procedure
- Providing just-in-time information, particularly at novice levels, when the player is not expected to know that information or when that information is not critical to achieving the learning objective
- Modeling or demonstrating procedures using gaming techniques or incorporating embedded videos
- Providing advance organizers, such as a schematic of the ship to aid navigation
- Providing characters to guide the player in decision making when a player has been unsuccessful after several attempts
- Providing access to "mini-games" as needed to increase proficiency in basic skills

Following a traditional game technique, the notion of a "score" was initially suggested as a form of implicit feedback, and the use of specific warning messages in response to student errors was suggested as a form of explicit feedback and guidance.

During this initial instructional design process, two key tensions were revealed:

- Directive instruction versus guided discovery
- Feedback for learning versus feedback for motivating gameplay

Assessment Strategy

To accomplish the task of embedding effective assessment in our game, our design process dictated that we begin by examining the damage control domain - a broad class of activities of which flooding control is a part. Working closely with the instructional design team, the assessment experts identified the key constructs and relationships that formed the sub-domain of flooding within the damage control domain and mapped these constructs and relationships into a Flooding Ontology (a visual representation of the constructs and that make up flooding and how those constructs are linked). The ontology was divided into three broad constructs—Situation Awareness, Communications, and Decision Making—as these encapsulated the cognitive and behavioral elements relevant to addressing a damage control situation aboard a ship. The goals and objectives of the training, along with the assessments, would be

Figure 5. Partial flooding ontology with mission 1 related elements (in gray)

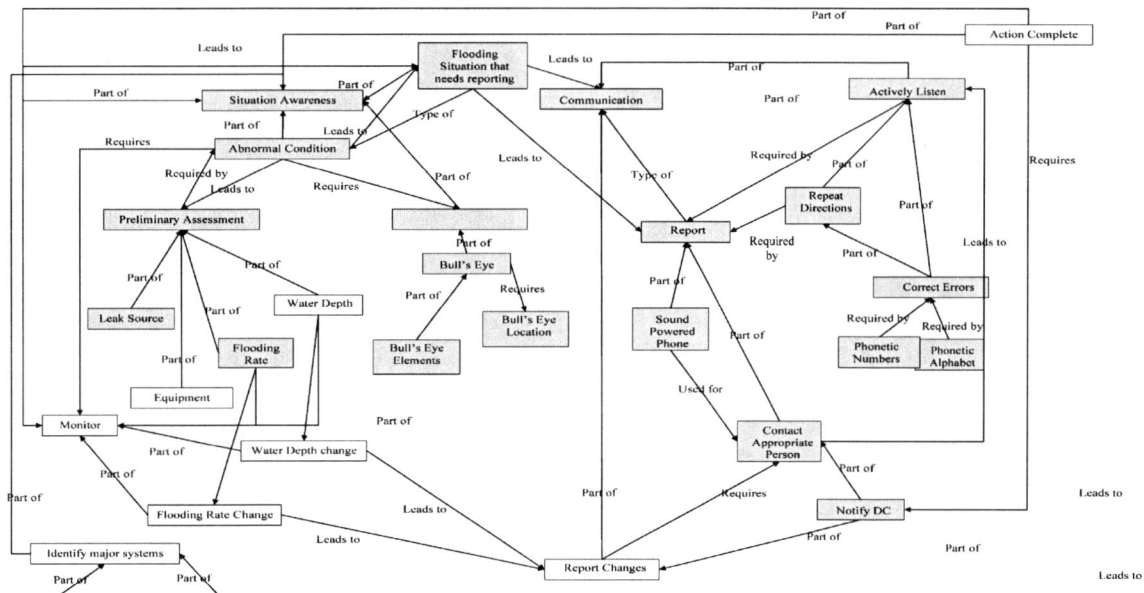

designed around these core concepts and relationships. Given this general Flooding Ontology, the specific elements relevant to the Mission 1 level were identified based on the initial instructional design (e.g., see Figure 5).

With this ontology completed, we drafted a preliminary set of functional game requirements necessary for player performance to be sufficiently assessed. These requirements were specific features that needed to be represented in the game to serve as indicators of performance. They included elements the player could use to communicate with Damage Control Central (to assess communication skills), opportunities for the player to choose from a variety of apparatus the appropriate one to contain the leak (to assess content knowledge and decision making), and opportunities for the player to interact with different types of doors under various material readiness conditions (to assess the player's understanding of readiness conditions and adherence to permission protocols).

Based on the initial instructional design and the general approach to gameplay, we proceeded to flesh out the detailed automated assessment strategy to be applied. The key learning objectives were mapped to key assessment requirements. We then identified, for each assessment requirement, the specific metrics that were possible to determine automatically in the game. These were tied as appropriate to specific scoring computations or feedback actions. The final assessment strategy incorporated the pedagogical objectives while respecting the need to have assessment fit in naturally with gameplay. The game score was defined to be monotonically increasing (i.e., total score could only increase in value, and would not decrease due to errors). The second iteration of instructional design, incorporating an assessment and feedback strategy for all learning objectives, was completed in mid-July.

In our assessment strategy, we attempted to identify behaviors that could lead to a failure of the mission. These serious errors would produce a demerit that interrupted gameplay to provide important feedback. Table 2 provides an example of the assessment strategy for a single learning objective (i.e., Follow safety protocol), showing

Table 2. Sample specification of assessment and feedback for a learning objective

Skill Area	Learning Objective	Assessment Requirements	Metrics	Positive indicator	Negative indicator	Game Score Feedback Effect - Success	Numerical/Aggregate Metric	Scoring/Feedback Effect - Negative
Decision-Making	Follow safety protocol	Did not enter the flooding compartment without permission.	Intent	Asked for permission to enter a room in which student had (incorrectly) reported flooding	Never ask for permission to enter a room	0		
			Accuracy	Asked for permission to enter the actual flooding room	Enter flooding compartment without asking permission	10 points if student enters the room after requesting and receiving permission	1 count per flooding compartment; 1 count per "first" entry into flooding compartment; 1 count per correct request to enter flooding compartment	Demerit: A demerit is issued if the student enters the flooded compartment without receiving permission.
			Multiple instances (flooding compartment)	N/A - only one flooding compartment in this level				
			Multiple occurrences (Over time)	Only keep track of requests prior to the first entry into the flooding compartment			Accuracy percentage: Total correct request / Total flooding compartment	

Figure 6. Preliminary guided-discovery instructional logic for obtaining appropriate permission before opening doors (light gray = direct instruction, dark gray = implied instruction, black = goal achieved)

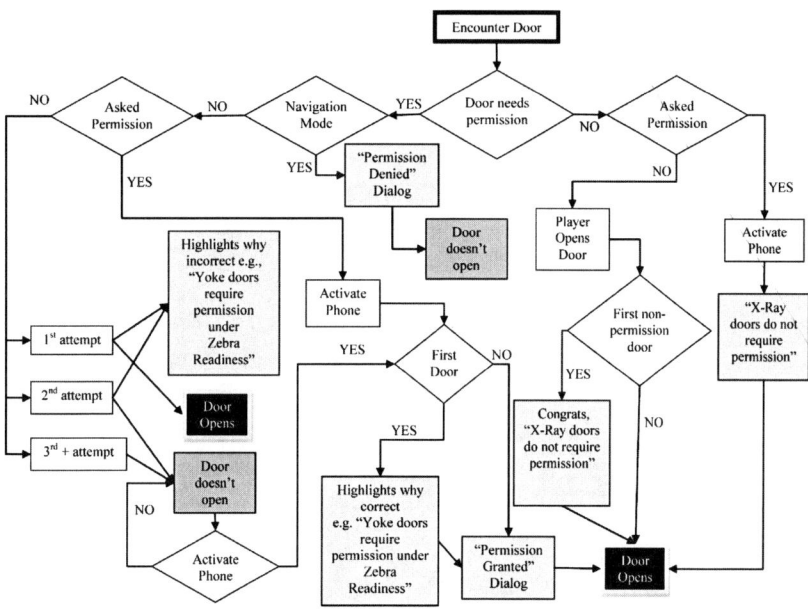

implicit feedback in response to accurate behavior via a change to the game score, as well as explicit "demerit" feedback in response to errors. Many of the details of the assessment strategy changed during subsequent refinements.

As the detailed assessment strategy was being developed, preliminary instructional logic was also created to identify when to capture data for assessment of player performance, as well as to identify possible situations for performance-based feedback opportunities. The logic for several of the behaviors involved in a flooding control mission was captured by mid-July using flowcharts such as that in Figure 6. The motivation behind these initial flowcharts was to capture an "ideal" trainer by working forward from the ontology and recommended functional assessment requirements to produce a guided discovery instructional strategy that would also support effective assessment. Consideration of gaming strategy, story and mission flow were left as future integration exercises. The logic allowed for multiple attempts to accomplish tasks while providing different guidance and constraining user actions depending upon the errors made (and amount of error repetition).

After mid-July, a collective effort was made to resolve the different representation formats preferred by the instructional designers, assessment experts and game designers and to integrate the various aspects of the product design. The most commonly agreed-upon format was the use of flowcharts. Using the preliminary instructional logic flowcharts (e.g., Figure 6) as a starting point, new flowcharts were defined for the key learning objectives of the mission 1 level. Figure 7 illustrates a flowchart specifying the instructional logic for "Following safety protocols", as determined by donning the appropriate personal protective equipment (PPE). Each flowchart provided a specific set of rules indicating under which circumstances

Figure 7. Complete instructional logic for learning objective of "following safety protocols: Don appropriate personal protective equipment (PPE)"

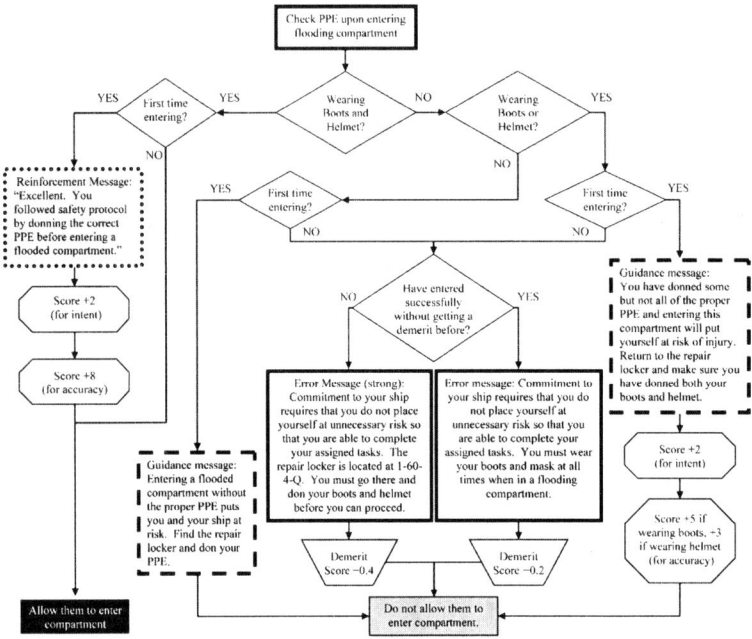

specific guidance messages (dashed border, light-gray), positive reinforcement messages (dotted border, light gray) or negative feedback messages (solid border, light gray) should be given. These flowcharts integrated the assessment strategy and basic character interactions to identify the instruction, scoring, and dialog to be implemented in the game. New flowcharts covering all key situations in Mission 1 were completed in mid-August. During this revision, a second score mechanism was added to keep track of penalties associated with errors. Specifically, when a negative feedback message was given in response to an error by the student, a negative demerit score value was decreased by some amount. The demerit score (trapezoid) and positive score (hexagon) were separate values. The possibility of mission failure due to a high total demerit score (e.g., −1.0) was also introduced.

During the development of the assessment strategy and the associated instructional design revisions, two key tensions arose:

- Developing an objective versus developing methods for assessing attainment of that objective
- The right documents for sharing knowledge

Software Development

By mid-June, a software requirements document and an initial software design document were completed that identified the key changes needed to the existing simulation in order to support our desired training game design. In addition to supporting the instructional levels we had designed, we adopted an additional requirement to adapt the gaming infrastructure so that pedagogical elements of the training would be specified as much as possible using a data-driven approach rather than implementing a hard-coded game. The goal of a data-driven approach was to facilitate rapid changes to instructional content.

An aggressive agile development schedule was determined with four end-of-month deliverables: the first at the end of June and the last for our final product at the end of September. The first deliverable would incorporate a basic mission flow (incorporating a briefing screen and a debriefing screen) and include some preliminary user interface elements (such as feedback and information windows). The second deliverable would provide an end-to-end walkthrough that exercised basic missions and basic forms of all interface elements. The third deliverable would include the data-driven infrastructure and all key graphics and animations needed to support the tutorial and main mission. The final deliverable would have a fully functional tutorial and the main mission.

In particular, the data-driven design adopted was as follows. A mission would be comprised of multiple tasks. For each task, specific trigger events would initiate it, and completion of the task would in turn initiate one or more other tasks. The specific tasks in a scenario and all details concerning each task, such as its triggers and its description, were specified in the scenario data file. Further, a mission could contain multiple "score" objects, each maintaining a distinct value representing user performance. A score object could be triggered by direct player actions, such as dialog choices or entering a room, as well as by indirect actions, such as completing a task. The score object would contain specific feedback messages to provide to the user upon being triggered (e.g., via a "demerit" message or an instructional guidance message), and would maintain internal state reflecting the number of times triggered. The message could vary depending upon the number of times triggered. The specific score objects and all details concerning them, such as their triggers and messages, were specified in the scenario data file.

The key tension of the software development were

- Hard-coded versus data-driven implementation

Introductory Movie

In late June, as we iteratively refined our story, we realized that our product would be particularly enhanced by the incorporation of an introductory movie scene that would present the backstory and lay the foundation for motivating the student. The use of introductory movies and cut-scenes within and between game levels is a powerful method used in commercial games to enhance the player's sense of immersion in the game. For our product, we determined that the introductory movie needed to:

- Motivate the desire to play
- Motivate the desire to serve in the Navy
- Introduce the backstory
- Promote the Navy core values of Honor, Courage, and Commitment
- Introduce the ultimate game objective
- Orient the student's approach to playing the game

The introductory movie design was captured in a table that divided the script into small one- or two-sentence segments to facilitate the association of dialog with the appropriate on-screen visual. The team met via phone conference to discuss what should be seen while the narration was spoken. In order to hasten development and save money it was decided that characters would only be seen when absolutely necessary to the story. Likewise, we decided that in order to allow the learner to relate to the game character, the player would

Figure 8. Final instructional objectives: Skill areas - terminal objectives - learning objectives

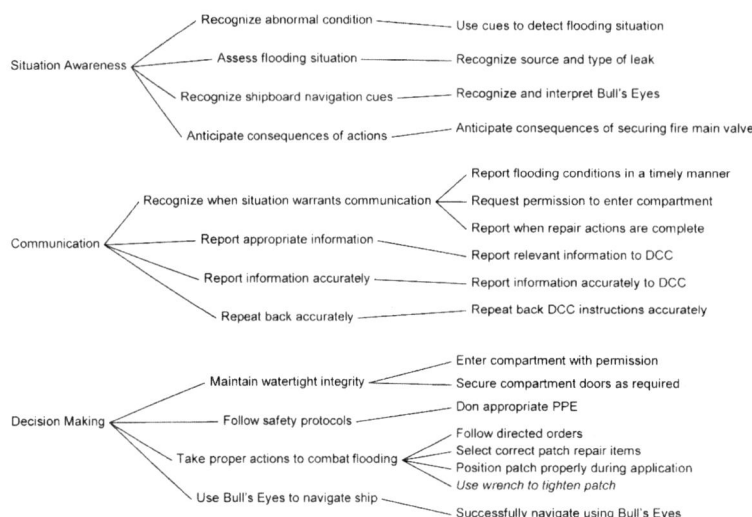

not be identified as male or female and that any time the learner had to "speak" it would be via text only. This meant that the learner would not ever see or hear himself or herself as a character in the game. This kept the player "generic" and allowed the learner's imagination to fill in the gaps. Thus, a player could be immersed in the game without the necessity of choices of character art or voicing.

An artist then created storyboards (sketches) for the intro movie. The sketches were scanned and placed into the draft of the script. This document was shared with the larger team (including the subject matter experts) and feedback was incorporated into successive drafts. To stay true to story and reality, we decided it was necessary to have one non-player character (the officer of the deck who welcomed the learner aboard). Once the storyboard and script were approved by the development team and SMEs, the artists began to create the art and animations for the movie. For maximum effect, a 3D modeling approach with realistic ships and naval scenes was used. After a number of iterations with frequent SME feedback on the accuracy and suitability of our models, the graphic rendering of the movie was completed.

In parallel, a narration and sound script for the scene was finalized with inputs from several team members and from our SMEs. This process also involved the creation of a "scratch track" of the narration of the introductory movie that was reviewed by our customer. Once the intro video was created, it was passed to our sound production team. Professional voice talent was used to record the narration for the movie, and the audio for the movie was finalized and enhanced with sound effects and music to add to the drama. The introductory movie was then incorporated into the game to be played before the tutorial mission.

There were minimal tensions raised during the development of the introductory movie. However, since the development of the movie was a substantial task, it was important to ensure that it was general enough to accommodate further story refinements in the game itself. During development, the specific wording used in the narration was varied slightly several times to reflect changes in the specific missions (particularly in the mission briefings).

Table 3. Assessment and feedback based on students' dialog choices

Dialog choice	Assessment	Feedback
"DCC, this is Seaman Taylor. I'm dressed out and ready to help."	Incorrect	Dialog response from DCC: "Seaman Taylor, DCC. I asked you to go to compartment 1-80-1-Q and report when ready to enter. Get on it!"
"DCC. I am at space 1-80-1-Q. Request permission to enter and inspect."	Error against 'Report Appropriate Information to DCC' learning objective	Dialog response from DCC: "Sailor, this is DCC. Who is this?" and Demerit (e.g., "A proper report should include your name and all relevant information about a situation." and 0.1 demerit score change)
"DCC, this is Seaman Taylor. I am dressed out and ready to enter 1-80-1-Q. Request permission to enter and inspect compartment."	Correct if student is actually reporting from a phone near 1-80-1-Q.	"Seaman Taylor, DCC. Aye. Permission to enter and inspect 1-80-1-Q granted. Report situation as soon as possible."
	If reporting from wrong location, then error against 'Report accurate information to DCC' learning objective	Dialog response from DCC: "I am not showing you near that space. Go find 1-80-1-Q and call me from the nearby phone." and Demerit (e.g., "You made a bad report. The DCC trusts you to be the eyes on the scene. Reassess the situation and give the DCC a better report." and 0.3 demerit score change)

Iterative Review, Refinement, and Testing

The instructional design, game level design, introductory movie, and software design were all refined in an iterative manner after the end of July. Efforts were conducted in parallel and discussions on one aspect of the product often involved team members working on other aspects. Customer design and product reviews were held bi-weekly (on August 4, August 17, September 4, September 17, and October 8). Three rounds of product testing with test subjects were held both before and after the initial product release. A pilot usability test was held on September 18, an end-user usability study on October 22-23, and a pilot validation study on November 17. Following each review and test, desired improvements were identified based on feedback and observations. These were then prioritized based on the importance of the changes and time/resource constraints.

In particular, four instructional elements were iterated upon and refined right up until the beta release in mid-September: instructional objectives, assessment methods, dialog content, and mission debriefing.

On-going refinement of the instructional objectives of Mission 1 led to the following final set of objectives being approved by the customer on September 15 (see Figure 8). For each broad skill category (Situation Awareness, Communication, Decision Making), several terminal objectives were defined. For each terminal objective at least one enabling objective was defined.

In the final implementation, the student's performance against every terminal objective was assessed automatically via the student's actions in the game and choices in dialogs. Dialog interactions formed a key method for assessing user performance against a variety of communication and situation awareness learning objectives. These assessments were context-sensitive (i.e., the same dialog choice may be correct or incorrect depending upon prior user actions). As shown in Table 3, a single dialog interaction could result in errors against different objectives (e.g., reporting appropriate versus accurate information), and different types of feedback (e.g., dialog responses versus demerits).

The mission was comprised of multiple tasks, and the student's actions were interpreted in the context of the current sub-task(s). At the end of the mission, the student was provided a debrief that summarized their performance against every terminal objective on a three-point traffic light scale (green, yellow, red). An analogous design was adopted for the tutorial mission, though with a much reduced set of terminal and learning objectives.

Tensions that arose during the iterative refinement process included:

- Designing for the novice while keeping the gamer happy
- Balancing revisions to instructional design with meeting software deadlines
- Gaming strategy versus instructional strategy

MULTIDISCIPLINARY TENSIONS AND LESSONS LEARNED

The Basis for Technology Decisions - Pedagogical or Technological

The tension revealed by the intertwined choice of training domain and gaming platform concerns the weight that should be given to pedagogical or technological factors when making decisions about which training technology to use. In our case, the decision was made primarily for technological reasons, and not for pedagogical ones. For instance, the view of the instructional designers and assessment experts was that the domain should have been chosen first, the cognitive task analysis performed next, and then, based on the key learning objectives identified, the technology choice should have been made that would best suit those objectives. Thus, choices such as 3D immersive versus 2D interactive environments, single-player versus multi-player, simulated characters versus no characters, and so on should, in principle, have been driven by pedagogy. The lesson learned is that the practical considerations of customer preferences, schedule, and available resources can significantly impact these choices.

Identifying all the Learning Objectives vs. Selecting the Objectives to be Addressed in Level 1 of the Game

The multi-disciplinary approach to instructional design had a clear effect upon the decisions made on how to sequence training in the game. The instructional designers, assessment experts, game designers and story developers all agreed upon a layered approach in which the training was broken up into multiple levels of increasing difficulty. The challenge was determining what was meant by "increasing difficulty." Everyone had a slightly different understanding of how the learning objectives might be linked to form a beginning level scenario that would be both engaging and prepare the target audience for a live simulation of the entire flooding process. Many interesting, collaborative discussions occurred in refining exactly what would occur in the earliest missions.

The instructional designers envisioned a layered approach focusing on the key steps in the flooding control process in all levels, but varying the complexity of the skills expected. In order to allow the students to experience the entire process in earliest levels, instructional designers advocated that some of the smaller steps in the process be provided to the learner or that the environment be structured in such a way that the player would not have to demonstrate more specific skills until higher levels of the game. Their primary goal was to build a foundation to ensure that players had a basic understanding of what happens in a flooding situation, and to reinforce all the key process steps in all the levels. The story developers encouraged the idea that not all information about the situation needed to be provided at the beginning, and that an incremental introduction of new story elements to provide context for new learning objectives over successive levels would be effective. The game designers wanted to keep the early levels simple and slowly build up the complexity of the gaming skills needed over multiple levels. Unlike the instructional designers, they did not feel that all aspects of the process needed to be present in every level (i.e., certain levels could focus upon certain elements of the process). The assessment experts were concerned primarily with cognitive overload and wanted to ensure that new skills and information were introduced in an incremental fashion over successive levels.

The decisions regarding which learning objectives to focus on in the first level were made with the intent of satisfying these goals as much as possible, and multiple refinements were needed over time to reach consensus. For example, there were several potential learning objectives pertaining to the rules for opening and closing doors under various readiness conditions. In our developed ontology all possible types of doors and hatches and appropriate entry rules were presented. In a classroom, these rules would likely be learned and tested with a paper and pencil test. An embedded tutorial addressing these rules was possible; however, we did not want this to be our primary instructional target for the beginning level of the game. To do so would burden the learner with information overload and require many permutations for mastering the distinctions governing the handling of different types of doors when in reality, given the readiness condition already set by the collision, most doors would require the same treatment. We resolved the issue by limiting the kinds of doors visible to the player so that the correct procedure would be technically correct while presenting limited cognitive load. Thereby, the learner could focus on the larger steps required to identify, report, and combat the flood. This solution managed the cognitive load appropriately, maintained cognitive and physical fidelity, kept the learner engaged in the story and prevented negative training.

A lesson learned from this key tension is to maintain communication and seek mechanisms for closer collaboration. Presenting the learning objectives and developing an ontology are critical to developing the appropriate story, rules, and consequences in the game. Instructional designers are accustomed to look at the relationships among objectives in terms of the development of student knowledge and skills and designing for a scaffolded progression based on skill level. Further research is needed to determine whether and when, for game-based training: every skill must be taught from the bottom up in level 1; levels should provide part-task training; or hybrid solutions will be effective. There is more than one way to develop skills, and gaming technology offers unique opportunities to develop skills that are not available through the use of other instructional media. We can select skills on which to focus in a specific scenario and manipulate the physical fidelity of the environment to meet the instructional design. In a game, this can mean restricting the environment or game space or sequence of events or having other characters perform tasks that are beyond the expectations for the level 1 player. There is a fine line between selecting tasks that would be fun but not expected to challenge the learner, and selecting tasks that would confuse or distract the learner at level 1.

Balancing Cognitive Elements with Experiential Elements

A difference in perspective between the instructional designers and the game designers was the basis for understanding and characterizing about the domain. During the knowledge acquisition process, the instructional designers gravitated towards a focus upon the cognitive elements of the domain (via a CTA) while the game designers gravitated towards a focus upon the context in which learning would occur in the domain (via a scenario element analysis). The goals of a CTA and a scenario element analysis are different. The former is focused on performance and associated knowledge states. The latter is focused on identifying the environmental contexts in which simulated activities occur, the steps of and interactions among those activities, and the possible variations in the details of those activities or contexts. Both analyses were conducted with collaborative discussions among team members. However, there were frequent misunderstandings of how to characterize the domain. Eventually, it became apparent that the two analyses were capturing complementary information. The exercise of merging them (e.g., Figure 3) resulted in several interesting discussions in which some previous misunderstandings were resolved. One benefit was that additional enabling objectives were identified. For example, in the context of the transit phase, an enabling objective for reporting a situation immediately is reporting any hazards encountered en route to the flooding area. On the downside, the merged document was difficult to follow for team members who had not directly been part of the knowledge acquisition process.

The lesson learned is that both the cognitive and context aspects of a domain are critical for game-based training and each leads to different ideas on how to structure the learning experience in a game. More investigation into effective methods for characterizing and refining the two domain aspects is needed.

Story First vs. Design First

In our effort, a lot happened very quickly. There was a general feeling that the story development (i.e., the StoryJam™ held on May 6-7) occurred too early in the process since the learning objectives from the CTA were not fully understood and had not yet been prioritized. However, without the forcing function of trying to develop a focused story to support the training, the necessary type of filtering and prioritization of the objectives may not have occurred in a timely manner. Throughout the successive refinements of the instructional design and game design, the basic story elements established at the early meeting provided positive guidance that encouraged convergence of ideas and consensus on decisions. The story decisions were also remarkably stable. For example, the supporting backstory and story elements used in making decisions in the May 29-30 game design meeting were largely present in the final design several months later. The participation of SMEs as well as the customer in the StoryJam was instrumental in establishment of a credible story.

Further research is needed to identify the best way to time and use the story development process to positive effect in game-based training development. However, a clear lesson learned is that a collaboratively developed story facilitates collaborative decision-making during subsequent development.

Gaming Strategy vs. Instructional Strategy

The key area of difficulty reflecting the multidisciplinary conflicts in the team was the relative emphasis of instructional elements and gaming elements. There are many different types of computer games and there are many different ways to structure instruction. In developing game-based training, open questions include what type of game is best suited to a particular instructional strategy as well as what type of instructional strategy is best for a particular type of game. When choosing a gaming strategy, one is typically choosing a particular suite of game mechanics (interactions the player may have within the game) and a particular type of event flow. Likewise, when choosing an instructional strategy, one is typically choosing a particular suite of instructional interactions with the student and a particular organization to those interactions.

A general methodology for determining how to provide game-based instruction for a particular domain is to map the enabling objectives to specific game mechanics and specific encounters within each level to ensure that the game player is learning what the designer has set out to instruct. During the knowledge acquisition process of our effort, we discussed a variety of possible game mechanics for various enabling objectives. In fact, the SMEs would occasionally proactively suggest a means for how they would "teach this in a game." In our case, these initial characterizations of the instruction led to an early choice of two "complementary" strategies - guided discovery instruction using adventure style gaming. Guided discovery instruction uses implicit and explicit interventions to encourage and focus a student's exploration of the training domain to achieve the learning goals. Adventure style gaming uses carefully placed hints and clues to encourage the player to continue to explore the game world and achieve the adventure's goals.

This choice led to a lot of synergy at the beginning of the design process. Instructional designers would suggest a need for an intervention, and the game developers could easily map this into an event furthering the adventure. However, as the instructional design process progressed, issues of how to embed new instructional elements often turned into discussions of how the "gameplay" might be adversely affected. Issues that caused the greatest tension were the instructional designers' and assessment experts' desires for increased guidance and feedback explaining all the student's mistakes as they were made, explicit didactic information on every element of relevance in the scenario (e.g., a help lookup facility), and increased scaffolding, such as the use of a compass-like aid to assist in navigating around the ship. Their goal was to reduce the student's cognitive load and to ensure that the student formed good mental models as early as possible during training. Countering this, the game designers wanted to minimize the amount of non-embedded information (i.e., not delivered as a natural part of interacting with the environment and other characters) to reduce negative impacts on the player's immersion and to maintain the sense of adventure. After many discussions and necessary compromises on everyone's part, we succeeded in resolving most issues to produce a final instructional design and game design that supported one another.

Hence, the two related lessons learned are that linking gaming strategy to instructional strategy can lead to good synergy of design and a compelling experience, but that an early choice of gaming strategy can lead to difficulties in trying to incorporate incompatible instructional elements later on.

Directive Instruction vs. Guided Discovery

Expectations for the design of directive instruction and discovery learning are not the same. Our task was to pull the best elements of each into an engaging game that fosters learning through guided discovery. Just how much guidance is required in guided discovery? Which are the elements that must be more closely guided? When is there no harm in letting learners play until they get it right versus when must one guide them so they are not bogged down by details that are insignificant and detract from learning (i.e., avoiding situations in which the learner becomes overwhelmed with less important details and misses the key learning opportunity)? A prime example was in reading the compartment identifiers, or "Bull's Eyes", which indicate the location and type of compartments. We purposefully focused the player's attention on two of the elements in the four element series of number and letters that defines a compartment identifier. The goal was for the player to learn to use the two elements well enough to navigate fore and aft and to recognize which side of the ship they were on. Once these tasks were mastered the additional elements could be addressed. In the game, however, how long should the player wander before being guided to information that will help him or her reach the intended destination?

The lesson learned is to use the cognitive task analysis to focus on identifying the actions a student must make to carry out a task as well as the typical wrong paths a student can take. In experiential learning, we include these actions in the environment and make some of them options

– deliberately selecting these to promote successful task development while not overwhelming the student.

Feedback for Learning vs. Feedback for Motivating Gameplay

Games offer opportunities for players to see the consequences of their actions in ways other instructional methods do not. Feedback in the form of natural consequences of actions can be powerful; the trick is to ensure that the feedback is driven by the learning objectives and emphasizes the cause and effect relationship between learners' actions and the consequences so they understand what has happened and why. Consequences of actions in games often provide the "wow" effect garnering the player's attention but not always making clear exactly what the player did to give rise to them. For example, one of the early feedback responses considered was to have a Chief non-player character yell at the player's character when certain key mistakes were made. This kind of feedback was not necessarily appropriate from a chain of command perspective and did not support the learning objective. However, from a gaming/story perspective, it added some emotional stress and made the experience more exciting. Providing the player with appropriate consequences resulting from his or her actions had to be carefully considered as part of the entire gaming/instructional strategy. The experiential aspect of the game environment opens possibilities that should continue to be explored while maintaining the instructional focus. Decisions about which type of corrective feedback to provide immediately within the game and which to provide in the post-game debrief remain important questions for research.

The lesson learned is that finding innovative ways to provide feedback that promotes reflection requires concerted effort, ongoing discussion, and continued research. The game design must examine the intended outcomes and the paths people take that compromise those outcomes. We can then devise strategies to promote awareness and reflection, such as by setting up distractions to entice players into making common errors and providing feedback to help them refine their thinking.

Developing an Objective vs. Developing Methods for Assessing Attainment of That Objective

A game story line must be developed so as to create situations that challenge learners with respect to the learning objectives and that provide opportunities for learners to react and receive feedback on their actions. Therefore, the objectives as well as the assessment strategy need to be developed first in the development of the game. In formal schooling, assessments often occur after a learning event (e.g., end of module test). As a team with such a short development cycle, it was easy to slip into this way of thinking. Also, team members had differing expectations about the role of assessment and even about the meaning associated with it.

Assessment is another area where gaming technology offers new possibilities for learning. Using both time to complete and path taken are possible options in addition to "correct" actions. One promising discussion focused on assessing intent and accuracy as one way to understand reasons for a player's actions and to tailor future trigger events so as to hone skills appropriate to that player's needs. This kind of assessment also has implications for the type of feedback provided. Van Merrienboer and Kirschner (2007) for example make the following recommendations for assessing intent and accuracy:

- If the learner makes an error that conveys an incorrect goal, the feedback should explain why the action leads to an incorrect goal and should provide a hint or suggestion as how to reach the correct goal. .

- If the learner makes an error that conveys a correct goal, the feedback should only provide a hint for the correct step or action to be taken and not simply give away the correct step.

The lesson learned is that objectives and assessments must be linked at the outset of the development process.

The Right Documents for Sharing Knowledge

A key challenge throughout the effort was conveying our thoughts and ideas to each other in an effective manner. Most stakeholders in the effort tended to think and operate at a different level of abstraction or with a different focus. Further, different stakeholders entered with different preconceptions about the motivations of other stakeholders. Exacerbating this quintessential inter-disciplinary communication issue was the fact that our team was widely distributed geographically. Despite roughly monthly face-to-face meetings and frequent discussions by telephone, numerous misunderstandings would arise and persist.

The knowledge sharing issue was particularly revealing in the documents that we generated to share our instructional design ideas. Every single performer had a different preferred format for capturing their ideas, even if the general type of document was similar. For instance, each of us had a different view about what a storyboard entailed. This confusion led to various persistent misconceptions. For instance, the game designers tended to conclude that the instructional designers were seeking to provide learning that was linear in nature due to the list or table formats they used to capture instructional events.

At several points during the effort, attempts were made to integrate the different perspectives and create a single format that met the needs of all performers. During knowledge acquisition, a table format merging the key CTA information and the scenario element analysis was created (see Figure 3). During development of instructional and assessment logic, a rich flowchart format was used to draw together inputs from several sources (see Figure 7). During the specification of the dialog, three different formats were explored (a dialog tree, a storyboard with optional paths, and a linear list with linked dialog fragments), but ultimately the actual dialog implemented in the game prevailed as our means of discussing and reviewing dialog content.

There is a strong need for effective collaboration mechanisms in a diverse game-based training development team. While we identified some collaborative document formats, further research and refinement is needed to find those communication means that are most effective for game-based training design.

Hard-Coded vs. Data-Driven Implementation

In the field of game-based training, there is always a tension between implementing exactly what is needed for the specific instruction and implementing general game mechanics which can be used to support a variety of instruction. In many cases, the former approach is used, leading to a system that requires significant effort to adapt. Due to our broader project goals, we adopted the latter approach. In addition to a data-driven task mechanism, several data-driven feedback mechanisms were implemented.

The software development iterations during July and August were focused on implementing a data-driven infrastructure and basic support for the game mechanics we wanted in our initial prototype. By the end of August, a preliminary implementation of the tutorial and Mission 1 levels had been implemented. However, further revisions to the instructional content continued right up the end of September. These revisions were easy to incorporate into the system, and we made numerous revisions to the instructional content without slipping our software deadlines.

A key lesson learned was that the focus on a data-driven approach to specifying the instructional logic in the game greatly reduced the stress of making changes to the game and avoided the need for any significant re-factoring of the code.

Designing for the Novice While Keeping the Gamer Happy

Aside from age, Navy recruits are the very embodiment of diversity – 40,000 recruits each year who come from every conceivable background. It was essential that our training system be accessible to everyone from super users to computer neophytes and be effective with everyone from expert gamers to non gamers. However, since our target population truly are novices in the domain of Navy skills, it was critical that the instruction be delivered at an appropriate level and not be overly challenging. In making our instructional design and game design decisions, there was an ongoing tension between designing for the novice and ensuring that the game was usable and appealing to a very broad population. This tension revealed several lessons learned.

The interface of the game needed to be simple to accommodate the broad range of users. Further, a simple interface is a common trait of many good games. However, in our effort, being simple was often at odds with being comprehensive. The more instruction, data, feedback, interventions, and details we added, the more complex the interface became and consequently, the less usable. In the end, we decided to postpone some instructional interventions in order to maintain a simple, clean interface.

Dynamic interaction is the most obvious quality of games. After all, if one can't interact, a game reduces to a spectator experience. However, what types of interactions are appropriate for novice game players, while retaining appeal for the experienced gamer? In our game, the player assumed the role of a Seaman recently assigned to his or her first ship. By making the role in the game reflect the near future for all recruits, issues of experience in gaming were partially avoided (e.g., all the recruits are new to being aboard a Navy ship, so navigating around a ship effectively requires diligent attention to details in the environment). Further, we made actions in the game reflect real-world actions. Since these actions (e.g., navigating within a ship, repairing a leak using correct procedures, communicating properly with a superior officer) are not typical of a commercial game, both the inexperienced and experienced gamer still needed substantial learning in order to do it right.

Each mistake made by a player creates a brief teachable moment where corrective guidance can make a huge difference. However, the right level of guidance to give to a novice can conflict with the degree of freedom expert gamers expect. Our approach was to provide both non-interruptive and interruptive feedback of varying detail depending on the nature of the mistake made while keeping the player immersed in the story as much as possible. Minor errors would result in some guidance provided in a small pop-up window that was not intrusive. Conversely, when the student made a critical error, a demerit would interrupt gameplay visually and aurally with a specific message about the error. For both hints and demerits, the initial message would be somewhat general. If the error was repeated, subsequent messages would be more detailed.

These decisions were borne out by results of product testing across all a variety of users. For instance, a student's first demerit stops mere button-clicking behavior and leads quickly to increased attention by both gamers and non-gamers.

Balancing Revisions to Instructional Design with Meeting Software Deadlines

In addition to our data-driven approach, another key contributor to our success was the use of an agile software development methodology. A particularly relevant aspect of agile development is frequent deliveries. One of the key tenants of agile development is getting working iterations of the software into the hands of stakeholders as early and as often as possible. With six geographically distributed groups, frequent releases helped mitigate communication issues and maintain a cohesive vision for the product.

As in many software products, the requirements provided by the customer may change over time. In our case, there was an increasing emphasis placed upon the depth of training to deliver. Initial requirements were for a skill practice environment with limited guidance. Later requirements were for a training environment with limited assessment that could produce some validated learning outcomes of certain skills. The final requirements, determined several months into the project, were for a training system with focused guidance, general assessment on all key skills, comprehensive feedback in response to errors, no negative training, and no irrelevant elements that would produce validated learning outcomes across most skills taught. Our agile approach to software development, iterative approach to instructional design and frequent interactions with the customer enabled us to accommodate these changes in requirements.

Although key to our success, the agile method was not without issues. In fact, it caused some interesting tensions and challenges among the team. In order to produce frequent releases, software development moves extremely quickly. Yesterday's ideas become tomorrow's implementation. This put an extra burden on quick turn arounds and the rapid iteration of ideas and forced the team to focus on what's important. Ideas that fell into the "could-be" or "might be nice" categories quickly got pushed aside by "must have" and "critical" requirements. Whenever an idea cropped up, it was immediately weighted for its instructional or gameplay value and then balanced against the rest of the to-do items. The most important items always got done and the rest got set aside for another day. The result was a game that was delivered on time and within budget, met all customer needs, and passed the usability tests with flying colors.

FUTURE RESEARCH DIRECTIONS

In the next two years of our project, our plan is to investigate further issues related to the development of game-based training and to build upon the lessons learned so far to identify design and development methods that support consistent and effective outcomes. We plan to create additional refinements of the flooding control training system, as well as to create additional training systems in other domains for the same customer. These development efforts will inform and be supported by efforts to create better authoring and assessment tools for game-based training (e.g., of which our data-driven infrastructure was an early step). To ensure that our advances are well-founded, we will conduct a variety of empirical investigations into the effectiveness of different gaming features for training, and the interaction between different gaming and instructional approaches.

The field of game-based training is at an interesting crossroads as it moves from a poorly understood cottage industry to a well-founded discipline of its own. The multifaceted nature of game-based training is both its strength and its weakness - no other medium can provide as rich and varied learning experiences, but few other instructional mediums are as difficult to get right. There is a need for increased communication across practitioners to share processes, mechanisms, and lessons learned. We believe our continuing efforts in this project will provide seminal contributions towards formalizing game-based development methodology.

CONCLUSION

Conducting research on instructional games begs the question, "How can games be made instructional *and* engaging?" A critical battleground for this debate is the learning objectives and how they are used during design and development of the game. Respecting the role of learning objectives was an easy point of agreement for the team in the abstract. However, it was difficult to reach a common understanding of what that meant in terms of game play, story, actions, dialog choices, feedback, guidance, and the game environment. Instructional design based on learning objectives can limit distractions that overload the player with extraneous information that interferes with learning. However, the objectives alone do not guarantee focus on relevant information. Structuring the story must work in concert with treating the consequences of errors to help the player develop a mental model of what actions lead to the desired outcomes. The interactions between the player and the game must maintain the player's sense of immersion while also ensuring that the experience remains focused on the skills being taught. Finding the balance between adhering to the learning objectives while maintaining the characteristics associated with a game was a persistent challenge that revealed itself in a variety of ways.

A second key battleground for this debate concerns the processes followed in producing an appropriately instructional and fun game. In general, the agile approach adopted in this effort resulted in ongoing priority conflicts between the instructional, gaming and story mindsets. The instructional designers by nature preferred a more traditional waterfall approach in which all key decisions regarding learning goals, content selection, and specification of instructional methods and strategies would be made prior to involving considerations about the story, the technology, and the gaming style. Their concern was that that key components, affordances, and experiences in the training system necessary from a pedagogical and assessment standpoint would be undermined or inadequately represented if instruction came second. The gaming developers by nature preferred clear direction on the basic interaction modes desired for instruction and the basic structure of a game level so that the associated software interfaces and graphics could be created in a timely manner, and were content to leave easy-to-change data details (e.g., the specific dialog and instructional messages, or sequencing of activities within a level) as part of an iterative refinement process led by the instructional designers. Finally, the story developers by nature preferred a strong early investment in coming to agreement on the basic story in order to ensure that successive iterative refinements of the instructional design or technology could be shaped to maintain a powerful story. From their vantage point, powerful stories create powerful experiences that invite participation while information and data, on the contrary, invite critique. The danger of ignoring the story aspect of a learning game at any stage of the process is that achieving a powerful story becomes difficult.

Of the many lessons learned in this effort, most involve the challenge of effective communication among professionals from diverse backgrounds. It is important to clearly articulate each team member's priorities as early as possible and work together to devise methods for sharing and integrating information that will accommodate any competing priorities. From this, specific critical paths and components must be identified that, if organized and prioritized correctly, can inform the overall design and development approach. Since the process of combining the craft of game design and the science of learning has yet to be established, the lessons learned from this project may serve to inform the process and lead to an efficient and effective model for the development of games for learning.

ACKNOWLEDGMENT

The research reported in this paper was conducted under the management of Dr. Ray Perez under the Office of Naval Research contract number N00014-08-C-0030. We would like to thank our customer Dr. Rodney Chapman, Chief Learning Officer of the Naval Service Training Command, and his staff for their active participation in refining and guiding our product design, and Lt. Greg Page from the Recruit Training Command for his assistance in providing subject matter expertise. We would like to thank Dr. Ray Perez and Mr. Paul Chatelier of the Potomac Institute of Policy Studies for identifying opportunities for putting our research efforts in context of the Navy's current needs. We would like to thank Steve Greenlaw of Intelligent Decision Systems, Inc. for providing subject matter expertise. Finally, we would like to thank the software designers and engineers who contributed to the development and testing of the product: Chris Rodgers and Brad Anderegg from Alion Science and Technology, Rachel Joyce and Julian Orrego of UCF, and Erin Panttaja, Todd Wright and John Ostwald of BBN Technologies.

REFERENCES

Bandura, A. (1977). Self-efficacy: Toward a unifying theory of behavioral change. *Psychological Review*, *84*, 191–215. doi:10.1037/0033-295X.84.2.191

Bandura, A. (1986). *Social foundations of thought and action: A social cognitive theory*. Englewood Cliffs, NJ: Prentice-Hall.

Baylor, A. L. (2001). Agent-based learning environments for investigating teaching and learning. *Journal of Educational Computing Research*, *26*, 249–270.

Bonk, C. J., & Dennen, V. P. (2005, March). *Massive Multiplayer Online Gaming: A Research Framework for Military Training and Education*. Office of the Undersecretary of Defense For Personnel and Readiness. Technical Report 2005-1.

Bransford, J. D., Brown, A. L., & Cocking, R. R. (Eds.). (1999). *How people learn: Brain, mind, experience, and school*. Washington, DC: National Academy Press.

Bransford, J. D., Sherwood, R. D., Hasselbring, T. S., Kinzer, C. K., & Williams, S. M. (1990). Anchored instruction: Why we need it and how technology can help. In D. Nix & R. J. Spiro (Eds.), *Cognition, education, and multimedia: Exploring ideas in high technology* (pp. 115-141). Hillsdale, NJ: Lawrence Erlbaum Associates.

Bruner, J. (1990). *Acts of meaning*. Cambridge, MA: Harvard University Press.

Cartier, J. L., & Stewart, J. (2000). Teaching the nature of inquiry: Further development in a high school genetics curriculum. *Science and Education*, *9*, 247–267. doi:10.1023/A:1008779126718

Chi, M. T. H. (2000). Self-explaining: The dual processes of generating inference and repairing mental models. In R. Glaser (Ed.), *Advances in instructional psychology: Educational design and cognitive science*, (Vol. 5, pp. 161-238). Mahwah, NJ: Lawrence Erlbaum Associates.

Chi, M. T. H., Glaser, R., & Farr, M. J. (Eds.). (1988). *The nature of expertise*. Hillsdale, NJ: Lawrence Erlbaum Associates.

Cognition and Technology Group at Vanderbilt (CGTV). (1992). Anchored instruction in science and mathematics: Theoretical basis, developmental projects, and initial research findings. In R. A. Duschl & R. J. Hamilton (Eds.), *Philosophy of science, cognitive psychology, and educational theory and practice. SUNY series in science education* (pp. 244-273). Albany, NY: State University of New York Press.

Cognition and Technology Group at Vanderbilt (CGTV). (1997). *The Jasper Project: Lessons in curriculum, instruction, assessment, and professional development*. Mahwah, NJ: Lawrence Erlbaum Associates.

Cognition and Technology Group at Vanderbilt (CTGV). (2000). Adventures in anchored instruction: Lessons from beyond the ivory tower. In R. Glaser (Ed.), *Advances in instructional psychology: Educational design and cognitive science*, (Vol. 5, pp. 35-99). Mahwah, NJ: Lawrence Erlbaum Associates.

Csikszentmihalyi, M. (1990). *Flow: The psychology of optical experience*. New York: Harper Perennial.

Diller, D., Roberts, B., Blankenship, S., & Nielsen, D. (2004). DARWARS Ambush! – Authoring lessons learned in a training game. In *Proceedings of the 2004 Interservice/Industry Training, Simulation and Education Conference (I/ITSEC)*, Orlando, FL.

Gentner, D. (1983). Structure-mapping: A theoretical framework for analogy. *Cognitive Science, 7*, 155–170.

Gerhard, M., Moore, D., & Hobbs, D. (2004). Embodiment and copresence in collaborative interfaces. *International Journal of Human-Computer Studies, 61*, 453–480. doi:10.1016/j.ijhcs.2003.12.014

Gist, M. E., Schwoerer, C., & Rosen, B. (1989). Effects of alternative training methods on self-efficacy and performance in computer software training. *The Journal of Applied Psychology, 74*, 884–891. doi:10.1037/0021-9010.74.6.884

Gist, M. E., Stevens, C. K., & Bavetta, A. G. (1991). Effects of self-efficacy and post-training intervention on the acquisition and maintenance of complex interpersonal skills. *Personnel Psychology, 44*, 837–861.

Glaser, R. (1989). Expertise in learning: How do we think about instructional processes now that we have discovered knowledge structure? In D. Klahr & D. Kotosfky (Eds.), *Complex information processing: The impact of Herbert A. Simon* (pp. 269-282). Hillsdale, NJ: Lawrence Erlbaum Associates.

Hill, R. W., Jr., Belanich, J., Lane, H. C., Core, M., Dixon, M., Forbell, E., et al. (2006*). Pedagogically Structured Game-Based Training: Development of the Elect BiLAT Simulation*. National Technical Information Service. Retrieved from http://handle.dtic.mil/100.2/ADA461575

Hussain, T. S., & Ferguson, W. (2005). Efficient development of large-scale military training environments using a multi-player game. In *2005 Fall Simulation Interoperability Workshop*, (pp. 421-431).

Hussain, T. S., Weil, S. A., Brunye, T., Sidman, J., Ferguson, W., & Alexander, A. L. (2008). Eliciting and evaluating teamwork within a multi-player game-based training environment. In H.F. O'Neil & R.S. Perez (Eds.), *Computer Games and Team and Individual Learning* (pp. 77-104). Amsterdam, The Netherlands: Elsevier.

Johnson, W. L., Wang, N., & Wu, S. (2007). Experience with serious games for learning foreign languages and cultures. In *Proceedings of the SimTecT Conference*, Australia.

Kanfer, R., & Ackerman, P. L. (1989). Motivation and cognitive abilities: an integrative/aptitude-treatment interaction approach to skill acquisition. *The Journal of Applied Psychology, 74*, 657–690. doi:10.1037/0021-9010.74.4.657

Kanfer, R., & McCombs, B. L. (2000). Motivation: Applying current theory to critical issues in training. In S. Tobias & J. D. Fletcher (Eds.), *Training and retraining: A handbook for business, industry, government, and the military* (pp. 85-108). New York: Macmillan.

Locke, E. A., & Latham, G. P. (1990). Work motivation: The high performance cycle. In U. Kleinbeck & H. Quast, (Eds.), *Work motivation* (pp. 3-25). Hillsdale, NJ: Lawrence Erlbaum Associates.

Locke, E. A., Shaw, K. N., & Saari, L. M. (1981). Goal setting and task performance: 1969-1980. *Psychological Bulletin, 90*, 125–152. doi:10.1037/0033-2909.90.1.125

Mayer, R. E. (2001). *Multimedia learning*. Cambridge, UK: Cambridge University Press.

Moreno, R., & Mayer, R. E. (2004). Personalized messages that promote science learning in virtual environments. *Journal of Educational Psychology, 96*, 165–173. doi:10.1037/0022-0663.96.1.165

O'Neil, H. F., & Perez, R. S. (2008). *Computer Games and Team and Individual Learning*. Amsterdam, The Netherlands: Elsevier.

O'Neil, H. F., Wainess, R., & Baker, E. L. (2005). Classification of learning outcomes: evidence from the computer games literature. *Curriculum Journal, 16*(4). doi:10.1080/09585170500384529

Pintrich, P. R., & Zusho, A. (2002). The development of academic self-regulation: The role of cognitive and motivational factors. In A. Wigfield & J. Eccles (Eds.), *Development of achievement motivation* (pp. 249-284). San Diego, CA: Academic Press.

Raghavan, K., Satoris, M. L., & Glaser, R. (1997). The impact of model-centered instrbction on student learning: The area and volume units. *Journal of Computers in Mathematics and Science Teaching, 16*, 363–404.

Raghavan, K., Satoris, M. L., & Glaser, R. (1998). The impact of MARS curriculum: The mass unit. *Science Education, 82*, 53–91. doi:10.1002/(SICI)1098-237X(199801)82:1<53::AID-SCE4>3.0.CO;2-#

Roberts, B., Diller, D., & Schmitt, D. (2006). Factors affecting the adoption of a training game. In *Proceedings of the 2006 Interservice/Industry Training, Simulation, and Education Conference (I/ITSEC)*.

Rothman, F., & Narum, J. L. (1999). *Then, Now, and In the Next Decade: A Commentary on Strengthening Undergraduate Science, Mathematics, Engineering and Technology Education*. Washington, DC: Project Kaleidoscope.

Salas, E., & Cannon-Bowers, J. A. (2000). The anatomy of team training. In S. Tobias & J. D. Fletcher (Eds.), *Training and retraining: A handbook for business, industry, government, and the military* (pp. 312-335). New York: Macmillan.

Schunk, D. H., & Ertmer, P. A. (1999). Self-regulatory processes during computer skill acquisition: Goal and self-evaluative influences. *Journal of Educational Psychology, 91*, 251–260. doi:10.1037/0022-0663.91.2.251

Schunk, D. H., & Zimmerman, B. J. (2003). Self-regulation in learning. In W.M. Reynolds, & G.E. Miller (Eds.), *Handbook of psychology: Educational psychology*, (Vol. 7, pp 59-78). New York: John Wiley.

Stewart, J., Cartier, J. L., & Passmore, C. M. (2005). Developing an understanding through model-based inquiry. In M.S. Donovan & J.D. Bransford (Eds.), *How Students Learn: History, Mathematics, and Science Inquiry in the Classroom*, (pp. 515-565). Washington, DC: National Academies Press.

van Merriënboer, J. J. G., & Kirschner, P. (2007). *Ten steps to complex learning: A systematic approach to four-component Instructional Design.* London: Lawrence Erlbaum Associates.

Vogel, J. J., Vogel, D. S., Cannon-Bowers, J. A., Bowers, C. A., Muse, K., & Wright, M. (2006). Computer gaming and interactive simulations for learning: a meta-analysis. *Journal of Educational Computing Research, 34*(3), 229–243. doi:10.2190/FLHV-K4WA-WPVQ-H0YM

Zimmerman, C., Raghavan, K., & Sartoris, M. L. (2003). The impact of MARS curriculum on students' ability to coordinate theory and evidence. *International Journal of Science Education, 25,* 1247–1271. doi:10.1080/0950069022000038303

Chapter 5
DAU CardSim:
Paper Prototyping an Acquisitions Card Game

David Metcalf
University of Central Florida, USA

Sara Raasch
42 Entertainment, USA

Clarissa Graffeo
University of Central Florida, USA

ABSTRACT

This chapter discusses DAU CardSim, a multiplayer card game for teaching defense acquisition strategies, and addresses the challenges in moving from a paper prototype of the game to a digital version. This post-mortem will break down the requirements and elements that went into the DAU CardSim design and the decision to adopt a card game system. The rapid development process used varying levels of simple prototypes for initial design and playtesting, as well as game balance and refinement. The culmination of the design process involved converting the physical card game to a digital version. This presented challenges in creation but lacked many of the inherent problems of developing a digital system from the ground up by streamlining the development cycle.

INTRODUCTION

DAU CardSim

An instance of UCF's CardSim card game framework, DAU CardSim is a multiplayer scenario-based card game designed to reinforce acquisitions skills and teamwork. A group of two to six players take on one or more acquisition team Roles such as Project Manager, Contracting Officer, or Systems Engineering lead. The players collaborate to most efficiently apply their knowledge, skills, and abilities (KSAs) toward a real world acquisition scenario while tackling various obstacles that arise during the course of the project.

Each game is played with one Scenario selected from the available Scenario deck. Scenarios have one or more possible solutions that introduce different completion requirements (varying numbers of required KSAs) and a limited number of allowed

DOI: 10.4018/978-1-61520-739-8.ch005

Figure 1. The major types of DAU CardSim cards

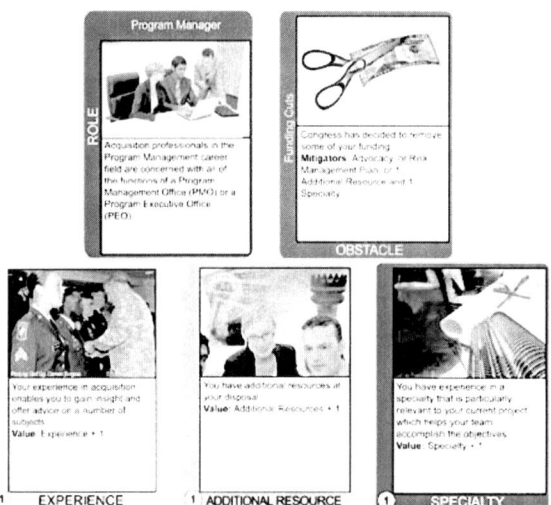

play rounds. In addition to the Scenarios, the game consists of several different color-coded card types that represent the players' Roles and KSAs (see Figure 1):

- Role cards (purple) – There are six of these, and players may take on one or more depending on the number of available players. These are used to identify players' team positions and to mitigate some obstacles.
- KSA cards – These are used to complete Scenario requirements and mitigate obstacles. They come in three different varieties and can be general or specific.
    ○ Experience (yellow) – These represent players' experience both in their overall field and in their specific role.
- Course Levels – These are a special variety of Experience card and represent education in the player's field at three levels: 200, 300 and 400 level courses. When used on a player's Role, this card grants the player additional moves each turn (i.e. the player can play two, three or four cards per turn rather than one).
    ○ Additional Resources (green) – These represent resources such as funding and personnel.
    ○ Specialty (blue) – These represent knowledge in specialized areas, such as Long-Term Contractual Management or Arbitration.
- Obstacles (red) – These represent project interrupters of varying difficulty and can be mitigated by specific KSA cards, multiple generic KSA cards and/or relevant Role cards. Obstacles come in two varieties:
    ○ Programmatic – These are general project changes or events, such as Funding Cuts. Players must divert some of their KSAs from the Scenario to address these obstacles.
    ○ Ethical – These are ethical conflicts or violations committed by team members, such as Nondisclosure violations. All ethical situations are prepared as resolved correctly or incorrectly, and either subtract from or add to the team's Scenario completion.

As a team, players select a Scenario and solution option. All players are assigned one or more Roles and dealt a hand of five KSA cards; all cards are kept face up on the table to encourage collaboration between players in planning out how to most efficiently complete all requirements. Varying Obstacle mitigation requirements, as well as the different Course Levels held by each player, make it valuable for players to assess all team members' hands and discuss who can perform each necessary action most efficiently. Any Obstacle cards drawn during the deal are immediately placed next to the Scenario card; Ethical Obstacles with positive results are applied to the appropriate Scenario requirement.

Each round begins with the Program Manager role and continues around to the left. Each player applies their cards to Scenario requirements or Obstacles as needed. When playing a card, players engage in a "talk aloud" component that involves providing a real world justification for their selection—for example, rather than simply playing an Additional Resource card to mitigate a Reallocation of Staff obstacle, the player explains "as Contracting Officer, to supplement our reduced staff I subcontract to a small business in order to complete a portion of the project work." This talk aloud component, in combination with each Role's use in mitigating certain Obstacles, reinforces all players' understanding of acquisition roles, both their own and those of other team mates. Play continues until either the allotted number of rounds has been exhausted or the team has completed the Scenario. In order to successfully complete the Scenario, players must not only provide all KSAs required by the Scenario but also completely mitigate all Obstacles in play.

After the game is concluded, Instructors guide players through an After Action Review. The AAR is a common element in military training exercises, in which the Instructor walks participants through the exercise to highlight and discuss all successes and failures (Morrison & Meliza, 1999). Generally the Program Manager for each game will act as a team scribe, taking notes during the game for use during the AAR phase to ensure that all relevant points can be addressed.

Development Requirements and Objectives

Most organizations and professions involve a large amount of work performed in teams. Defense Acquisition University, the primary training entity for the acquisition, training, and logistics (AT&L) workforce, is no exception and found themselves in need of new training methods to modify their individual training organization into a more appropriate team-based approach. They also had an interest in implementing gaming and simulation into their courses to increase learner engagement and provide more powerful practice exercises for their students.

The non-tactical nature of their work combined with limitations of their training classrooms necessitated that a low technology gaming intervention be used for intact team training. Many of the DAU classrooms would not have a large number of computers or particularly high-end machines available to run technologically demanding simulations, and given the lack of tactical operations these sorts of simulations would not necessarily be required to communicate accurate scenarios. In addition, DAU's learners tend to be older individuals who may not have a high amount of computer literacy or experience with gaming, meaning that a simpler game that avoided too many hardcore gaming conventions or skill sets would be preferable. Finally, most DAU courses are on a short schedule of no more than a week, meaning that any game or simulation should ideally be playable in a fairly short time frame and without too much training needed beforehand.

The last overall requirement, and one of the most important, was to purchase or develop simulations and games that would be easily modifiable. This would enable DAU to leverage simulations designed for other subjects and modify or replace the assets and scenarios to reflect acquisition situations and information. It would also provide a cost benefit for DAU by allowing them to use the same simulation throughout a large portion of their curriculum, rather than necessitating development or modification of different simulations for each course or subject area.

In addition to DAU's general requirements for games and simulations, specific learning objectives were identified with the help of DAU instructors for DAU CardSim:

1. Apply management skills and strategic thinking within a team setting.
2. Assimilate quickly and get to know team members quickly.
3. Exhibit improved skills in resource allocation, risk management, strategic and tactical planning and execution, teamwork, and problem solving.
4. Gain additional knowledge of acquisition processes, resources, products, and situations.
5. Apply teamwork skills within IPT and other defense acquisition focused teams.
6. Gain additional knowledge about roles and responsibilities of DAU career specialties.
7. Apply performance monitoring and backup behavior within real world team settings.
8. Communicate and coordinate with team members in IPT and other defense acquisition focused teams.
9. Self-diagnose and self-correct as a team.

The overall acquisitions process, DAU's general training priorities and limitations, as well as the identified learning objectives guided development of DAU CardSim. We identified cards as the best game type in order to fit easily into DAU classrooms, regardless of technical limitations, and to minimize issues with audience familiarity and learning curve due to the commonality of card games. To satisfy DAU's need for team-specific training, we prioritized cooperative team play within the game framework. The basic elements of the game: acquisition roles, real world project scenarios, obstacles and KSAs, were included to satisfy these learning objectives and tie the game firmly to learners' job needs.

During this development process, we benchmarked prior research on serious games for training and team training techniques, and studied various card games to help design the game play and ensure alignment with the subject matter and learning objectives. To accommodate quick development on a low budget and the need for both internal and external playtesting we also studied and implemented rapid prototyping techniques.

BACKGROUND

Card Game Structure and Educational Effectiveness

The research foundation of DAU CardSim was partially based on evaluation of the probable familiarity of card games among the target audience to promote rapid uptake, along with the use and effectiveness of card games in educational contexts.

It is difficult to obtain comprehensive statistics on card game play, considering the approximately 3,000 year history of card games and the existence of anywhere from 1,000 to 10,000 distinct games depending on where lines of distinction are drawn (International Playing Card Society, 2006), but we can estimate that most people are familiar with at least one game involving the use of cards. The collectible card game *Magic: The Gathering* alone has over six million players spread across 75 countries (Wizards of the Coast, 2004) and this can be considered to represent only a subsection of card game players, as this game is common primarily among younger individuals and heavy gamers. As such, leveraging a card game framework may result in a greater ease of adoption and reduced learning curve among a wide-ranging audience, especially one that leans more heavily towards older adults.

A sizeable amount of research is continually being conducted on the use of games for educational purposes. As an interactive medium with the possibility of simulating experience and providing instant feedback, games have demonstrated accelerated learning times and greater success rates at completing tasks and synthesizing information (de Freitas, 2006). The ability of games to provide dynamic, naturally increasing difficulty and a balance of new and repeated challenges can supply a flexibility and responsive experience that traditional learning methods often struggle to match (Gee, 2003). These aspects of interactivity, instant feedback and adjustable learning curves/difficulty contribute to the creation of a flow experience through games; this flow experience is what produces the sense of engagement and "fun" that instructors wish to harness by using games. Flow Theory describes optimal experience, where individuals experience a feeling of self-control, control over their environment and enjoyment. In order for a flow experience to exist, certain conditions must be met: "There must be a goal in a symbolic domain; there have to be rules, a goal, and a way of obtaining feedback. One must be able to concentrate and interact with the opportunities at a level commensurate with one's skills" (Csikszentmihalyi, 1990).

Research on serious games and simulations generally focuses on video games of various types (e.g. role playing games, puzzle games, shooting games, etc.) rather than card games, but a number of card games have been developed and evaluated as teaching tools in subject areas including Instructional Design processes, software engineering, and basic biological systems (Betrus, 2001; Odenweller, Hsu, & DiCarlo, 1998; Steinman & Blastos, 2002). These games have seen use in business environments as well, including for management training at Intel Corporation (Clifton, 2006).

When evaluated using not only student satisfaction surveys but also pre- and post-game tests for subject matter knowledge, these games have displayed promising results. Steinman and Blastos (2002) conducted a study of learning outcomes from their immune system card game with three student groups of various ages—a group of 25 eighth graders, a second group of 18 tenth graders and a third group of 8 first-year medical students—and found that the majority of students in all three groups significantly improved their test scores after use of the game. No significant increase was seen on information outside the game's content as a result of re-testing. When game rules were structured to correspond with conceptual foundations and processes within the subject area, students' absorption of game play strategies and feedback related to the subject matter indicate that card games facilitated a greater understanding and synthesis of these concepts and processes (Baker, Navarro, & van der Hoek, 2005; Steinman & Blastos, 2002). Overall student reactions to the use of card games for learning are also promising. Students reported that games were "thought provoking, promoted discussion of important….concepts, and were effective in reviewing the material" (Odenweller et al., 1998, p. S82) and that they liked "the various strategies you can employ" (Baker et al., 2005, p. 12).

Paper Prototyping and Iterative Development

Originating in the fields of engineering and software development, the process of iterative development or rapid evolutionary development describes a development process that makes use of continuous rapid prototyping and testing. Through this repeated process, prototypes give "so much feedback that you can't help but come up with better products" (Arthur, 1992, p. 112). Receiving this feedback and adjusting to it at early stages of development—rather than developing based solely on extensive front-end documentation and conducting testing at the end as in the traditional "waterfall" model—creates flexible growth in the project and reduces development costs and cycle times by an estimated factor of four or more; this is accomplished by shifting most radical changes to points in the development schedule where they require less alteration to the existing product, and therefore are less time consuming and expensive (Arthur, 1992; Royce, 1987; Fullerton, 2008). When the customer is involved throughout the process, this also increases customer satisfaction. The prototype "puts something in front of the client up front—really early—and helps them see what their final project product is going to look like" (Jones & Richey, 2000, p. 75), an especially useful feature because customers often may not fully understand their requirements until they see an implementation (Tripp & Bichelmeyer, 1990).

One method for creating these evaluation prototypes is by assembling paper mockups of various complexities; by using paper stand-ins for system components the development team and any external testers brought in can perform usability testing and then quickly revise the prototype for the next round of evaluations (Snyder, 2003). Designer Eric Zimmerman (2003) describes using paper prototyping in the development of his game SiSSYFiGHT, where the first iteration of the game was "played with post-it notes around a conference table" (p. 2). Tracy Fullerton (2008) breaks the paper prototyping process for games down into four major phases:

1. Foundation – Construct representations of core game play component(s); play just with these, and record questions that arise.
2. Structure – Gradually add specific elements in order to flesh out or accomplish essential game priorities (e.g. add point values or costs associated with the basic game play actions).
3. Formal details – Gradually add rules and procedures to expand and codify game play. If things begin to seem imbalanced, examine these carefully. If necessary, strip down to basics and reintroduce details one at a time to locate the stress points.
4. Refinement – The game is a playable system at this point; question fine details, perform minor clarifications, add flavor, and continue testing for optimal balance and flow within the game play.

This process is especially suited to game design and development, as game design presents a creative endeavor as well as a second-order design problem. Designers create core mechanics and systems, but the real object is the crafting of play—enjoyment, meaning and social interaction—that arises from players enacting the game's systems. This "delicate interaction of rule and play is something too subtle and too complex to script out in advance, requiring the improvisational balancing that only testing and prototyping can provide" (Zimmerman, 2003, p. 12). Even simple paper prototypes can be invaluable in evaluating the core mechanics or entire structure of your game, giving the design team an opportunity to test drive the concept, find the fun in the game—or realize that it is lacking in fun—and also to begin the process of game balancing early on when making adjustment is simple. In this way, prototypes are parts of the design process but also become tools for the designer's understanding of systems and design processes, intermediary products that feed back into and impact the creative process (Rathbun, Saito, & Goodrum, 1997; Briskman, 1981).

These benefits are leading more game developers to opt for iterative processes—often through agile project management methodologies like SCRUM—with rapid, sometimes paper, prototyping. Rapid prototyping, including early paper prototypes, are used routinely by Carnegie Mellon's Experimental Gameplay Project to develop games on one-week schedules (Gabler, Gray, Kucic, & Shodhan, 2005). Paper prototyping and iteration are also a key component in the development process for activist game development group Values at Play (Flanagan & Nissenbaum, 2007). Some large-scale AAA publishers have also turned to agile and iterative methods, including High Moon Studios, Bungie, Epic, and Valve (Buscaglia, 2007; Keith, 2007).

Simulation and Training for Teams

A sizeable proportion of common tasks and projects encountered in the workplace are too large and complex to be resolved by individual efforts. As a result, teams form a cornerstone of the work environment in industry, government, and military settings. These teams pose a challenge for training and simulation because teams have different needs compared to individual students. Due to the widespread need for team training, much research has been devoted to identifying these specific needs and determining best practices for training to meet them.

Team performance can be linked to two overall types of skills. The first type consists of skills related to the technical portions of the job. The second type consists of knowledge, skills and attitudes (KSAs) needed to be an effective team member. These KSAs include interpersonal skills (e.g. communication and conflict resolution), an understanding of the work environment and necessary tasks, knowledge about other team members (e.g. strengths, weaknesses and preferences), interpositional knowledge—an understanding of various team roles and responsibilities—and adaptability (Cannon-Bowers & Salas, 1998b).

One strategy suggested for team training is to ensure that team members train together as an intact unit. By training as they work, team members can rehearse working together and the transfer of knowledge from simulation to real work is smooth. When the team participates in training exercises as a unit, it demands that they use and therefore refine many of the skills necessary to their cooperation.

For greater efficiency, team training should support a guided team self-correction process, usually by including a robust feedback component; this allows team members to share expectations, thoughts, and problems with one another. Furthermore, the instructor can assist students in walking through successes and failures, ensuring that team members locate and discuss the most important aspects of the exercise and helping to mediate team conflicts if absolutely necessary (Cannon-Bowers & Salas, 1998a). Military training exercises often implement such a feedback component using the After-Action Review process, in which instructors and participants review an exercise in detail; feedback gained from these AAR sessions are used not only to improve future team performance, but to develop a knowledge base for other teams and potential revisions of the training exercise (Morrison & Meliza, 1999).

Another strategy that has been implemented for team training is cross-training. During cross-training, team members change roles with one another; as each team member gains experience with the tasks and responsibilities of other team members, they develop better communication skills and a greater degree of interpositional knowledge (Blickensderfer, Cannon-Bowers, & Salas, 1998). Some research has demonstrated that teams with cross-training experience outperform teams who have not been cross-trained (Volpe, Cannon-Bowers, Salas, & Spector, 1996).

CREATING DAU CARDSIM

The beginnings of the CardSim framework began to form in 2005 and 2006 upon examination of the casual game space, along with research showing that casual games could be more inclusive than traditional video games, extending the traditional play base by covering a broader age range and greater gender parity. Casual games such as card and board games have proven quite successful, particularly with companies such as Wizards of the Coast and their *Magic: The Gathering* game, Steve Jackson's *Munchkin*, the *Lord of the Rings* trading card game and a host of others; these games provided enough strategy, complexity and extensibility to support a larger knowledge domain that could be used to promote learning.

As we began examining these ideas, we found that very few of these games had been studied empirically for corporate or government use. We did find several studies, including one substantial pilot of a custom game used at Intel that was primarily a card-driven game with a board as a placeholder (Clifton, 2006). We examined the results of this 500 person study and presented evidence of its success for classroom use to DAU. The idea was compelling based on its portability, scalability and extensibility, which allowed us to map the same game framework to other knowledge bases while using the common infrastructure of cards, which was more broadly accepted within the typical demographic of the military acquisition community.

With the help of John Gilbert, a student at Duke University and graduate of their game design academy, we began playtesting various games to determine play style, game design, and the potential to apply our learning rubric to fit that framework. We also did a patent search and examined what play characteristics had already been patented to ensure that we did not use these techniques or in any way violate intellectual property rights.

Based on a blend of simple card structures and dynamics of group play, no one card game completely fit the early military user requirements. We chose to follow a model that combined the simplicity of the five card hand, found in traditional card games like poker and some collectible card games like *Magic: The Gathering*, with the notion of obstacles found in Mille Bourne, scenario based learning techniques and a team play dynamic. We wrote up the game design, the instructional treatment and the overall programming benefits, including the ability to expand to other delivery platforms (i.e. classroom, online, perhaps mobile and virtual worlds in the future) and presented this to DAU as part of a multi-year plan. They accepted our proposal and we were able to begin work on the actual game design document.

DAU CardSim Design Process Overview

Initial Design

Prior to creation of DAU CardSim, we benchmarked and playtested collectable, trading and traditional card games to see what game play conventions were common in the card game genre, as well as to see what strengths and weaknesses different methods of play offered. These games included well-known commercial games such as *Magic: The Gathering* as well as other training games including the Intel management card game and independent games. The game play and usability of these games was catalogued for use in design of DAU CardSim.

Commercial trading and collectable card games tended to use competition and require outside work to play the game, which we wanted to avoid in DAU CardSim's design in order to provide team training. *Magic: The Gathering* requires players to consider strategies when building their own deck outside of actual game play. We considered many of these strategies for play when building the game, but aimed to make the game playable without outside work like deck building or customization by the player. Many of these games also required large time commitments to learn, but DAU CardSim's design specifications necessitated that this could not be the case.

In examining the cards in commercial games, we noticed a common style of design for the cards: utilizing a picture to represent the card, a text box with either flavor text or information on how to play the card, and color-coding on the border of the card to identify the card type or associations.

The color-coding and graphical representation is a style taken from traditional playing cards. Due to the familiarity of this style the DAU CardSim cards were designed in a similar way. The use of text on the card, giving information about the card and directions on how to play the card was also used, as it eliminated the need to look at a rule book every time a card is played.

To familiarize ourselves with the DAU course subject matter, members of the design team combed through the DAU course guide as a method of identifying possible directions for the games. The course guide was also used to identify what roles the players should be able to take in the game. This analysis also gave a breakdown of what skills and abilities different DAU specialties needed to progress in course levels. This information, combined with a thorough review of the defense acquisition lifestyle, provided the early framework for the game. Milestone B on the Acquisition Lifecycle was pinpointed as the focus of the game.

Initial design was undertaken with a team from METIL and subject matter experts from DAU. Large scale brainstorming meetings, as well as scenario solicitations were used. The game design team then identified scenarios to generate the first cards used in the game.

The subject matter experts from DAU met with METIL team members both in person and through electronic means. The SMEs were solicited to write sample scenarios for incidents a typical acquisition team might face. These scenarios gave a baseline of the types of scenarios and skills that would need to be incorporated into the game play. Game designers from the lab worked to identify key words and commonalities between the scenario proposals to use as the first prototype cards in the game. These common elements were broken down into categories, becoming the basis for the KSA cards. During this time frame, the basics of the game play were also being determined. The requirements of the game necessitated a cooperative rather than competitive framework. Working from this information, as well as the scenarios submitted by the SMEs, the team decided that the best way to have to the team work together was to have them work cooperatively to solve a single scenario.

The skills identified by the team were put into a spreadsheet designed to categorize them as knowledge, skills or abilities. The scenarios submitted tended to have their own built-in obstacles, so in the first version of the game there were no Obstacle cards. Of the scenarios submitted, three were chosen for use in the initial paper prototype. These scenarios were given win conditions based on the best outcomes; they each had very specific requirements to meet win conditions, generally consisting of an assortment of unique cards that players were required to play.

Internal Playtesting with Early Paper Prototypes

The first paper prototype was a printout of the card classification spreadsheet. The cells, which each represented a card, were taped onto standard playing cards to create the feel of a card deck while playtesting. This initial version of the prototype was used to test early game play. The cells were colored on the edges to show their category. The team used these cards to play the three scenarios selected for playtesting. We quickly discovered that having many required cards to complete the scenario made the game take too long to complete. When a round limit was imposed to shorten play, the scenarios became impossible to beat. Additionally, the roles at this point had little to do with actual game play. This early version of the game showed many flaws in the design and core scenarios.

Working with the notes from the early prototype sessions, the team went back to refine game play and scenario requirements. The obstacles were removed from the scenarios and added as a distinct card type in the deck to increase both the challenge of the game and the sense of realism

Figure 2. Early prototypes using regular playing cards

(since real work obstacles tend to arise randomly throughout the project). Scenarios were given a set number of rounds to be completed in and given a certain number of KSAs required for completion, rather than an assortment of specific cards. Instead, obstacle cards were given specific cards required to mitigate them. Certain cards were specified to only be played by certain roles and cards could be traded to other players to assist in meeting play requirements.

The second round of paper prototypes consisted of a new version of the printouts with more information taped onto the cards. During this additional playtesting, more problems were identified. The ability to only play one card during each turn still made the game last too long and the scenarios were rarely completed. The specific mitigator cards for Obstacles often did not come into play, further complicating the ability to succeed at the scenarios.

After this second round of playtesting, Course Level cards were added to the deck; these allowed players to play additional cards during their turn. Course Level cards had to be played in order—for example, you could not use a course level 3 card before playing a course level 2 card. This allowed for faster rounds and increased the rate of solution for scenarios. Generic KSA cards were also added to the deck during this iteration to supplement scenario completion. Role cards were also given the ability to mitigate specific Obstacles without playing a card, to show that different roles were more effective at different tasks.

During the third round of playtesting the deck seemed to work more efficiently, but half of the Course Level cards were unable to be played due to the limitation of playing them in order. The generic KSA cards added more cards to the deck, thus decreasing the likelihood of drawing an Obstacle card and taking some of the challenge out of the game. During this round of playtesting, DAU asked for the addition of ethical obstacles to the game. These were a special obstacle type that stopped game play while players were asked to read and respond to a special ethical scenario. These scenarios were developed from the DAU ethics guidelines and curriculum. The cards were given several scenarios and solutions, which would affect the rest of the game play by adding or subtracting cards or turns. Regular obstacle cards were also tweaked to have specific, non-specific and role mitigators.

The fourth round of playtesting consisted of full card mockups printed on card stock. These

DAU CardSim

Figure 3. Card stock prototypes with one adjusted number card, used in playtesting our SYS203 expansion

cards contained all the text, graphics and color-coding that would be included on the final cards, but were easily and inexpensively created. This portion of playtesting mostly consisted of deck balancing. Cards were added and removed from the deck during different playtesting sessions to see what deck configuration was optimal. The Course Level cards were changed to be playable by anyone at any time.

This fourth prototype deck was used for an extended period of time for balancing, usability and game play testing. The rest of the scenarios received from the SMEs were converted to playable scenarios. Each of these scenarios was playtested multiple times with varying numbers of players to see the likelihood of scenario completion. The deck was adjusted and playtested several times during this period until a final deck configuration was determined.

External Playtesting with DAU Personnel

After SME approval and our internal playtesting of the game were completed, a provisional card deck was finalized. The next step was to demonstrate the game to other DAU instructors and some students in order to test our thoughts about the game play and learning while getting a chance to observe players who understood the acquisitions subject matter but were not already familiar with the game. This deck was produced in house using blank standard poker-sized playing cards and labels. Labels were printed for the front and back of each card then affixed to each side of the blank card. This method allowed the METIL team to create a functional deck at minimal cost. Due to the double labels, these cards were somewhat thicker than standard playing cards, making the deck rather large, but due to their sturdy structure they were easier to shuffle and held up better to

Figure 4. Sticker prototypes

repeated playtesting than the simple card stock versions.

These card decks were used in several playtesting sessions conducted at multiple sites, including the DAU Western Division, the Capitol Northeast at Fort Belvoir, DAU South, and several other locations. Our first testing session was held with a group of DAU instructors and one student. These sessions were useful to acculturate the instructors and students to the game play dynamics and learning outcomes.

We saw multiple useful results from the playtesting sessions. The first major return was that the instructors and players found the game interesting and thought it could be a valuable addition to the classroom. We were also able to see directly, both from practice at the playtesting themselves and commentary by the testers, how the implementation dynamics for the game have a bearing on game play; we noticed that the physical logistics of the room and the time allotted for game play were all factors that could greatly affect use of the game and needed to be taken into consideration.

Early on we discovered that many DAU personnel had different ideas about the best length of play; with the tight schedules for DAU courses, it was determined that a 1-2 hour block would be an optimal length for typical play sessions. The various limited number of rounds for scenarios, as well as the ability to adjust the fidelity required in the talk-aloud portion, would allow for either short play or long play (defined as ranging from one hour up to a full day). We discovered, however, that our initial setup and practice round was significantly longer than initially proposed or anticipated. Playtesting revealed techniques to shorten this process by introducing a demo movie detailing the game play procedure, easier student instructions, and a standardized initial game round delivered using a "starter deck" that would take everyone through the same set of cards in a specific order and demonstrate the key basic game elements. From there each game session took 15-30 minutes depending on the length of the Scenario, the number of Obstacles that appeared at random as players drew cards from the deck

DAU CardSim

Figure 5. The online DAU CardSim interface

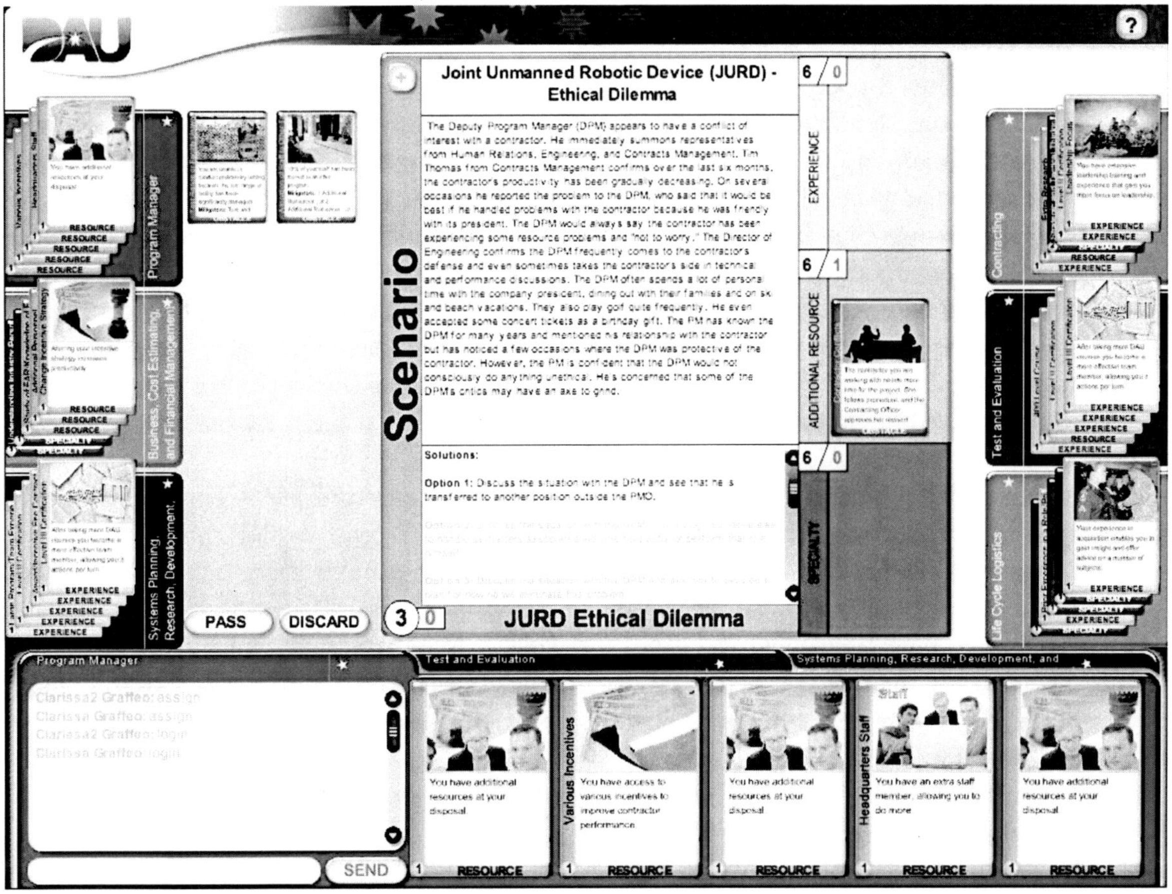

and the amount of talk-aloud technique to promote learning that was achieved during the sessions. The shorter time frames proved optimal and gave more time for AAR and feedback.

We also confirmed that as course objectives and requirements changed, cards and game play structures needed to be adapted. Certain changes would benefit the game overall in terms of implementation logistics (amount of time needed to play, complexity, etc). While the core game play structure remained the same, certain game play elements such as ethical obstacles needed to be further optimized for quicker game play. The sturdier sticker prototypes used during these external playtesting and demonstration sessions were more resilient and usable than our internal prototypes, but the stickers still allowed us to revise cards on-the-fly to work out these new adjustments.

Flash Version Development

In order to support distance learning courses and play outside the classroom, the team partnered with an outside team of developers to create an online Flash version of DAU CardSim. A Flash game played through a standard web browser was selected for the digital format in order to be easily used across any computer and to avoid any security issues when accessed through military computers and networks, since it would not install on the local machine or require any active ports other than port 80 for standard web traffic.

Once the developer was selected, we supplied them with all information about the game rules and deck structure, including a hands-on demonstration session with their team where we played the game with them and answered questions. Having the game rules and deck structure already established freed us up in the early development phases to focus on usability issues related to the interface. All assets required for play and the aspects we wished to prioritize were lined up, enabling us to immediately begin organizing these assets clearly on the screen and gearing the arrangement and functionality to reinforce important game elements (e.g. integrating a text chat to support team communication and organizing Roles around the edge to simulate the team sitting together at a table playing).

Once the layout and supporting visual design elements were established, the development team was able to immediately begin coding the game to existing game specifications. When testing was conducted during development, there were only two areas that required checking: the game behavior compared to the rules for the physical game and functional bugs in the code that would result in crashes, display errors, etc. Code revisions to correct these issues were much easier, faster, and cheaper to implement than if we created a digital game from the ground up, as we had no need to alter entire basic game functions in order to deal with balance problems.

Due to using an outside freelance developer unfamiliar with the game and with other project commitments, in combination with our own largely part-time team structure and other work, did result in some loss of time. In a project developed digitally from the ground up, this most likely would have resulted in an extremely long project and multiple problems related to ironing out game balance and playability. By taking care of all game system development up front with the physical version, however, we were able to reduce a large amount of initial development time and revisions, balancing schedule stretch from other factors.

Expansion Deck Development and Rebalancing

As discussed earlier in the chapter, one of DAU's core requirements for simulations was the ability to easily customize and expand them, both to ensure the scenarios would always reflect real world acquisition information and to enable extensive reuse of the same simulations for many scenarios and courses. Since we determined that card games could be easily modified and expanded through swappable expansion decks, we built in plans to eventually create decks for other DAU courses outside of ACQ 201B.

The first expansion deck constructed to move DAU CardSim into another course was a deck for a Systems Engineering course (SYS 203). Due to the necessity for realism in scenarios and subject information, as well as alignment with course learning objectives, cards were initially developed by a Subject Matter Expert based off of existing cards. A compendium of the full ACQ 201B deck was supplied to provide a framework for deck proportions and scoring values (e.g. the staggered difficulty levels for Obstacle mitigators) and to allow the instructor to retain general cards from the previous deck that would still be applicable for the new course.

Upon delivery of the edited compendium, we discovered an issue resulting from the large amount of information conveyed in the Systems Engineering course. The suggested deck we received did not match the original ACQ 201B deck in the overall number or internal ratio of cards. With a larger deck, meaning a wider distribution of cards, and an altered ratio of Obstacles to KSA cards, we engaged in another internal playtesting cycle to ensure the deck would be balanced. The overall playtesting procedure was similar to the final rounds of our initial development playtesting.

Available Illustrator source files from the previous deck were populated with the images we had already pulled along with the new card text and printed onto card stock using the sticker template. Our team (anywhere from two to four people on any given play through) repeatedly played several Scenarios, especially the most difficult one, to test win/lose ratios with the new deck. Balancing ideas were implemented and tested on the fly by simply crossing out and rewriting the card text. To maintain and examine the overall number balance of the deck, we used our previously developed playtesting spreadsheets to record which Scenario we played, the number of rounds we got through, whether we won or lost the game and by how many points, and the number of Obstacles we encountered. We also labeled each game or sheet by what revision version of the deck we were using once we began making changes to card numbers or scoring.

During the course of playtesting we attempted different solutions for the balance. In order to maintain all important information, we initially tried to maintain the same number of cards delivered by the SME by only altering KSA card values and Obstacle mitigation requirements. We still experienced a markedly increased difficulty due to the larger deck size; however, Course Level cards were not drawn as frequently because they were more widely dispersed throughout the deck and might not appear until too late in the game, if at all. The increased number and altered proportion of programmatic obstacles to ethical dilemmas also caused problems. We found ourselves encountering far too many Obstacles, and there were almost no ethical dilemmas to add to the Scenario requirements; many more Obstacles now had Role mitigators as well, paired with KSA mitigators of extreme difficulty, which skewed the amount of cards drawn by each player (since Role mitigation requires no cards to be used) and the amount of KSA cards diverted away from Scenario requirements. Eventually, with approval from the SME, we implemented a combined solution: reduction of the deck back to original numbers, with revised Obstacle mitigation requirements and increased point values of some KSA cards to balance the revised internal ratios (especially of programmatic to ethical Obstacles).

This rapid implementation and testing procedure enabled us to repair the deck very quickly. All testing and balancing was accomplished in less than one week by a group of no more than four people working part time. The ability to change rules on the fly and alter the cards immediately by writing on them, rather than having to recreate more complex prototypes or consult a complicated rules addendum sheet, definitely saved us valuable testing time.

LESSONS LEARNED

The DAU CardSim development team had game design experience going into this project, but still faced challenges in development. Traditional game design tends to focus on digital prototypes, and while institutions like USC are working to change this, paper prototypes are less used in game development than digital. The DAU team was willing to work with us to create a paper version of the game first then move to a digital version.

This game, like most, started as an idea on a whiteboard. DAU CardSim moved from the whiteboard to notes, and these became the game. In the design of a typical digital game, at this point the designers would plan all aspects of game play and all game elements such as characters and items, and then put them in a game design document. To test certain methods of game play, like puzzles or driving mechanics, the designers might create prototypes. By using an iterative design method that focused on paper prototyping, most of the game design was done in creation of these prototypes instead of being developed through the documentation for the project. This method allowed for more flexibility in the game and more agile changes in game play. Nothing in the game was set in stone, and any aspects of game play or game assets could be change at any time during the development. In traditional digital games, once

Figure 6. Interface mockup for DAU CardSim mobile

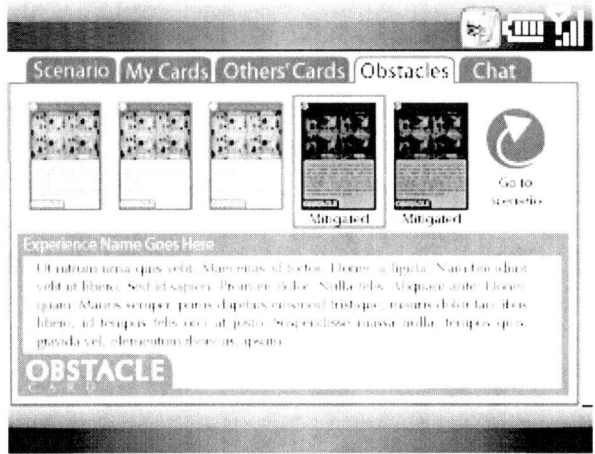

the game production has started it is difficult to change any aspect of the game, especially if they involve core components or functions. Changing out one card for another in the deck or adjusting the scoring numbers was much easier when writing on or printing cards than reprogramming a game engine.

Paper prototyping can be used in a variety of situations, even in games that are solely developed in digital format. User interface design is greatly aided by paper prototyping. Designers can create a mock-up of the interface on paper and record user reactions and timing of use to see how efficient their design is. The elements of the UI can be made individually so designers can rearrange the layout and see how the users feel about each version. Paper prototyping or numerical modeling can also be very good for testing combat systems; they can be easily balanced before programming in the game engine by performing calculations on paper of how hit points and attacks react in specific scenarios. Puzzle game elements can be played completely on paper to test mechanics as well as usability, so that during programming of the game only small elements need to be changed. Role playing games and games based in numerical systems tend to be the most appropriate for paper prototyping, especially since many of them derive from paper and dice heavy tabletop games like *Dungeons & Dragons*.

While there are many elements that can take advantage of paper prototyping and rapid iteration as a design method, there are just as many that must be made in a digital format. Games involving complex physics models and large amounts of digital fidelity, such as first person shooters and racing games, do not lend themselves well to paper prototyping; there is no good way to use a paper prototype to test gun recoil in a FPS or to test the way a car drifts around a curve in a racing game. For some elements, however, these types of games can use numerical calculations as a method of paper prototyping. Calculations can determine how much damage a gun can do to a specific enemy, or how much acceleration versus braking a car should apply on a race track. This sort of numeric balancing can be useful as a method of paper prototyping in many games that otherwise would not be able to take advantage of paper prototyping techniques.

DAU CardSim also took advantage of rapid iteration as a design method. By utilizing a paper prototype model, we were able to make changes to the game play and game elements on the fly. It was possible to make a change in the middle of a testing session if something was not working. During one testing session, when a scenario became unbeatable, we changed the number and types of cards in the deck to see if that made a difference. If a game is created solely in digital, changing game elements or systems can be time consuming from both a programming and an asset creation standpoint. Rapid iteration also allowed game play to be tested without having completed assets.

In addition to rapid prototyping, using a card game model offered several advantages for DAU CardSim. The needs of the client could have been met in different ways, but developing a card game framework allowed for an easier design process with a smaller team. While a fully digital simulation could have been created from the start, the objectives of the project, specifically intact team training, could be met without a high technology intervention. Using a card game framework allowed for easy adoption, based on the previous game experiences of the students, and also took less time to learn than the game play of a fully digital intervention.

FUTURE RESEARCH DIRECTIONS

Given the encouraging results from our current deployments, we plan to continue expanding DAU CardSim and refining the online component. Much of our future research direction involves keeping with the ideals of flexibility, easy modification, and scenario reuse that has proved successful and valuable thus far.

Immediate next steps include additional expansion decks for courses on subjects like Acquisition Leadership, Space Acquisitions and Lean Six Sigma. We also intend to develop an authoring tool that will expedite the process of gaining subject matter expert suggestions for new cards, as well as populating those cards into the online version of the game. By having a tool to streamline the initial information gathering and content creation for the cards, it will enable both our team and DAU to be extremely agile in addressing key subject areas or changes to course requirements.

We also hope to continue extending DAU CardSim and its core scenarios into other delivery modalities in order to best serve the entire acquisitions community, as we did in developing an online version to suit distance-learning courses. Multiple delivery options will ensure that all students and instructors can use the content in a way best suited to their course organization. In concert with DAU interest in these technologies and platforms, we have been examining usability issues for delivering DAU CardSim through mobile devices and integrated with courses offered in virtual worlds such as Second Life (see Figure 6 and Figure 7).

As we continue to work with DAU to deploy DAU CardSim into additional courses, we look forward to examining the resulting course metrics and student feedback. Beyond our development playtesting, these student comments continue to identify remaining problem points in the game and its integration into courses. With this data, we hope to further refine DAU CardSim and determine the optimal way to position it within overall programs and individual courses, including best practices for overcoming student learning curve, most effective motivations (e.g. whether to introduce competition between teams for completion efficiency) and how to time the game's use in the course to bolster student attention and interest without distracting from other course material or procedures. We feel that this data will not only assist our team and DAU in DAU CardSim's deployment, but prove valuable to fellow serious game developers and instructors implementing game-based solutions.

Since DAU CardSim's core framework was developed for expansion and specialization, we also hope to extend DAU CardSim beyond the acquisitions framework and into other subject areas. Many business environments and processes use similar skill sets and require the same team training concepts at the core of this game system,

Figure 7. Test virtual world integration of DAU CardSim using Second Life

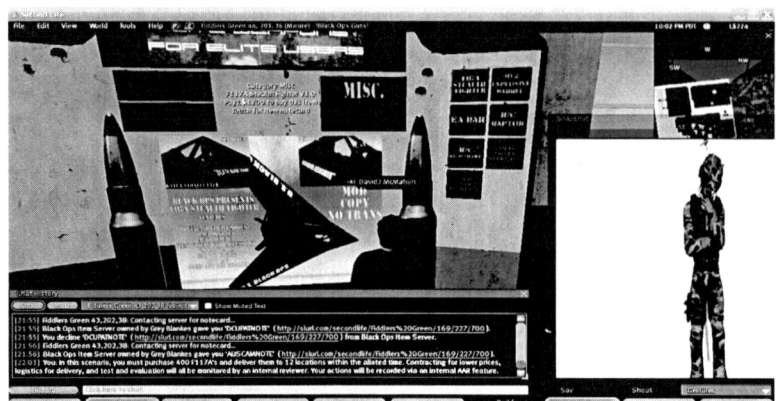

so we hope to extend this framework out to meet those needs.

CONCLUSION

Paper prototyping is a quick and flexible way to ensure that an initial game design is functional and enjoyable while still retaining important information in the case of serious games. The ability to test and refine game systems early on, especially when it can be done as quickly and easily as printing out or writing on a piece of paper, saves time and money while ensuring good game design and alignment with the customer's needs. This process can be vital when other elements of the project, such as a lack of full-time personnel or a requirement for outsourcing to other teams, introduce delays and complications into the development process.

Paper prototypes may not be able to completely encompass all games as well as they do a card game like DAU CardSim, but even very complex video games can benefit from this design process. These simple mockups can model many of the core game play actions and numerical systems involved in the game, and should be kept in mind as the designer's first line of defense against imbalanced or boring game play and the costs associated with fixing it later in the production process.

REFERENCES

Arthur, L. J. (1992). Quick & dirty. *Computerworld, 26*(50), 109–112.

Baker, A., Navarro, E. O., & van der Hoek, A. (2005). An experimental card game for teaching software engineering processes. *Journal of Systems and Software, 75*(1-2), 3–16. doi:10.1016/j.jss.2004.02.033

Betrus, A. (2001). *The many hats of an instructional designer card game – overview*. Retrieved November 10, 2008 from http://www2.potsdam.edu/betrusak/idcardgame/idcardgameoverview.html

Blickensderfer, E. L., Cannon-Bowers, J. A., & Salas, E. (1998). Cross training and team performance. In J.A. Cannon-Bowers & E. Salas (Eds.), *Making decisions under stress: Implications for individual and team training* (pp. 299-312). Washington, DC: American Psychological Association.

Briskman, L. (1981). Creative product and creative process in science and art. In D. Dutton and M. Krausz (Eds.), *The Concept of Creativity in Science and Art*. Boston: Martinus Nijhoff.

Buscaglia, T. (2007). Game law: Scrum deals – good, bad or ugly! *Gamasutra: The Art & Science of Making Games*. Retrieved November 17, 2008 from http://www.gamasutra.com/view/feature/1662/game_law_scrum_deals__good_bad_.php

Cannon-Bowers, J. A., & Salas, E. (1998a). *Making decisions under stress: Implications for individual and team training*. Washington, DC: American Psychological Association.

Cannon-Bowers, J. A., & Salas, E. (1998b). Team performance and training in complex environments: Recent findings from applied research. *Current Directions in Psychological Science, 7*(3), 83–87. doi:10.1111/1467-8721.ep10773005

Clifton, T. (2006). *Training Developer Forum Use of Business Simulations to Grow General Manager Skills*. PowerPoint, Palm Desert, CA.

Csikszentmihalyi, M. (1991). *Flow: The psychology of optimal experience*. New York: Harper Perennial.

de Freitas, S. I. (2006). Using games and simulations for supporting learning. *Learning, Media and Technology, 31*(4), 343–358. doi:10.1080/17439880601021967

Flanagan, M., & Nissenbaum, H. (2007). A game design methodology to incorporate social activist themes. In *CHI '07: Proceedings of the SIGCHI conference on Human factors in computing systems* (pp. 181-190). New York: ACM Press.

Fullerton, T., Swain, C., & Hoffman, S. (2008). *Game design workshop: A playcentric approach to creating innovative games,* (2nd ed.). Burlington, MA: Morgan Kaufmann Publishers.

Gabler, K., Gray, K., Kucic, M., & Shodhan, S. (2005). How to prototype a game in under 7 days: Tips and tricks from 4 grad students who made over 50 games in 1 semester. *Gamasutra: The Art & Science of Making Games*. Retrieved November 2, 2008 from http://www.gamasutra.com/features/20051026/gabler_01.shtml

Gee, J. P. (2003). What video games have to teach us about learning and literacy. *ACM Computers in Entertainment, 1*(1).

International Playing-Card Society. (2006). Playing-card games. *The International Playing-Card Society*. Retrieved November 15, 2008 from http://www.cs.man.ac.uk/~daf/i-p-c-s.org/faq/games.php

Jones, T., & Richey, R. (2000). Rapid prototyping methodology in action: A developmental study. *Educational Technology Research and Development, 48*(2), 63–80. doi:10.1007/BF02313401

Keith, C. (2007). Scrum and long term project planning for video games. *Gamasutra: The Art & Science of Making Games*. Retrieved November 14, 2008 from http://www.gamasutra.com/view/feature/3142/scrum_and_long_term_project_.php

Morrison, J. E., & Meliza, L. L. (1999). *Foundations of the after action review process* (Special Report 42). Alexandria, VA: U.S. Army Research Institute for the Behavioral and Social Sciences.

Odenweller, C. M., Hsu, C. T., & DiCarlo, S. E. (1998). Educational card games for understanding gastrointestinal physiology. *The American Journal of Physiology, 275*(6 Pt 2).

Rathbun, G. A., Saito, R. S., & Goodrum, D. A. (1997). Reconceiving isd: Three perspectives on rapid prototyping as a paradigm shift. In O. Abel, N.J. Maushak & K.E. Wright (Eds.), *19th Annual Proceedings of Selected Research and Development Presentations at the 1997 National Convention of the Association for Educational Communications and Technology* (pp. 291-296). Ames, IA: Iowa State University.

Royce, W. W. (1987). Managing the development of large software systems: concepts and techniques. In *ICSE '87: Proceedings of the 9th international conference on Software Engineering* (pp. 328-338). Los Alamitos, CA: IEEE Computer Society Press.

Steinman, R. A., & Blastos, M. T. (2002). A trading-card game teaching about host defence. *Medical Education, 36*(12), 1201–1208. doi:10.1046/j.1365-2923.2002.01384.x

Tripp, S., & Bichelmeyer, B. (1990). Rapid prototyping: An alternative instructional design strategy. *Educational Technology Research and Development, 38*(1), 31–44. doi:10.1007/BF02298246

Volpe, C. E., Cannon-Bowers, J. A., Salas, E., & Spector, P. E. (1996). The impact of cross-training on team functioning: An empirical investigation. *Human Factors: The Journal of the Human Factors and Ergonomics Society*, 87-100.

Wizards of the Coast. (2004). *Magic: The gathering: Noteworthy facts about the premier trading card game*. Available online at http://ww2.wizards.com/Company/Press/

Zimmerman, E. (2003). Play as research: The iterative design process. *Ericzimmerman.com*. Retrieved November 5, 2008 from http://www.ericzimmerman.com/texts/Iterative_Design.htm

ADDITIONAL READING

Aldrich, C. (2003). Simulations and the learning revolution: An interview with Clark Aldrich. Retrieved from http://technologysource.org/article/simulations_and_the_learning_revolution/.

Aldrich, C. (2005). *Learning by doing: A comprehensive guide to simulations, computer games, and pedagogy in e-learning and other educational experiences*. San Francisco, CA: Pfeiffer.

Bogost, I. (2007). *Persuasive games: The expressive power of videogames*. Cambridge, MA: The MIT Press.

Bogost, I. (2008). *Unit operations: An approach to videogame criticism*. Cambridge, MA: The MIT Press.

Cannon-Bowers, J. A., & Salas, E. (1998). *Making decisions under stress: Implications for individual & team training* (1st ed.). Washington, DC: American Psychological Association (APA).

Csikszentmihalyi, M. (1991). *Flow: The psychology of optimal experience*. New York, NY: Harper Perennial.

Fullerton, T., Swain, C., & Hoffman, S. (2008). *Game design workshop: A playcentric approach to creating innovative games* (2nd ed.). Burlington, MA: Morgan Kaufmann Publishers.

Gee, J. P. (2007). *What video games have to teach us about learning and literacy* (2nd ed.). New York, NY: Palgrave Macmillan.

Gee, J. P. (2007). *Good video games and good learning: Collected essays on video games, learning and literacy* (1st ed.). New York, NY: Peter Lang Publishing.

Guzzo, R., & Salas, E. (Eds.). (1995). *Team effectiveness & decision making in organizations*. San Francisco, CA: Jossey-Bass.

Hargrave, C. P. (2001). *A history of playing cards and a bibliography of cards and gaming*. Minneola, NY: Dover.

Jenkins, H. (2008). *Convergence culture: Where old and new media collide* (Revised.). New York, NY: NYU Press.

Koster, R. (2004). *A theory of fun for game design*. Phoenix, AZ: Paraglyph.

Michael, D., & Chen, S. (2005). *Serious games: Games that educate, train, and inform*. Florence, KY: Cengage Learning, Inc.

Moronta, L. (2003). *Game development with actionscript*. Florence, KY: Cengage Learning, Inc.

Parlett, D. (1991). *A history of card games*. New York, NY: Oxford University Press.

Prensky, M. (2007). *Digital game-based learning*. St. Paul, MI: Paragon House Publishers.

Rosenzweig, G. (2007). *Actionscript 3.0 game programming university*. Que Publishing.

Rouse, R. (2001). *Game design: Theory and practice* (2nd ed.). Wordware Publishing, Inc.

Salen, K., & Zimmerman, E. (2003). *Rules of play: Game design fundamentals*. Cambridge, MA: The MIT Press.

Salen, K., & Zimmerman, E. (2005). *The Game Design Reader: A Rules of Play Anthology*. Cambridge, MA: The MIT Press.

Sigman, T. (2005). The siren song of the paper cutter: Tips and tricks from the trenches of paper prototyping. *Gamasutra: The Art & Science of Making Games*. Retrieved November 2, 2008 from http://www.gamasutra.com/features/20050913/sigman_01.shtml.

Snyder, C. (2003). *Paper prototyping: The fast and easy way to design and refine user interfaces (The morgan kaufmann series in interactive technologies)*. Morgan Kaufmann.

Chapter 6
Kinesthetic Communication for Learning in Immersive Worlds

Christopher Ault
The College of New Jersey, USA

Ann Warner-Ault
The College of New Jersey, USA

Ursula Wolz
The College of New Jersey, USA

Teresa Marrin Nakra
The College of New Jersey, USA

ABSTRACT

Despite the maturation of the video games medium, most self-identified learning games take the traditional but flawed approach of transmitting fact-based content to the user, frequently through the superimposition of "drill and practice" quizzes on top of interactive game-play that has little inherent relationship to the subject matter. A model is described for a Spanish-learning video game that adopts a different approach, through a close integration of the learning content and the game world context, and through the application of a motion-based controller that provides the user with an innovative and pedagogically potent mechanism for communicating with the learning system. Foundational research is discussed pertaining to kinesthetic learning techniques and their potential for language acquisition. A proof-of-concept is detailed, in which the user demonstrates learning by executing appropriate gestural responses to commands or questions spoken by non-player characters. Language mastery is essential to the user's success in the immediate game environment, and also to resolving the game's underlying narrative.

INTRODUCTION

The question of whether video games can facilitate real learning dates to the very beginning of interactive computing technology in the late 1960s. Early experiments in Artificial Intelligence focused on the learning potential of the computer by developing game-playing agents for classic board games such as checkers. Researchers soon observed, however, that such interactions often led to demonstrable learning on the part of the human user, inspiring the hypothesis that interactive computer interfaces could foster rich learning environments for people.

Almost half a century later the promise still appears unfulfilled, despite significant technological advances including immersive three-dimensional worlds and sophisticated input devices, and despite the maturation of the medium to include a variety of game styles and to appeal to a wide spectrum of game players. An underlying theory of how learning can take place in such environments remains in its infancy, with most self-proclaimed learning games resembling either a multimedia, screen-based version of a printed textbook, or an awkward hybrid of game-play glued on top of rote instruction or vice-versa. Such games do little to encourage larger abstract problem-solving skills, nor do they accommodate likely variations in learning styles on the part of the players/students.

In this chapter we posit a game-design architecture that fully exploits the pedagogical potential of a rich graphical environment, coupled with an intuitive and sophisticated kinesthetic interface. While many educational games are structured around their particular content — the information to be imparted to the user — our approach is grounded in the game's context, with the assumption that genuine learning takes place in context, and that the richer and more compelling that context, the more thoroughly concept and skill mastery will occur. In this way, our work complements that of Janet Murray (1998) and other theorists who point toward the need for richer, contextualized environments to improve the educational applications of interactivity. Building on the Constructionist approach of Seymour Papert, we endeavor to move beyond games that demand the single "correct" answer to a multiple choice question in favor of broader processes of discovery and problem-solving. By creating a Spanish-learning game in which students are able to exercise considerable control over their environment, and fulfill learning goals by means of the innovative and intuitive physical interface, the students' journey through the story is intimately tied to their learning process.

The development of kinesthetic interfaces such as the Nintendo Wii controller opens up countless new possibilities for computerized foreign-language teaching. Using a motion-sensitive controller has allowed us, for instance, to assimilate well-tested pedagogical principles that link sensory-motor experience and second-language learning, such as James Asher's Total Physical Response (TPR), first introduced in the 1960s. In Asher's movement-based method, teachers give students verbal commands to act out without having to speak. During the first few hours of classroom time, students perform simple commands such as jumping, sitting, and standing, though instructions quickly become more complex (Asher, 2003). Studies by Asher suggest that in beginning foreign language classrooms, students taught using Total Physical Response outperform those schooled using more traditional methods (Asher, 1965, 1966, 1969, 1977; Asher, Kusudo, & De la Torre, 1974). Other researchers have supported the idea that bodily movement aids in vocabulary retention (Swaffar & Woodruff, 1978; Thiele & Schneiber-Herzig, 1983; Wolfe & Jones, 1982). Most recently, Seth Lindstromberg and Frank Boers (2005) suggest that TPR may aid in more complex forms of language processing.

While we have always encouraged the consideration of alternate, specialized controllers in our game development courses at The College of New Jersey, the ubiquity and easy interoperability of the Wii controller inspired students and faculty to create a game in which motion-sensitive controllers allow players to interact with a language-learning game in an entirely new way. In itself, the controller sets this game apart from any other language-learning game on the market. Yet our decision to use the Wii-Mote to interact with the game is not driven by no novelty, but the battery of studies suggesting Total Physical Response to be one of the most effective methods for teaching beginning language students. Our resulting proof-of concept is a single-player Spanish language-learning game that is loosely modeled on the "first-person-shooter" genre, although there is no shooting or violent fighting. In the game, the player must traverse a complex story, solving immediate problems in order to achieve an ultimate outcome, all the while listening and responding appropriately to Spanish utterances that increase in sophistication at each game level. The player demonstrates language mastery through gestures, by executing actions that support h h his or her immediate survival — such as swiping an ID card to enter a building — while also advancing the game-play. Using the Wii-Mote allows the player to interact with the game though gestures alone, a means of response that is backed up by decades of research into language-learning: most importantly the notion that comprehension precedes production — that students are able to understand far more than they can produce verbally or in written form, and that it is through listening to comprehensible utterances in the target language, and not through speaking or writing, that competency emerges.[1] In addition, this dynamic means of interaction promises to appeal to kinesthetic learners who learn best while reenacting real-world experience, a group of learners whose needs are too often ignored in the traditional classroom setting, as well as in traditional learning games.

INTERACTIVITY AND LEARNING

In the early days of the video games medium, landmark games such as *The Oregon Trail* (1971) and *Lemonade Stand* (1973) attempted to capitalize on both the allure and the processing power of computers in order to impart facts and simulate real-world experiences. As with any new medium, most early examples of educational or "edutainment" software were simply digital versions of printed workbooks and other traditional teaching materials, similar to the "shovelware" of the mid 1990s, when companies and individuals simply "shoveled" their existing printed content onto the web. Papert — a pioneer of computer-empowered learning with the Logo programming environment in the late 1960s — was an early and vocal critic of this approach, reflecting on the typical "edutainment" game in a 1998 essay:

The player gets into situations that require an appropriate action in order to get on to the next situation along the road to the final goal. So far, this sounds like "tainment." The "edu" part comes from the fact that the actions are schoolish exercises such as those little addition or multiplication sums that schools are so fond of boring kids with. (p. 88)

Much of the early efforts in educational games — as well as much of the criticism — focused on the content of the games, an approach consistent with a behaviorist approach to learning. From the behaviorist perspective, there exists a body of knowledge that is objectively and universally true. The facts are the facts, whether it is a multiplication problem, the date of a Greek battle, the chemical formula for dichlorinedioxde or the correct spelling of *insouciant*. The job of the educator — or the educational game — is to impart that body of knowledge to students, usually through repetition and reinforcement, both positive (you get a gold star) or negative (you get a red X). Games like those described by Papert took the accepted, objective facts about a subject and imparted them to players by means of text and simple graphics on a computer screen. Each player was presented with the same information, and each answered the same questions to demonstrate their comprehension of that information. The answer was either right or wrong — right and player moved on to the next parcel of knowledge, wrong and the player got another look at the previous one. Such early educational computer games did little to leverage the expressive or processing power of the new medium, with little or no difference between the content on the screen and the content in a textbook. (In fact, a book with photographs and illustrations presented allowed for more information and context than a single-color computer monitor.)

More recently, James Paul Gee (2003) has argued for a more constructivist view of learning in video games. While he acknowledges that games such as *Civilization* may impart objective facts about historical events, he stresses that the more valuable learning takes place as a result of ongoing experimentation and exploration on the part of the player, a sentiment which echoes the following statement from Papert's essay:

Another factor is that games are designed so that the learner can take charge of the process of learning, thus making it very different than school learning, where the teacher (or the curriculum designer) has made the important decisions and the "learners" are expected to do what they are told—which is no way to learn to be a good learner. (p. 88)

Games such as the *Myst* series or *Fable* confront the player with a new world, just as a student may be confronted with a new subject, and the player has to map out the geometry of that world, to learn its customs, its vocabulary and its grammar, and most importantly to learn what qualifies as success and failure — what it means to win or lose. One learns these things by playing the game. Steven Johnson (2006) refers to this idea as "probing the physics" of the game world, something he likens to the scientific method:

1. The player must *probe* the virtual world (which involves looking around the current environment, clicking on something, or engaging in a certain action).
2. Based on reflection while probing and afterward, the player must form a *hypothesis* about what something (a text, object, artifact, event, or action) might mean in a usefully situated way.
3. The player *reprobes* the world with that hypothesis in mind, seeing what effect he or she gets.
4. The player treats this effect as feedback from the world and accepts or *rethinks* his or her original hypothesis. (p. 45)

This line of thinking is frequently carried beyond the classroom to the workplace. While it is true that being a corporate lawyer, a pastry chef or a chemical engineer requires a firm grasp of some objective knowledge, excelling at any of those vocations also requires a great deal of thinking on one's feet, an ability to quickly divine the rules of a new environment and recognize what qualifies as good work and bad work in that domain. Ian Bogost (2007) points out that Gee largely sidesteps the particular content or context of a particular game, arguing more for the educational potential of the medium as a whole. In a sense, any game that encourages the player to explore, adapt, and develop his or her own understanding of a context creates an opportunity for learning. From this perspective, a game such as *Grand Theft Auto: San Andreas*, which presents several easy targets for critics in terms of objectionable content, is in fact rich in educational potential based on the wide-open, non-linear, flexible structure of the game.

Content aside, applying Gee's argument to a game like *Grand Theft Auto* — or something less controversial like a recent *Legend of Zelda* game — is relatively easy, as the games' physics allow for that sort of free-form exploration. But applying Gee's view retroactively, to games such as *Lemonade Stand*, becomes more problematic. The potential for Gee's brand of learning has evolved just as the medium itself has evolved, including the development of more capable hardware and more sophisticated authoring techniques. What might qualify as teaching and learning in the text-only games of the seventies is demonstrably different from the teaching and learning that take place in an immersive, 3D game of today — a trajectory that parallels, in some respects, the emergence of constructivism in contrast to behaviorism. Critics of the behaviorist approach often call to mind the stereotypical college lecture, with a professor leaning on the lectern, reading aloud pages prepared many semesters ago, imparting their knowledge to a group of students who absorb this knowledge and accept its veracity based on the authority and the experience of the professor. Each student is an empty (and largely passive) vessel for the teacher to fill. From a constructivist perspective, effective teaching and learning in a classroom require an ongoing dialogue — some sort of back and forth between the student and the teacher, between the student and the text or whatever form the material happens to take. The student is encouraged not to placidly accept but to actively query the teacher, the material, the environment, and with each round of this dialogue, the student's perspective evolves.

Chris Crawford — creator of landmark games such as *Balance of Power* (1985) — uses similar language to describe effective interaction design, whether the design object is a game or a word processor (Crawford, 2002). He approaches interactivity as a conversation between the computer (or the game console) and the user (or player). The computer speaks to the user in various ways — visually, as green text on a black screen in the case of the early text-based computer games, or as a collection of thousands of carefully shaded polygons arranged in 3D space. The computer also speaks audibly, through headphones or speakers. And many game consoles also speak to the player through the hand-held controller, vibrating when a player collides with an object, for instance. Users receive the computer's output — they hear what the computer is saying — and they react appropriately, sometimes mashing buttons in a split-second reflex, or taking the time to deliberate over a complex puzzle. Then the user has a turn to speak, using a game controller, a keyboard, a mouse, or in some cases a microphone. Finally, the computer considers this input, processes it, and responds in kind, continuing the dialogue. Crawford argues that the most satisfying interaction is a well-balanced conversation, where both parties hold up their end, and the more means of expression the better (Crawford, 2002). After all, much of the richness of a conversation depends not only on the words being spoken, but also how they are spoken, and the expressions, gestures and other "body language" that give the words greater context.

Applying this model of a conversation to the question of teaching and learning in games, one would expect that the richer the dialogue — the more means each party has to speak, think and listen — the greater the potential for learning. Unlike early games, today's games allow both the computer and the user more vocabulary and more means of contributing to the conversation. In "real life," conversation skills vary — some people may be smooth talkers while others stutter, some may punctuate every statement with a hand gesture while others may keep their hands rooted deep in their pockets. Still others may prefer to pass notes or send text messages, even when their conversation partner is within earshot. Building on the constructivist model of learning as a dialogue, today's games give players more ways to participate in that dialogue, and in doing so provide the player with more ways — different ways — to learn.

KINESTHETIC LEARNING AND DEVELOPMENTAL THEORY

Teaching to Students' Strengths

In the fields of gaming and educational technology, the notion of interactivity as a conversation has allowed theorists to rethink the relationship between the learner and the educational game. By similarly imagining the classroom to be a conversation between teacher and students, we can also begin to consider new and different ways to allow students to participate in that conversation. Prior to 1970, few studies catalogued the wide array of factors that determine how people learn. Now, it is commonly accepted that individuals focus on, process and retain information in varying and sometimes contradictory ways. In 1972, Rita and Kenneth Dunn developed the Learning Styles Index (LSI), which initially included 12 factors; in 1990 they added nine additional variables (Dunn & Dunn, 1993). Of all the environmental, physiological, psychological and sociological preferences the LSI measures, Dunn and Dunn suggest that the perceptual channel through which students begin to "concentrate on, process and retain new and difficult information" is of great importance (Dunn & Dunn, 1993, p. 2). In fact, researchers often label a student as a specific kind of learner based on their preferred perceptual channel. While auditory learners are most capable of learning through aural input, visual learners prefer visual input, tactual learners excel when presented with tactile input, and kinesthetic learners learn best when reenacting real-world experiences.

Studies have shown that students' perceptual preferences are not static, but that they evolve as they get older. While children are predominantly kinesthetic learners, the secondary school population is comprised of 20-30% auditory learners, 40% visual learners and 30-40% tactual and kinesthetic learners (Dunn & Dunn, 1993). However, it is unclear if if students' perceptual preferences change due to biological or environmental factors, as primary school classrooms often abound with kinesthetic and tactual experiences and activities, while these are rarely present at the secondary level. Though tactual and kinesthetic learners make up 30-40% of the secondary school population, Dunn and Dunn suggest that high school teachers rarely cater to the sensory needs of these adolescents. For instance, teachers often present new material through assigning readings. They review content by writing notes on the blackboard, by lecturing or engaging the class in discussion. While these methods may be effective for visual and auditory learners, who process and retain information based on what they see and hear, they are rarely effective for kinesthetic and tactual learners who learn by doing and by manipulating objects (Dunn & Dunn, 1993).

Language Learning and Sensory-Motor Experience

As many studies have suggested, kinesthetic learners will benefit when teachers present content according to their sensory strengths. But beyond reaching out to kinesthetic learners, researchers have suggested inherent connections between language acquisition and sensory-motor experience. For instance, Jean Piaget (1963) theorized that young children acquire language during their sensory-motor period of development (their first two years of life). According to the Piagetian model, pre-verbal children begin to conceptualize their worlds based on their own bodily movement as well as through repetitive physical interactions with the people and objects that surround them. In this way, Piaget suggests that it is through "motor-concepts" and not speech that children begin to form concepts about their surroundings.

While Piaget's theory linking sensory-motor experience and language development applies to very young children, Asher theorizes that adults wanting to learn a second language will experience greater success through techniques that approximate the way in which they learned their first language (through sensory-motor experience, active-listening and mimicry). In the 1960s, Asher (a developmental psychologist) introduced the concept of the "Total Physical Response" (TPR). In Asher's movement-based method, teachers are to give students verbal commands to silently act-out. During the first few hours of classroom time, students perform simple commands such as jumping, sitting, and standing, though instructions quickly become more complex (Asher & Adamski, 2003). Asher and other researchers suggest that with a little creativity, teachers can use TPR to for even complex grammar structures.[2] By continually recombining the order of commands, instructors can assure that students are not merely memorizing fixed sequences of behavior. In addition, Asher, Kusudo, and De la Torre suggest that introducing surprising, zany, and bizarre commands will maintain a high level of student interest (Asher et al., 1974).

Two research findings highlight the need for more kinesthetic and tactual experiences in the secondary school classroom. First, studies have shown that almost all so-called underachieving or problem students at the secondary level are tactual or kinesthetic learners. Second, a battery of studies of secondary school students have revealed that "adolescents achieved more, behaved better, and liked learning better" when teachers present information in a way that takes their perceptual strengths into account. Studies of primary-school children and of adult continuing-education students have documented similar findings (Dunn & Dunn, 1993, pp. 11-18).

Asher's own studies suggest that TPR surpasses other methods for teaching vocabulary to beginning students (Asher, 1965, 1966, 1969; 1977; Asher et al., 1974). Similarly, studies by other researchers find that physical movement helps foreign-language students to retain vocabulary (Swaffar & Woodruff, 1978; Thiele & Schneiber-Herzig, 1983; Wolfe & Jones, 1982). In 2005, Lindstromberg and Boers found that TPR contributes not only to vocabulary retention, but to understanding more complex forms of language. For instance, these researchers suggest that motoric enactment, the cornerstone of Asher's TPR, not only helped Dutch learners of English "better acquire English manner-of-movement verbs in their common literal senses," but that it aided in their interpretation of previously unknown metaphorical uses of those verbs (Lindstromberg & Boers, 2005, p. 241). In addition to suggesting that vocabulary retention and understanding are increased as a result of acting out the meaning of a verb, these researchers also suggest that observing someone else do the action produces better recall than listening to an aural definition of the verb (Lindstromberg & Boers, 2005, p. 254).

Research into the correlation between sensory-motor experience and language learning is consistent with the idea that teachers should provide classroom instruction beyond the typical "drill-and-kill" methods in which students memorize dialogues, repeat verb conjugations and complete fill-in-the-blank activities. Stephen Krashen and Tracy Terrell (1983) suggest that the principal means by which students develop competency in a second language is through understanding contextualized messages in the target language, a phenomenon they call the reception of "comprehensible input." Moving away from the tenets of most twentieth-century language teaching methodologies, Krashen and Terrell advocate eliminating explicit grammar instruction, drills and rote memorization. Through messages delivered in the target language that students find interesting and challenging, Krashen and Terrell suggest that students will "pick-up" or acquire language in a way similar to the way in which they learned their first language.

Krashen and Terrell's focus on comprehension over production (or input over output) has been so widely adopted that Bill VanPatten, a leading theorist in second-language acquisition, recently affirmed that "few discount a role for input and acquisition is generally characterized in just about any theory as being input dependent in some way" (VanPatten, Williams, Rott, & Overstreet 2004, p. 29). VanPatten and many others propose a theory of second-language learning called "Connectionism," a context-based model that takes ideas from Piaget, Krashen, and Terrell. Connectionists suggest that second-language acquisition occurs by means of a series of "form-meaning connections," consisting of "(1) making the initial connection, (2) subsequent processing of the connection, and

(3) accessing the connection for use" (VanPatten et al., 2004, p. 5).

While some foreign-language textbooks take current research into account, many books, companion web-sites, ancillary materials and so-called Spanish-learning video games continue to drill verb forms and vocabulary in a way that does not help students to make connections and which in no way appeals to kinesthetic and tactual learners. Similar to the classrooms that Dunn and Dunn criticize, these types of games and activities rarely cater to non-visual and auditory learners and rarely exploit the possibilities the computer offers for students who learn by doing, manipulating objects and by re-enacting real-world experiences. As far as we know, there is no foreign-language learning game that explicitly stimulates the connection between sensory-motor experience and language acquisition.

NEW INTERFACES AND MEANS OF PARTICIPATION

Our game development courses have always included a discussion of the evolution of physical game interfaces and the forces driving that evolution. As we have developed our games, an ongoing design concern is how the hardware interface might contribute to or detract from the immersiveness of the game experience. Even before the Wii-Mote was available, the students and faculty recognized the potential of such a device to open new methods of game-play for experienced players, but more importantly to break down barriers of entry for people new to games, and others who may have felt comfortable with the relatively simple controls of early games, but who lost interest in games when joysticks started to give way to 12-button control pads. The Wii-Mote combines a number of sophisticated sensors in an inexpensive, robust device that communicates easily with a PC using the Bluetooth wireless data transmission protocol. Its combination of positional and accelerational gestural information, along with more standard push-button input, has made the Wii-Mote a powerful tool for simulating real-world interactions not possible with a standard game controller. As the market success of the Wii console has since confirmed, the ability for players to pick up the controller and do more-or-less what they would expect to do — swing a racket across one's body or point at a target on the screen and pull the trigger — has increased curiosity and eased intimidation on the part of many casual game players. Industry analysis, personal experience, as well as our own user testing, indicate that the Wii-Mote encourages more rapid and persistent engagement with the game world. As Murray (1998) suggests,

The same phenomenon occurs when a child rocks a teddy bear or says "Bang!" when pointing a toy gun. Our successful engagement with these enticing objects makes for a little feedback loop that urges us on to more engagement, which leads to more belief. (p. 112)

The Nintendo Wii-Mote, like any game controller, is a liminal object, existing at the threshold between the physical world and the game world, and facilitating the transition from one world to the other. Similar to a child's teddy bear as described by Murray, the Wii-Mote exists as a physical object with a presence independent of its manifestation in a particular game. And, like a teddy bear, this physical object is often treated with considerable affection on the part of the user. Wii players protect the object with colorful "skins" and decorate it with stickers. They carry the controllers in special cases. The "Wii-Mote" moniker is itself a term of affection conceived by users, not the company. Nintendo has, however, encouraged a view of the controller as a prized object by allowing players to store their personalized Mii avatars inside the Wii-Mote, and thus transport the characters from one Wii console to another — one physical location to another — by means of the object. Simultaneous to its physical manifestation, the

Wii-Mote is represented in various ways in various game worlds, ranging from the concrete — such as a sword — to the abstract, such as the ability to levitate. The effectiveness of any object as a liminal device can be judged by the ease of that transition between worlds. Is the illusion of the game world violated when a player has to pause and think about which combination of buttons to press on a controller, or even look away from the screen and down at her hands? Conversely, is the controller so unobtrusive that a player neglects to notice when the object is on a collision course with an antique vase? The success of the Wii and its controller has been due, in large part, to lowering that barrier between the physical world and the game world, making it easy for players to move quickly between the two. By integrating this controller into our game, we exploit a technology that is readily available, reliable and provides a sufficiently flat learning curve, in contrast to the speech recognition technology employed by other language learning software, that still requires significant training with each individual player's voice before valuable game-play can begin.

This notion of a "barrier to entry" with regard to playing games is analogous to a pivotal concept in theories of second-language acquisition. Theorists describe the "affective filter" as an invisible barrier that prevents students from being able to absorb input in a language class. Various attitudinal variables determine whether a given student's affective filter is high or low. Factors that raise the affective filter include low self-esteem, a poor concept of the culture that speaks the target language, or teaching techniques that make students feel self-conscious. On the other hand, if students have positive self-images, if they think well of the target culture and if they genuinely enjoy classroom activities, the affective filter is said to be low (Krashen, 1983, p. 37-38).

Because TPR does not require students to produce speech, but merely to enact sequences of commands, theorists have commented on the propensity of TPR for diminishing performance anxiety in the classroom and therefore lowering the affective filter. The type of playful, bizarre, and zany commands that Asher, Kusudo, and De la Torre recommend are meant to develop a playful environment and sense of camaraderie in the foreign-language classroom.

1. "When Henry runs to the blackboard and draws a funny picture of Molly, Molly will throw her purse at Henry."
2. "Henry, would you prefer to serve a cold drink to Molly or would you rather Eugene kick you in the leg?"
3. "Rosemary, dance with Samuel, and stick your tongue out at Hilda. Hilda, run to Rosemary, hit her in the arm, pull her to her chair and you dance with Samuel!" (1974, p. 27-28).

The Wii-Mote has a similar, easily observable effect on the social atmosphere surrounding video games. The prevailing stereotype of a "gamer" is still the lone male teenager in a dark basement, and while there are plenty of images to counter this one — including multiplayer online games that are profoundly social and collaborative — much of the social interaction between players happens on screen. Even if people are playing with (or against) each other in the same physical environment, the social dynamic plays out in the virtual environment, to the extent that someone in the same room not actively playing the game might easily lose interest. With the Wii and similar kinesthetic interfaces like a dance pad, however, those bystanders can also watch the players, not just the screen, as they wave their arms and punch the air and hop on one foot — a more engaging social activity, similar to the experience of students observing their classmates in a kinesthetic learning exercise. Much of the time, the Wii game player or the gesturing student look a little silly, which has the effect of lowering the anxiety level of the other participants.

KINESTHETIC LEARNING IN A VIDEO GAME

Our two-semester game development sequence emphasizes cross-disciplinary collaboration, with faculty and students contributing from a range of fields. The first two games to emerge from the course, for example, featured original scripts from English students, scores from Music students, graphics from Art students and code from Computer Science students. In addition to its current emphasis on a broad, liberal education, The College of New Jersey has a long tradition of teacher training. This fact, combined with a renewed interest in so-called "serious games" within the academic and games communities, as well as the perceived cynicism (or at best ambivalence) on the part of the students toward self-proclaimed learning games, led the faculty to consider developing just such a game as part of the next course sequence. Students and faculty would have an arena in which to formulate and test their ideas about how learning does or does not take place in video games, based on both personal experience and established theories of learning.

Furthermore, the simultaneous emergence of the Wii led us to consider how we might leverage the affinity between the Wii-Mote and gesture-based, kinesthetic learning techniques such as TPR. Similar to such exercises in a real-world classroom, we devised a system in which players were expected to respond to Spanish commands by executing an appropriate gesture, such as sweeping one's hand rapidly downward in order to slam the snooze button on an annoying alarm clock, or to hold the controller sideways and shake it quickly up and down to brush one's teeth. As we began to develop and test the system, feedback from undergraduates indicated that the novelty of the physical interface was sufficient to overcome their reflexive reluctance to develop a self-proclaimed "learning game." Our hope was that would also hold true for the game's target audience, college or high-school Spanish students, many of whom may be skeptical of the value of learning a foreign language in the first place, let alone playing a language-learning video game. By means of the game, we sought to overcome preconceptions about educational games being the "drill and kill" variety derided by Papert, and similarly to move beyond from the rote memorization and repetition exercises traditionally associated with foreign language instruction.

The narrative frame of the game targets the same students who may benefit from kinesthetic exercises in their language classes, but relocates those exercises from the classroom to an immersive 3D world. The game's subject matter suggested several likely settings, ranging from the fantastic — complete with Spanish-speaking pirates — to the foreign, featuring a digital simulation of life in Spain or Latin America. Ultimately, the students and faculty chose to situate the game in a world that would resonate with the target audience, and also highlight the challenge of accomplishing relatively mundane tasks in a foreign language. Our game takes place in a faithfully recreated, 3D version of our own campus. College students expressed enthusiasm for the prospect of encountering a virtual version of the man working the cash register in the dining hall, and we surmised high school students would appreciate playing the role of a college student on an actual (though virtual) college campus. In a nutshell, the story is that the main player and his (or her) friends emerge from a science lab on campus — a lab shielded from an unexpected and significant electromagnetic event that occurs while the group is inside — to discover that the entire campus is suddenly speaking Spanish and only Spanish. The player must manage the everyday tasks of a college student, from finding books in the library to ordering food in the dining hall, while at the same time trying to solve the mystery of this startling transformation. The premise strikes a balance between familiar and fantasy. Players take pleasure in finding the details of their day-to-day lives represented in the computer, and to a degree this process of discovery is their motivation to play and keep playing. The academic setting also clearly grounds the game in the context of education. Unlike many educational games, we do not try to disguise our intention by wrapping the learning inside an outer space adventure or some other fantastic scenario. The fact that this familiar campus is speaking a foreign language, however, is enough to remind the player that this is a game, not just a simulation, and that there is a puzzle to be solved.

In the Spring 2008 semester, we implemented a proof-of-concept of this approach. We chose to work with the *Source* (*Half Life 2*) engine because we already had extensive experience with its tools and processes, and therefore could focus our efforts not on custom modeling or programming, but instead on mapping out the story and testing the implementation and effectiveness of the Wii-Mote. The player responds to spoken Spanish commands (recorded by College conversation hour leaders) that increase in complexity as the game unfolds, demonstrating comprehension with physical gestures while holding the controller. We built three basic game levels. The first was a typical campus dormitory room, in which the player had to follow his friend's spoken commands in order to ultimately open the door and leave — *apague el despertador* (turn off the alarm) and *lávese los dientes* (brush your teeth). The second level was a typical classroom, in which the player had to follow his professor's commands to *entregue la tarea* (turn in the homework) and *borre la pizarra* (erase the blackboard). Finally, the third level was the exterior campus, allowing the player to move from one locus of game-play to another. *Source* allowed us to implement these levels relatively quickly and gauge their effectiveness with the intended audience. The results were encouraging: players already comfortable with Spanish enjoyed demonstrating their knowledge in an unconventional way, and those less familiar with the language were intrigued by the game's premise and interface to the point that they *wanted* to understand Spanish in order to progress to other levels and locations.

DECENTRALIZED LEARNING AND ASSESSMENT

Many educational video games fall into "tutor" category of the "Tutor, Tool, Tutee" model proposed by Robert Taylor (1980): "The computer presents some subject material, the student responds, the computer evaluates the response, and, from the results of the evaluation, determines what to present next" (Taylor, 1980, p. 3). Teaching and assessment are embodied in an omniscient teacher, be it a nameless narrator asking multiple-choice questions, or a clearly defined personality such as Mavis Beacon from the popular typing tutorial software. In these games, this teacher is the source of all the questions and all the answers, all of the game's learning content. Consistent with our emphasis on learning from context as well as content, our design distributes teaching and assessment throughout the game system, embodied in the behavior of preprogrammed agents — other students, faculty, dining hall employees — with whom the player interacts.

The player's success and survival depends on both short- and long-term measurements, with the short-term resembling the so-called "health meter" ubiquitous in today's video games. The player takes instructions, clues, hints and advice for various Spanish-speaking characters, and the player's "health" depends on executing a response appropriate to the social context. For example, heeding a roommate's advice to brush one's teeth, by gesturing up-and-down in front of the bathroom sink, increases the health meter, while ignoring a suggestion to eat lunch decreases it. Repeatedly failing to respond appropriately will decrease the player's health to the point that he or she has to start over within that context, be it a dorm room, a classroom, or the dining hall.

With a wider scope, the system also gauges the player's level of Spanish proficiency, prompting the game's non-player characters to use language that grows increasingly complicated in terms of syntax, vocabulary, colloquialisms, etc.

Also tied to this broader measurement is the ability to acquire assets, ranging from the material—clothing, a skateboard—to the immaterial, such as songs for an in-game music player. As evidenced by popular games such as *The Sims*, this basic consumerist behavior provides further incentive for players to continue playing, continue exploring, and therefore continue learning.

Our approach distributes the pedagogy among the characters and situations in our game, rather than adopting the traditional scope and sequence curriculum that is often implicit if not outright explicit in computer-based tutors. We make no demands regarding the order in which language is learned, but simply require the user to survive and progress from one situation to the next. The vocabulary, syntax and idiomatic sophistication of each situation depends on the demonstrated sophistication of the player, not on the player's position within a pre-determined linear sequence, providing opportunities for unanticipated situations and emergent behavior that could not be prescribed or predicted by us, the developers. This sort of richness is common in entertainment games, but frequently lacking in educational games.

FUTURE RESEARCH DIRECTIONS AND CONCLUSION

Our approach seeks to build on established theories from such diverse fields as foreign language learning, computer science, interaction design, and ludology, and to combine those ideas in the form of a video game that is innovative, engaging, and educational. Our proof-of-concept — consisting of a dormitory room and a classroom — demonstrates progress in this regard. Unlike many games that simply layer content-based learning on top of tenuously related game-play, ours allows learning to emerge naturally from the exploration of the game's geography, the exposition of the game's narrative, and the execution of the game's gesture-based mechanics. Having successfully implemented this system on a limited scale, we hope to use it as a means of pursuing further research, particularly in the area of kinesthetic learning.

An obvious application of our system is to further gauge the benefits of kinesthetic learning techniques. As discussed earlier, many studies have suggested ways in which bodily movement aids in vocabulary retention as well as in more complex forms of language processing. However, receiving input by means of a video-game interface and responding via a game controller creates variables that have not been explored with regard to TPR. We have the flexibility to create two different versions of our game, for instance, one responding to gestural input from the Wii-Mote, and another responding to input from a conventional push-button game controller. Will user testing demonstrate that one approach is significantly more successful than the other with regard to either vocabulary retention or to more nuanced forms of understanding in a foreign language? We can also explore the considerable gradation between those two methods of user input. The fact is pushing buttons is still kinesthetic, though markedly less so than waving a Wii-Mote through the air. The question arises: At what point is the player's physical action sufficiently kinesthetic to trigger the benefits of TPR as described by Asher and others (Asher, 1965, 1966, 1969, 1977; Asher, Kusudo, & De la Torre, 1974; Lindstromberg & Boers, 2005; Swaffar & Woodruff, 1978; Thiele & Schneiber-Herzig, 1983; Wolfe & Jones, 1982)? The boundary between the virtual and the physical also becomes an issue for exploration. If a relatively low-kinesthetic action on the part of the user — such as pushing a button — triggers a high-kinesthetic action from a realistic on-screen avatar — a kick, for example — does that virtual world gesture confer any of kinesthetic learning benefits of the identical gesture in the physical world? Eventually, we hope to allow for more complicated gestural input and variable tracking than allowed by our current game engine, *Source*, including perhaps input from additional kinesthetic devices, such as dance pads or the Wii balance-board.[3] As we continue to refine our system's pedagogical parameters and establish its technical framework, we are also considering the potential of our design for learning other languages and ultimately other disciplines.

REFERENCES

Asher, J. (1965). The Strategy of the Total Physical Response: an Application to Learning Russian. *International Review of Applied Linguistics, 3*(4), 291–300. doi:10.1515/iral.1965.3.4.291

Asher, J. (1966). The Learning Strategy of The Total Physical Response: A Review. *Modern Language Journal, 50*(2), 79–84. doi:10.2307/323182

Asher, J. (1969). The Total Physical Response Approach to the Second Language Learning. *Modern Language Journal, 53*(1), 3–17. doi:10.2307/322091

Asher, J. (1977). Children Learning Another Language: A Developmental Hypothesis. *Child Development, 58*(1-2), 24–32.

Asher, J., & Adamski, J. C. (2003). *Learning another language through actions*. Los Gatos, CA: Sky Oaks Productions.

Asher, J., Kusudo, J., & De la Torre, R. (1974). Learning a Second Language through Commands: The Second Field Test. *Modern Language Journal, 58*(1/2), 24–32. doi:10.2307/323986

Bogost, I. (2007). *Persuasive Games: The Expressive Power of Videogames*. Cambridge, MA: MIT Press.

Crawford, C. (2002). *The Art of Interactive Design: A Euphonious and Illuminating Guide to Building Successful Software*. San Francisco, CA: No Starch Press.

Dunn, R. S., & Dunn, K. J. (1993). *Teaching Secondary Students Through Their Individual Learning Styles: Practical Approaches for Grades 7-12*. Boston: Allyn and Bacon.

Gee, J. P. (2003). *What Video Games Have to Teach Us about Learning and Literacy*. New York: Palgrave Macmillan.

Johnson, S. (2006). *Everything Bad is Good for You: How Popular Culture is Making Us Smarter*. London: Penguin.

Krashen, S. D., & Terrell, T. D. (1983). *The Natural Approach: Language Acquisition in the Classroom*. Englewood Cliffs, NJ: Alemany Press/Regents/Prentice Hall.

Lindstromberg, S., & Boers, F. (2005). From Movement to Metaphor with Manner-of-Movement Verbs. *Applied Linguistics, 26*(2), 241–261. doi:10.1093/applin/ami002

Murray, J. H. (1998). *Hamlet on the Holodeck*. Cambridge, MA: The MIT Press.

Papert, S. (1993). *The Children's Machine: Rethinking School in the Age of the Computer*. New York: Basic Books.

Papert, S. (June, 1998). Does Easy Do It? Children, Games, and Learning. *Game Developer Magazine*, 88.

Piaget, J. (1963). *The Origins of Intelligence in Children*, (M. Cook, Trans.). New York: Norton.

Swaffar, J., & Woodruff, M. (1978). Language for Comprehension: Focus on Reading a Report on the University of Texas German Program. *Modern Language Journal, 62*(1/2), 27–32. doi:10.2307/324112

Taylor, R. (1980). *The Computer in the School: Tutor, Tool, Tutee*. New York: Teachers College Press.

Thiele, A., & Schneibner-Herzig, G. (1983). Listening Comprehension Training in Teaching English to Beginners. *System, 11*, 277–286. doi:10.1016/0346-251X(83)90045-3

VanPatten, B., Williams, J., Rott, S., & Overstreet, M. (Eds.). (2004). *Form-Meaning Connections in Second Language Acquisition*. Mahwah, NJ: Lawrence Erlbaum Associates, Inc.

Wolfe, D., & Jones, G. (1982). Integrating Total Physical Response Strategy in a Level I Spanish Class. *Foreign Language Annals, 14*(4), 273–280. doi:10.1111/j.1944-9720.1982.tb00258.x

ENDNOTES

[1] For summaries of current research on the importance of comprehension over production as well as studies about the importance of a contextualized approach to the teaching of a foreign language, see VanPatten, Williams, Rott, & Overstreet (2004).

[2] "For example, the *future tense* can be imbedded into a command as, 'When Luke walks to the window, Marie *will* write Luke's name on the blackboard!' The *past tense* can be incorporated into the command structure. For instance, say: 'Abner, run to the blackboard!' After Abner has completed the action, say: 'Josephine, if *Abner ran* to the blackboard, run after him and hit him with your book.'" (emphasis in the original) (Asher et al., 1974, p. 26)

[3] Typical of a first-person-shooter game, *Source* allows the player one active weapon at any given time; and by pressing the designated action key, that weapon performs its one function, be it smashing something with a crowbar or shooting something with a crossbow. The action is binary — the gun is either firing, or not — there is no gradation. The Wii-Mote is more advanced in this regard, with accelerometers able to sense continuous acceleration along three axes. By having to translate this sensor data into the one-at-a-time, on-or-off logic of *Source*, we neglect the opportunity to incorporate more complex and perhaps sequential gestures into the game. The engine is also linear in nature, intended for games in which a player completes on level then advances to the next, with no motive or opportunity to return to the previous level. The concept for our game is less constrained, with players being able to travel from one location to several other locations and back again. The characters, information, and tasks encountered in a particular place at a particular time depend on several factors, including current social dynamic between the player and other non-player characters in the space, his or her demonstrated language aptitude, what locations the player has already visited and what tasks he or she has already accomplished. This kind of complex variable tracking and analysis is also not inherent to *Source*, which was built to track linear and largely independent variables such as how many bullets the player has remaining and how many hits he or she can sustain. *Source* is a comprehensive, empowering tool for building realistic shooter games along the lines of *Half Life 2*, and its extensive model, sound and behavior library saved us considerable time developing a prototype, but the inconsistencies between the engine's fundamental mechanics and our goals for the game have led us to seek other tools as we move beyond the initial proof-of-concept stage.

Section 2
Applications of Serious Games

Chapter 7
How Games and Simulations can Help Meet America's Challenges in Science Mathematics and Technology Education

Henry Kelly
Federation of American Scientists, USA

ABSTRACT

The quality of the U.S. workforce is critical for a competitive economy in today's fast-paced, tightly connected global economy. Given current trends the quality of the U.S. workforce, measured in educational attainment and knowledge of key concepts in science, mathematics, and engineering will actually decline in coming years. Conventional methods will simply not be adequate to the enormous task of improving this situation. The technology and management tools that have been used so successfully to increase both quality and productivity in other service enterprises can play a vital role in improving science and mathematics education and training.

INTRODUCTION

There is universal agreement that a U.S. economy that successfully recovers from the current deep recession must be built around businesses able to innovate and adapt quickly to innovations. Innovation is particularly critical in the operation of the nation's system of education and training that will be forced to meet dramatically increased demand – both in the quantity of what must be learned and in the number and the diversity of people needing to learn. New information technologies, particularly a new generation of simulations and computer-based games, provide a powerful set of tools for meeting these goals at an acceptable price. The discussion below reviews the nature of the challenge, the ways simulation and game technology can help meet the challenge (with an emphasis on science, mathematics, technology, and engineering education), and outline a path forward that can accelerate development and, where appropriate, rapid adoption of the new technologies.

DOI: 10.4018/978-1-61520-739-8.ch007

WHAT'S AT STAKE

The jobs in this new economy will demand a wide range of adaptive skills including ability to: master new concepts quickly, gather information and make decisions under conditions of uncertainty, work comfortably with people with very different backgrounds and skills, and communicate effectively with peers, specialists in other fields, and novices. Most jobs will also require a solid foundation in the basics of science, engineering, and mathematics. Jobs in everything from trucking to surgery will be imbedded in sophisticated production networks involving information systems, sensor networks, data management, sophisticated labs and testing equipment, and constant series of innovations. U.S. leadership will depend on research labs, staffed by world class scientists and engineers able to push the frontiers of scientific knowledge, convert discoveries into marketable products and services, and build businesses around them.

None of this can happen without a U.S. workforce able to contribute to technologically sophisticated enterprises that are in a continuous process of innovation and renewal. Our ability to compete will depend in essential ways on the quality of the U.S. workforce. Federal Reserve Chairman Ben Bernanke points out that "taking full advantage of new information and communication technologies may require extensive reorganization of work practices, the reassignment and retraining of workers, and ultimately some reallocation of labor among firms and industries" (Bernanke, 2006).

Given this reality, it's unsettling that the U.S. workforce ranks 13th among high income countries in the quantitative skills of its workforce, and 14th in document literacy (National Commission on Adult Literacy, 2008). It's even more disconcerting that the educational quality of the U.S. workforce is likely to decline during the coming generation because of large numbers of poorly educated workers and the fact that a growing share of population growth will come in minority groups most poorly served by the existing educational system (Kirsch, 2007). The cohort with the largest population growth in the U.S. from 2000 to 2020 will be people with less than a high school education. U.S. high school graduation rates reached a peak of 77% in 1969 but are now 70%. Shockingly, only half of many minority groups graduate. The overall achievement rates of U.S. students are consistently behind other affluent nations and are essentially unchanged since the 1980s. And while the average education level in nations like China and India remain far below those in the U.S., measured in absolute numbers these countries are becoming very large. China's colleges are producing more than 6 million graduates annually while the U.S. produces only about 1.5 million (U.S. Department of Education, 2008; People's Daily Online, 2009). There is undoubtedly still a significant difference in the quality of the degrees granted but it's obvious that a lot of talented people are engaged in the global marketplace who will be in direct competition with U.S. employees in a tightly coupled global marketplace.

The gap separating the U.S. from the countries with the best educational records is so substantial that it cannot be closed in the next two decades even if we solve our problems with college enrollment and high school graduation rates (Jones, 2007). An aggressive program in adult education will be needed if the U.S. workforce is to be competitive in terms of educational quality. While many high school drop outs succeed in getting GEDs, only 27% of GED graduates enroll in post secondary education and 85% of those who do must take at least one remedial course (National Commission on Adult Literacy, 2008). Adult education is, however, largely used only by people who already have a reasonable education. In a recent survey people were asked "did you get any training" in the previous five years. 19% of people lacking a high school degree had gotten some training, 33% of high school graduates, and

Figure 1. From Kirsch (2007)

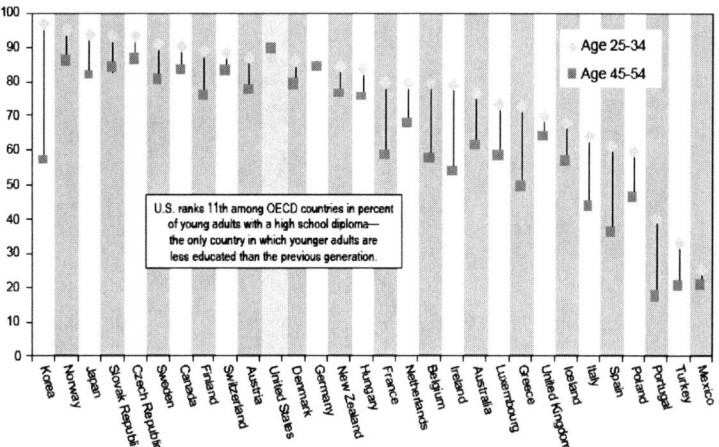

50% of people with college degrees had obtained training of some kind (National Commission on Adult Literacy, 2008).

People with poor educations will face growing problems given the clear direction of the U.S. economy. While people doing routine manufacturing and clerical occupations were able to maintain middle-class incomes a generation ago, they are unlikely to do so in the future. One symptom of this is the rapidly growing gap separating the earnings of people with different levels of educational achievement. The lifetime earnings of males with BAs were 51% higher than high school graduates in 1979, 96% higher in 2004 (National Commission on Adult Literacy, 2008). Manufacturing jobs that provided so many middle class jobs for high school graduates are all but disappearing. Manufacturing employment as a fraction of nonagricultural employment fell from 32% in 1960 to 10% in 2007 (Morrison, 2008). More than 85% of U.S. jobs are in the service sector, but this sector is achieving surprising productivity growth based on skillful use of advanced information technology tools. In fact much of the productivity growth in the last decade has come from the seemingly impossible – productivity growth in services resulting from intensive investment in both hardware and software Bosworth, 2006). The average productivity of 22 service industries (weighted by value added) which had grown 1.1%/year between 1977 and 1995 grew 3.0%/year between 1995 and 2000 (Triplett, 2003). This means that large numbers of routine clerical tasks were automated and only the people able to design and operate sophisticated networks were benefiting from the innovations.

Between 1984 and 2000, demand for occupations with high literacy increased 57% (from 34 to 54 million), while demand for jobs with moderate literacy increased 14.3% (from 70-80 million). The Bureau of Labor Statistics (BLS) forecasts suggest that this trend will continue. Between 2004 and 2014 predict 18% growth (9 million jobs) in high literacy and 10% growth in moderate literacy (10 million jobs) (Kirsch, 2007). Yet, as Figure 1 above indicates, the overall rates of achievement in prose proficiency, as defined in the Adult Literacy Survey, are declining.

Taken together, growing demand for high levels of education and the declining levels of

Figure 2. From National Commission on Adult Literacy (2008), used with permission

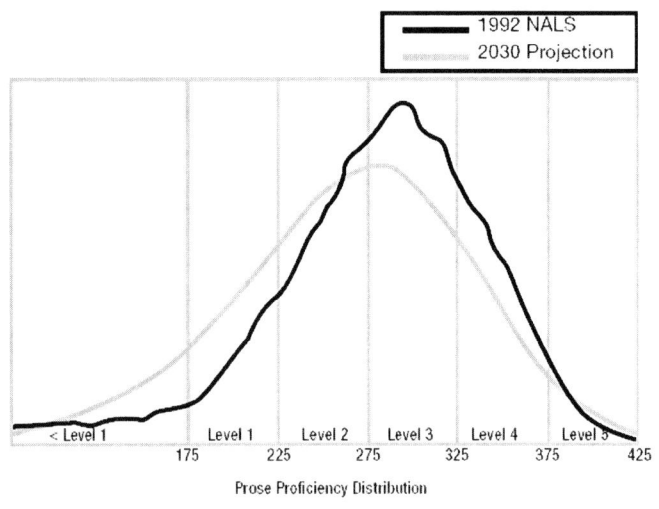

education in the U.S. workforce, point to a perilous future for the U.S. economy. Either the U.S. faces difficulty competing or it manages to build competitive enterprise in ways that reward only the shrinking fraction of the workforce with excellent educational skills, accelerating the growth of income inequality that is already painfully large. In a recent survey of wealthy nations, the U.S. had the second highest spread in educational outcomes (National Commission on Adult Literacy, 2008). This gap is very likely to grow.

All of these sobering statistics and indicators point to a stark need to strengthen the U.S. educational system at all levels – both for people being prepared to enter the workforce and for people already employed. We must find a system adaptable enough to provide people from all educational and cultural backgrounds a sophisticated education – including an ability to operate efficiently in a technology-rich work environment. And we must find a way to do this at an acceptable cost.

ASPIRATIONS

The tools that have been used to achieve huge productivity gains in service industries throughout the economy clearly have the potential to make learning more productive, more engaging, and more closely tailored to the interests and backgrounds of individual learners. Computer games provide a particularly spectacular example of what can be achieved. Many of these games require players to master complex skills to advance in the game. Players can interact with complex equipment and natural landscapes that might include many species of trees and flowing water displayed with a fidelity that is increasingly difficult to distinguish from nature. These simulations are, in effect, a higher order of scientific and engineering visualization that has always been a powerful element of instruction. These simulations provide much more than the clearest graphic since the "viewer" is free to explore a dynamic environment and manipulate it at will.

These simulation tools, coupled with other tools available from modern software, raise fascinating opportunities to address the core challenges just discussed: providing inexpensive access to first rate instruction to all Americans and systems of instruction adaptable to and appealing to a hugely diverse population – including people with very poor educational backgrounds.

There's a catch, of course. While the opportunity clearly exists, it has proved extraordinarily difficult to seize. There are three key issues: (1) designing the course of instruction so that it is both rigorously correct and constantly engaging, (2) ensuring that the system adapts to the background and interests of individual learners, and (3) evaluating the expertise of learners in ways that make sense to them and to future employers. In discussing these topics I'll explore the design challenges involved and provide some concrete examples of the decisions made as our group built and tested the instructional game "Immune Attack" (Kelly, 2007). This game is designed to help students master key concepts in immunology in what is, in effect, a war game. Invaders are bacteria and viruses and the player must train the autonomous agents of the immune system (macrophages, neutrophils, and the like) to tell friends from foes, track invaders, and signal for assistance.

INSTRUCTIONAL DESIGN

New information technologies make it possible to imagine affordable ways of actually doing some of the things that have been recommended for years by cognitive scientists. For example, a National Academy of Sciences (NAS) study of student assessments emphasizes that the challenge of continuously gathering and evaluating complex information about students probably cannot be achieved without new information technology. The NAS study noted that "New capabilities enabled by technology include directly assessing problem-solving skills, making visible sequences of actions taken by learners in solving problems, and modeling and simulating complex reasoning tasks"(NAS, 2002, p. 268). The report states that computer simulations can test a much more sophisticated range of expertise

Gaining expertise appears to require both exposure to the logical structure of a body of information and the basic facts, and an opportunity to translate this raw material into memory patters that are permanent and accessible to each individual (Pliske, 2001). It probably also requires a balance between letting learners extract information from trying to solve practical problems (forming their own associations based on discovery) and formal instruction that provides them clues about what to look for and how to structure the observations. Debates about the best way to acquire expertise have undoubtedly been going on for a millennia but a growing research base is providing increasingly powerful insights into what works both for education and for training (Donovan, 1999). The challenges include the practical difficulties associated with mixing real-world experience with formal instruction and understanding how the optimum mix varies with the student's characteristics or the subject matter being taught. Should the lessons be given before, after, or during the explorations? Should the lessons be open-ended games or projects with fixed objectives? Would the explorations be improved if an expert participates, giving constant advice, or should the learner be allowed to fail and seek advice (or debrief the mission) after the failure? How effective are paper and pencil exercises in comparison with experiences gained in more realistic settings. Since the cost of actually providing different kinds of field experience is prohibitive in most learning situations more research is needed to address these questions. Games and simulations introduce a fascinating new dimension since the challenge presented is often a very practical one – defeating bacteria in the case of *Immune Attack* – that might be associated with training.

Figure 3.

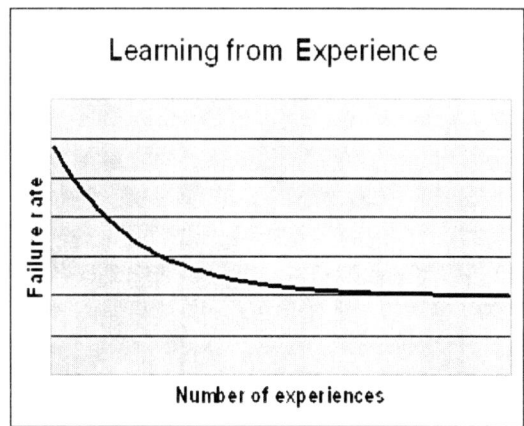

But the practical goal provides the motivation to acquire basic knowledge and understanding that clearly qualifies as education.

The value of experience and practice in practical settings is clearly documented in the few places where data is available. An experience or learning curve for acquiring expertise applies for acquiring expertise in many different fields is shown generically in Figure 3. This experience curve applies to solving geometry problems (Anderson, 1990), training surgeons (Gibbs, 2001), training pilots (Defense Science Board, 2001), and others attempting to transfer expertise from a classroom to a practical environment. It appears that this kind of experience based learning is important for virtually any kind of learning – from mastering simple mechanical skills to developing sophisticated expertise. It's reasonable to expect that similar curves would apply to learning everything from repairing computers to searching for a rare plant in the jungle. However, the experience curve implies that theoretical understanding of the information does not automatically transfer into performance. The learners are gaining something from experience that their initial grasp of the basic information didn't equip them to do.

There also appears to be something akin to a "forgetting curve" where memories that are not used decay. The few studies that have examined this in detail show a great, and not well understood, range of decay rates. A year after mastering a series of skills, after a year only about 25% of the students performed to criteria in cardiopulmonary resuscitation after 12 months and only about 5% passed the "don gas mask" test satisfactorily (Montague, 1993).

A reasonable hypothesis is that practical experience forces learners to restructure the way information is represented in their memories – building patterns that allow rapid recognition of problems and intelligent reactions. The experiences also force learners to extract essential information from the flood of information that's always present in real situations – including the dynamics of working with other people. The ability to experiment may also help organize the information so that it is efficiently embedded in the learner's memory. Presumably theory learned in the classroom provides critical clues that make this process more efficient.

A few essential rules seem to emerge from this brief examination of acquiring expertise (Donovan, 1999). These are:

- **The learner should be provided clues to how to organize information into a**

logical structure and lots of practical experience. *Immune Attack* invites players to explore and acquire information from a variety of sources but provides many clues about how to organize information so that it can be useful in meeting game challenges. The challenges were selected so that they can only be met by displaying authentic expertise.
- **The learner should be provided an opportunity to apply facts and theories in practical situations.** Fact and theories are unlikely to be retained as long-term memory if the learner is forced to move to new material without such opportunities.
- **The broadest possible set of experiences should be provided so that the learner doesn't mistakenly form and record false associations.**
- **Experience and practice should continue as long as it is challenging and reinforces expertise**. It should stop when expertise has been reached or the problems become uninteresting.

In the construction of *Immune Attack* we struggled to create challenges that would be motivating but require that players master key subject matter in order to win a level. In the first level of the game, for example, a simple bacterial infection can be cleared by training macrophages flowing in the blood stream to recognize the signs inside blood vessels indicating that an infection is near, escape from the vessel, follow a chemical signal to the site of the infection, and kill the bacteria without injuring host cells. Several attempts were required to get this right. One failed because smart gamers could figure out how to win the game without actually learning any biology by finding inadvertent hints left by the designers. Another failed because it provided too little guidance at the beginning; players got bored and confused attempting to find all the information needed. Still another failed because the presentation was too prescriptive leaving little room for exploration.

Reproducing the Fun of Discovery, Playing Around

While U.S. education plainly faces major challenges, there is little enthusiasm for adopting systems from Europe or Asia that produce students with higher test scores. The American system seems to be able to produce people with unique skills in creativity and flexibility (Zhao, 2008). Freed from the soul-crushing university entrance examination systems that dominate many other national programs, the U.S. system continues to turn out people capable of adaptability, entrepreneurial activity, and ingenuity. But even here there's reason to be concerned. The openness and diversity of the U.S. education system makes it difficult to make poorly performing systems accountable. The *No Child Left Behind* statute is a clear response but, unless carefully implemented, runs the risk of forcing instruction into narrowly focused "teach to the test" instruction.

Equally distressing is the shrinking possibilities for unrestricted exploration and experimentation that has been the driving motivation for generations of scientists and engineers. Granted that much work at the frontiers of science and engineering involves abstractions that have little to do with a common sense understanding of the physical world (the phenomena of quantum mechanics for example), most scientists were attracted to the field by a keen interest in understanding how the things around them work – natural curiosity that is all too often deadened by conventional instruction.

Most students today, of course, live in urban settings where they don't have the opportunity to play with fields, farm implements, farm animals, fields, and streams. The problem is particularly severe for low income children who may face a closed "latch-key" life where there is virtually no time to run free in an outdoor environment.

With manufacturing employment falling sharply, young Americans don't grow up in a world where conversations involve discussions of tools and materials. Two generations ago a "shop" crammed with tools and odd pieces of wood and metal, was a routine part of nearly every home. The experience of helping a parent build or fix something was a natural part of home experiences. But shops are never in the listings of modern homes. This means that children have no routine personal exposure to the natural environment – what happens when you throw a stick, or toss a stone into the water, or try to make vines into a rope – and therefore lack the instincts for physical and biological processes that an earlier generation took for granted.

The classroom laboratories that could provide something of a substitute for this experience provide a pale substitute. Well equipped laboratories are expensive and few schools can afford to keep their equipment up to date. Concerns about liability have greatly limited the scope of what can be offered and experiments are rigidly prescribed. Playing around is strictly discouraged. Even playgrounds have been stripped of the equipment most likely to convey a sense of the physical world: teeter totters, hand pushed merry-go-rounds, have been removed (Vogel, 2005). Picking up sticks and stones during recess is forbidden in many jurisdictions. The idea of home chemistry sets and home-made rockets would make legal advisors go catatonic.

While no substitute for direct physical experience, computer games and simulations may provide at least a partial solution. They invite play and experimentation – even dangerous experiments – without fear of liability. They open up the possibility of using tools inaccessible to any school (want your own mass spectrometer or gene sequencer – no problem). And they can do things impossible in even the best equipped laboratories because they allow travel to places no field trip is likely to go (travel to remote wetlands, explore the chemistry of methane lakes on Europa, navigate the inside of a pancreas). Importantly, they can also be environments that encourage play, exploration, construction, and creativity; conveying sophisticated information in an efficient and appealing way; and, providing a platform to continuously evaluate each learner and measure their increasing expertise using sophisticated metrics.

Learning that occurs in the process of approaching a specific task or challenge can have enormous advantages over formal instruction. Strategies include anchored instruction, cognitive apprenticeships, cognitive flexibility theory, constructivism, exploratory learning, functional context education, situated action, situated cognition, and situated learning (Resnick, 1989). Case-based lessons are used extensively in professional schools, such as law and businesses, and surveys suggest that they enjoy strong faculty support because the "faculty believe the case-methods approach promotes problem identification, decision-making, and specific concept learning, as well as more general educational benefits such as reasoning ability, oral communication skills, self confidence, and professional self image" (Allen, 2000, p. 443).

Games have always played a central role in education but are usually viewed with considerable suspicion. Only the Department of Defense (DoD) has the courage to call its critical training exercises "games." In all societies, games let inexperienced people imitate the tasks of experts, experiment with tools and concepts, and develop styles for collaborating with peers (Collada, 2009). The process can be extraordinarily efficient, but games also have the potential to be colossal wastes of time.

A particularly difficult challenge for the design of Immune Attack was to find a way to let players manipulate the immune system, explore, and play without allowing activities that were not biologically correct. Proposals letting players shoot and kill bacteria, for example, were rejected because this is not the way kills are made in real systems. The solution chosen was to give the players equipment that is clearly not biologically correct (and visually marked in ways to prevent any confusion). These include a miniature submarine the player pilots and small probes that are used to inspect phenomena at a molecular level. "Minigames," clearly not part of the biological simulation, are used to let players learn how to identify key marker molecules and acquire other skills. This knowledge is then use to "train" the macrophages and other immune cell types.

Theory of Flow

The most distinctive feature of problem-based learning and games is that they provide their own motivation. Enormous amounts of learning and practice may be involved in succeeding, but the rewards are intrinsic and powerful (Rieber, 1996; Papert, 1980; Papert, 1993). Simulations and games strip away the artifice of learning, connecting the act of understanding directly to compelling goals. Participants can be completely swept into the process, spending hours perfecting physical skills and mastering intellectual challenges needed for success. The test of these skills is immediate and provides clear motivation to try harder to reach new levels and seek more challenging tasks or opponents; the distress and confusion associated with mastering new concepts is happily endured because of the potential rewards. And participants can spend hours discussing strategies, tactics, and telling the equivalent of "war stories" about their triumphs and defeats.

Many of the most successful games implicitly implement a concept that the psychologist Mihaly Csikszentmihalyi developed as he attempted to understand "the process of achieving happiness." He calls the concept "flow" which he defines as (Csikszentmihalyi, 1990) "—the state in which people are so involved in an activity that nothing else seems to matter; the experience itself is so enjoyable that people will do it even at great cost, for the sheer sake of doing it" (p. 4). People are at their happiest when they are "invested in realistic goals, and when skills match the opportunities for action... These periods of struggling to overcome challenges are what people find to be the most enjoyable times of their lives" (p. 63).

A key issue in pedagogy, and in using games and other tools to incorporate good pedagogy, is how to get learners into a "flow" where they are constantly able to build skills that match the opportunity for action. This is an instinctive talent of brilliant tutors, able to present intellectual challenges in a sequence that constantly pushes the limits of what the learner can achieve, and motivates the learner to achieve new skills to overcome these limits. The order of presentation need not follow a rigid taxonomy of knowledge driven by the logic of the material, but may well begin with a complex but interesting task whose completion requires mastery of a number of more elementary skills which will be acquired quickly once the learner understands their relevance and the power they give in solving the problem at hand. A key design issue is to advance the instruction at a rate consistent with the revealed expertise of the player. If the game moves to new challenges too quickly, players are confused and frustrated. If it moves too slowly, they are bored.

In *Immune Attack* players are given challenges of increasing difficulty. The first level is designed simply to familiarize the players with the controls of the miniature submarine and understand how to navigate the veins, capillaries, and arteries. This proved challenge enough for beginners. Subsequent levels confront players with increasingly devious infections. The first involves bacteria that can be attacked and destroyed with one immune cell (the macrophage). The second is multiplying too fast to be cleared by the macrophage and requires the player to train macrophages to signal for help and train neutrophils (another immune cell type) to recognize the signal and navigate to the site of the infection. The design challenge is to have each infection present new dilemmas for the player that can be addressed only by mastering new skills.

The Role of Instructors

The issue of how best to combine human and simulated instruction remains largely unresolved. In most school settings, software is used as an "enrichment" tool rather than a substitute for traditional text and lecture methods. This is the way most teachers use *Immune Attack*. Systems like the Carnegie *Algebra Tutor* let students learn entirely at their own pace alone on computers. Teachers introduce concepts and help students with problems that can't be resolved through materials presented by the software. Many advanced Defense Department training simulations also employ this method with instructors providing initial lectures, providing assistance, and in some cases holding "debriefing" sessions where the instructor can replay the game just played and discuss the decisions made by the student.

It is likely that successful instructional games will involve a wide range of mixtures of human instruction, automated help, and simulations. *Immune Attack* combines a system of active and passive instructors –some provide context-sensitive information to help the player succeed where he is failing, while others provide more general hints, clues, and support to players as they make their way through the levels. Live instructors can also appear as avatars in games or simulations, using the virtual world as a platform for instruction – leading tours, demonstrating how equipment works, pointing out geological features on an expedition. Networked computers make it possible for an instructor to monitor each student's progress in real time and automated systems can be developed to trigger a call for instructor help when certain conditions are met that suggest a student needs human intervention. Student questions can be parsed much like commercial "help desks" so that frequently asked questions receive automated answers, but options remain to ask for human help when the automated systems fail. The software can also allow instructors to adjust game play in real time --adding complexity or interest appropriate to individual students. Many large scale DoD simulations have teams of instructors who operate "red teams" that react intelligently to decisions made by teams of students operating the "blue team." Given the power of networking there is no reason why all participating instructors need be in the same physical space as students. Systems can be designed that connect specialists as needed for answering specific questions, providing counseling, or other services. This will be particularly important for helping groups of students with very different backgrounds and who speak many different languages.

Multi-Player Participation, Collaboration

Simulated environments offer another unique opportunity by allowing groups of students to work together on the same challenge. The potential can be seen in commercial multi-player games like the World of Warcraft, large scale military war game simulations, and simulations of medical procedures where a surgical team or an emergency room team must learn to work together. The ability to work as an effective member of a team, and to learn how to enter an unfamiliar team, is a key skill in the emerging economy. But it has only begun to be explored as an intentional element of formal instruction. Little work has been done to understand where and when group instruction is most effective – DoD has taken the lead. While there has not been a formal evaluation of this point, anecdotal data suggests that players of *Immune Attack* prefer to play as pairs.

Continuous Testing and Evaluation

Technology also opens new options for testing a learner's expertise in many (but obviously not all) fields (Edys, 2009). A well designed game provides the player with continuous information about their level of skill and expertise. A game that gets a player into a "flow" state has them constantly aware of what they can and cannot do and highly motivated to master the skills to move to the next level (something very close to continuous formative assessment). In most cases skills acquired in early parts of the game are constantly put to use and refreshed even as new ones are acquired (Kelly, 2005).

The simulations can test quite sophisticated skills including mastery of science and mathematics content as well as skills much in demand in the modern workplace such as an ability to work effectively in a team, communicate with others, acquire information from diverse sources, and making decisions under uncertainty. The DoD now places considerable value on skills demonstrated in realistic simulations (Fletcher, 2009).

One of the more intriguing features of game-based testing is that players expect to fail, and learn from failure exactly what they need to master to succeed in the next effort. This is in stark contrast to the "high stakes testing" at fixed intervals that dominates education. Shank argues that one of the great values of problem based instruction is the very process of failure (Shank, 1997).

Standard testing mechanisms also force all learners to face testing at precisely the same time – breaking the key design feature of "flow." But experience reveals a large range in the speed with which different people master different subjects. Forcing all students to proceed, and be tested, at the same rate means that a large fraction will either be bored because of the slow pace or frustrated and confused because of the rapid pace. Recently the U.S. Navy conducted an experiment in which they converted a standard 8-week course to a course where students would be given certificates as soon as they had demonstrated mastery in a performance based test. Most students demonstrated mastery in 4 weeks, some in less than a week (K. Moran, personal interview, 2008). The Navy students had roughly equal backgrounds. The diversity of people in an average U.S. classroom – particularly one that includes adults – is likely to be much higher.

In Immune Attack the "win condition" is set by clearing the patient of the infection – failure means that the patient dies. This proves to be highly motivating. It does, however, create a tension because players rushing to beat the clock (the bacteria keep multiplying if they aren't controlled by the immune cells) don't have adequate time to explore and learn more detail about subjects that may interest them. Our compromise was to "stop the clock" whenever the player wanted to read into a topic.

GETTING PRACTICAL

Challenges

Even if it's accepted that simulations and game based instruction provide powerful learning experiences, three enormous problems must be overcome before the methods can be used at a significant scale: (1) they are very expensive to build, (2) they are expensive to maintain since updated versions must be created to match changes in computational technology, and (3) there is a proliferation of "one of a kind" software – each with it's unique interfaces and conventions – no two of which interoperate.

Costs

A modern, high-end computer game today costs on the order of $50 million dollars. While a significant fraction of this is undoubtedly marketing costs, it's clear that development and testing costs are far beyond what is typically spent on textbooks (textbooks can cost several millions to produce). To take an extreme hypothetical case, it would cost about $72 billion a year to produce high quality game-like materials that could cover all the hours now spent by students in K-12. This is a huge number and it plainly makes no sense to have students using computers all the time, but it still represents only about 18% of the total cost of K-12 education. This cost, of course, does not include any of the research, development, testing, and evaluation needed to garner the knowledge needed to undertake a major investment in instructional simulations and games.

While schools may well be able to save costs by reducing textbook purchasing and other costs, it seems unlikely that anything approaching these sums would be available. One of the most successful instructional technology systems, the Carnegie Algebra Tutor and Geometry Tutor, took more than a decade to develop and received over $15 million from various federal sources. But it was by far the exception. *Immune Attack* received $1.2 million from the National Science Foundation (NSF) and a roughly equal amount from other sources. But a good fraction of this must be considered a first of a kind research investment. Grants available from NSF and foundations to build instructional software are typically in the range of 500K or less.

What this means is that instructional technology can't use the development techniques that successful game developers feel are essential. One of the most important aspects of successful game designs are repeated tests with real players. Commercial developers pay players for thousands of hours of game testing for every feature of the game. These tests, and modifications, are expensive. The revisions to *Immune Attack* described earlier were based on testing on a comparatively small number of students and many suggestions and ideas for improvements could not be undertaken for reasons of cost. But Immune Attack probably went through more cycles of testing than most instructional games. A number of highly promising and creative games, such as a genetics course developed by the Concord Consortium, failed to reach its potential because the funds needed to go through cycles of play and evaluation were simply not available.

Interoperability

One of the factors driving the high costs of commercial games and instructional games is the absence of interoperability. Materials developed for one game are seldom reused in another. Each project starts virtually from scratch building virtual landscapes, interface tools, score or trophy rooms, evaluations, and other features that often are unique. This not only increases costs, it means that every tool requires a significant amount of investment in startup. Immune Attack is no exception. It takes on the order of 15-20 minutes before players are familiar with how to control the devices in the game, where to find information, how to navigate, the iconography, and other features. They, and their instructors, would need to go through this learning cycle for each game used.

Recognizing this problem, the gaming industry is struggling to develop standards that would allow the reuse of 3D elements designed for games and possibly even scripts through projects like the open-source Collada project (Collada, 2009).

Technical Obsolescence

A related issue involves the technical obsolescence of the instructional materials developed. Work on instructional technology has been underway for decades but much of the material is completely inaccessible because it remains on obsolete media (remember floppy disks) or has not been upgraded to run on modern operating systems. Most of this is permanently lost.

This problem has largely been solved for text, images, and video available over the web since with few exceptions each generation of new browsers designed for advanced hardware and software will be designed to ensure compatibility. It has not been solved for simulations and games – there is no html for these technologies.

Solutions

Fortunately a new generation of software tools is emerging that makes it possible to address the practical issues of cost, interoperability, and sustainability. Web 2.0 strategies have revolutionized the way groups of people can collaborate in design, testing, and use of elaborate products. These tools can greatly reduce the cost of building and testing new instructional simulations and games, and they can introduce a revolutionary new set of capabilities: a marketplace in instructional approaches. Given the enormous commercial interest in these tools, virtually everything needed to create applications for education is already available. What is needed is a way to make them easily available to the education community.

Based on our experience with building and testing *Immune Attack* and several other education games and simulations, the Federations of American Scientists (FAS) has begun to do exactly this in a project called Medulla (FAS, 2009a). The project involves locating and combining tools for two key "use cases:" (1) collaborative construction of simulations and virtual environments, and (2) easy use of these environments to create educational experiences and evaluations.

Use Case 1: Design Simulated Environments

The first step in creating a simulation or game based on a simulation is creating the environment in which the action will take place. This can be an ancient city where players walk through streets, interact with "non-player characters" – avatars controlled by artificial intelligence, and undertake other activities. Or it could be the human circulatory system, a simulated organ, a simulated swamp with AI controlled species and a sophisticated environmental model to track impacts of different interventions. It could be an engineering laboratory or the surface of Jupiter's exotic moon Europe. Whatever the environment, a number of key tools are essential:

- Project management tools (building, reviewing, commenting, rating, upgrading)
- Identifying team leaders, contributors, reviewers
- Scheduling and notification
- Managing payments for services as needed
- Connecting to object repositories with recognized meta-data
- APIs for scripts, AI, physics engines, simulations
- APIs for object creation and adding metadata
- Version control

Open source or commercial tools suitable for performing each of these functions is available. Identity management and project management tools allowing careful tracking of services like peer review are available in open source formats.

A key unresolved problem, however, is the tool for viewing 3D worlds online. These tools are needed to create the equivalent of html for 3D interactive worlds. While a number of systems are available (the most actively used being Second Life), none are completely satisfactory for the task at hand (FAS 2009b). Given the considerable commercial interest in these systems for entertainment and other applications, significant progress is likely in the next few years. The key is to represent the components in standardized formats (e.g. standard 3D graphic formats) with metadata suitable to the domain. Library objects, for example, should use the Dublin Core metadata already vetted by this community.

The difficulty of developing standards for scripts was discussed earlier. The advantage of using these tools is that artists, subject matter experts, graphics specialists, and others can collaborate in building and reviewing extremely complex sites.

And since the material will be built in an enduring format, it can be continuously reviewed and improved as new knowledge is acquired.

Use Case 2: Collaborative Use

Once a functional set of simulations are available, a separate set of challenges involves using the environments to build powerful learning tools. In most instructional software developments (including *Immune Attack*) a single team was responsible both for the design of the simulations and visualizations, and for the design of the pedagogy. There is no reason why this should be the case. A team capable of building a powerful and up-to-date simulation of an ancient city may not be the team best qualified to invent games and experiences in the environment best suited to instruction. A variety of different uses may be made of the same environment. In a virtual city, for example, experiences could include: free exploration, tours led by an expert, simple challenge tasks such as finding a hidden object, or highly elaborate game play. The events could be used by individuals exploring alone or in groups and they could be facilitated by instructors and "role players" who would control avatars or other activities in the scene.

The tools needed to create these experiences include:

- Design tools for projects, games, assignments, performance tests
- Storing and rating these experiences
- identifying students and connecting to student records (including private records)
- Identifying (and paying) instructors, tutors, counselors, guides, role players
- Scheduling and notification

The ability to design different experiences in the same environment opens the possibility of a market for instructional experiences. Students and teachers could rate different approaches creating a situation where powerful, creative approaches to instruction could propagate worldwide quickly. A standardized format will be needed to encapsulate and preserve different instructional strategies.

NEXT STEPS

There is a clear and pressing need to develop affordable instructional tools that will let all Americans participate in a rapidly changing, technology-rich economy. The software tools that have achieved spectacular results in most of our service businesses create an opportunity to meet this goal if applied in education and training. Education in the U.S. has always been a complex mixture of federal funding, state and local funding, and individual investment. There is little likelihood that the potential of the new technology can be realized absent creative federal leadership. This emphatically does not mean imposing federal standards on content, but it does mean leadership in establishing technical standards and other standards that will permit interoperability, collaboration, and competition on a national if not international scale. A new national media infrastructure that allows all instructors and all students easy access to rich simulated environments could be national media resource able to make contributions to the nation's education as powerful – if not more powerful – than the contributions made by public radio and television in an earlier generation. The new infrastructure will provide greater flexibility in adapting curricula to local needs – skills required, student backgrounds – than is possible using existing technology.

Building elaborate simulations of biological systems that would be the logical extension of *Immune Attack* may require inputs from hundreds of specialists over a period of years. A system allowing groups to collaborate in building, and constantly improving, such a simulation is entirely possible given current technology. What is lacking is (a) a major and sustained investment in research, development, testing, and evaluation, (b) a set of agreed standards allowing the construction and use of interoperable, persistent simulations and games, and (c) a concerted effort to address the complex issues of intellectual property rights, financing, and other practical problems that must be resolved. Some progress is being made. Congress recently authorized a *National Center for Research in Advanced Information and Digital Technologies* (110th Congress, 2008) which, if funded could provide research infrastructure. PBS commissioned an exploration of its "digital future" (PBS, 2005). It's time to put the pieces together and craft a national program consistent with the magnitude of the challenges we face.

REFERENCES

Anonymous. (2009). China moves to solve graduate unemployment issue. *People's Daily Online*. Retrieved January 16, 2009 from http://english.peopledaily.com.cn/90001/90781/90879/6571767.html

Avedon, E. M., & Sutton-Smith, B. (Eds.). (1971). *The Study of Games*. New York: John Wiley & Sons, Inc.

Barksdale, J., & Hundt, R. (2005). *Digital future initiative: Challenges and opportunities for public service media in the digital age*. Retrieved on June 22, 2009 from http://www.newamerica.net/publications/policy/digital_future_initiative_report

Bernanke, B. S. (2006). *Before Leadership South Carolina, Greenville South Carolina* [PDF]. Retrieved from http://www.federalreserve.gov/newsevents/speech/Bernanke20060831a.htm

Bosworth, B., & Triplett, J. (2006). *Is the 21st Century Productivity Expansion Still in Services? And What Should Be Done About It?* Paper from the 2006 Summer Institute sponsored by the National Bureau of Economic Research and the Conference on Research in Income and Wealth, Cambridge, MA. Retrieved from http://www.nber.org/confer/2006/si2006/prcr/bosworth.pdf

Centers for Disease Control and Prevention. (2009). Retrieved June 22, 2009 from http://www.cdc.gov/HomeandRecreationalSafety/Playground-Injuries/index.html.

Chatham, R., & Braddock, J. (2001). Training Superiority & Training Surprise, Final Report, *Defense Science Board*. Retrieved from http://www.acq.osd.mil/dsb/reports/trainingsuperiority.pdf

Collada (2009). *Collada* [computer software]. www.collada.org.

Csikszentmihalyi, M. (1990). *Flow: The Psychology of Optimal Experience*. New York: Harper & Row.

Donovan, M., Bransford, J., & Pellegrino, J. (Eds.). (1999). *How People Learn*. Washington, DC: National Academy Press.

FAS. (2009b). Retrieved June 22, 2009 from http://vworld.fas.org/wiki/FAS:Introduction.

FAS. (2009a). Retrieved June 22, 2009, from http://www.fas.org/programs/ltp/games/medulla.html

Federation of American Scientists. (FAS). (2008). *A small survey taken by FAS asked students whether they were interested in biology before and after being given conventional instruction in the subject. A majority were less interested after completing the course.*

Fletcher, J. D. (2009). Education and Training Technology in the Military. *Science, 323*(5910), 72–75. doi:10.1126/science.1167778

Gibbs, V., & Auerbach, A. (2001, July). Learning Curves for New Procedures—the Case of Laparoscopic Cholecystectomy. In *University of California at San Francisco (UCSF) –Stanford University, Evidence-Based Practice Center, Making Health Care Safer: A Critical, Analysis of Patient Safety Practices, AHRQ* (Publication 01-E058). Retrieved June 22, 2001 from www.ahcpr.gov/clinic/ptsafety/pdf/front.pdf.

Hagman, J., & Rose, A. (1983). Retention of military skills: A review. *Human Factors, 25*(2), 199–213.

Herron, R. E., & Sutton-Smith, B. (Eds.). (1971). *Child's Play*. New York: John Wiley & Sons, Inc.

Jones, D. (2007). *Mounting pressures facing the U.S. workforce and the increasing need for adult education and literacy.* National Commission on Adult Literacy. Retrieved June 22, 2009 from http://www.nationalcommissiononadultliteracy.org/content/nchemspresentation.pdf

Kelly, H. (2005). Games, cookies, and the future of education. *Issues in Science and Technology,* (21): 33–40.

Kelly, H., Howell, K., Glinert, E., Holding, L., Swain, C., & Burrowbridge, A. (2007). How to Build Serious Games. *Communications of the ACM, 50*(7), 45. doi:10.1145/1272516.1272538

Kirsch, I., Braun, H., & Yamamoto, K. (2007). *America's Perfect Storm: Three Forces Changing Our Nation's Future.* Princeton, NJ: Educational Testing Service. Retrieved June 22, 2009 from http://www.ets.org/Media/Education_Topics/pdf/AmericasPerfectStorm.pdf

Morrison, W., & Labonte, M. (2008, November). *China's currency: A summary of the economic issues* (Congressional Research Service Report RS21625). Retrieved June 22, 2009 from http://www.fas.org/sgp/crs/row/RS21625.pdf.

National Commission on Adult Literacy. (2008). *Reach higher America: Overcoming crisis in the U.S. workforce.* Retrieved June 22, 2009 from http://www.nationalcommissiononadultliteracy.org/ReachHigherAmerica/ReachHigher.pdf

Papert, S. (1993). *The children's machine: Rethinking school in the age of the computer.* New York: Basic Books.

Papert, S. (1993). *Mindstorms: Children, computers, and powerful ideas.* New York: Basic Books.

Pliske, R., McCloskey, M., & Klein, G. (2001). Decision Skills Training: Facilitating Learning From Experience. In E. Salas & G. Klein (Eds.), *Linking Expertise and Naturalistic Decision Making* (pp. 37-42). Retrieved June 22, 2009 from http://books.google.com/books?hl=en&lr=&id=lAo5kOgrdRoC&oi=fnd&pg=PA37&dq=Skills+Training:+Facilitating+Learning+From+Experience&ots=4mpqoAVod-&sig=sobMZht5N51sPU1hTFL-Jf5ssL8

Quellmalz, E. S., & Pellegrino, J. W. (2009). Technology and Testing. *Science, 323*(5910), 75–79. doi:10.1126/science.1168046

Resnick, L. (Ed.). (1989). *Knowing, learning, and instruction: Essays in honor of Robert Glaser.* Hillsdale, NJ: Lawrence Erlbaum Associates.

Rieber, L. P. (1996). Seriously considering play: Designing interactive learning environments based on the blending of microworlds, simulations, and games. *Educational Technology Research & Development, 44*(2), 43-58. Retrieved June 22, 2009 from http://it.coe.uga.edu/~lrieber/play.html

Shank, R. (1997). *Virtual learning: A revolutionary approach to building a highly skilled workforce.* New York: McGraw Hill.

Sutton-Smith, B. (1997). *The Ambiguity of Play.* Cambridge, MA: Harvard University Press.

110th Congress (2008). *Higher Education Opportunity Act Conference Report to accompany H.R. 4137* (House Report 110-803). Retrieved June 22, 2009, from http://www.congress.gov/cgi-bin/cpquery/?sel=DOC&&item=&r_n=hr803.110&&&sid=cp110pOPWa&&refer=&&&db_id=cp110&&hd_count=& Allen, B., Otto, R., & Hoffman, B. (2000). Case-based learning: contexts and communities of practice. In S. Tobias & J. Fletcher (Eds.), *Training and Retraining* (pp. 453-472). New York: Macmillan Reference.

Triplett, J., & Bosworth, B. (2003). Productivity measurement issues in services industries: "Baumol's disease" has been cured. *Economic Policy review, Sept*, 23-33. Retrieved June 22, 2009 from http://econpapers.repec.org/article/fipfednep/y_3a2003_3ai_3asep_3ap_3a23-33_3an_3av.9no.3.htm

U.S. Department of Education, Institute of Educational Sciences. (2008). *Fast Facts*. Retrieved January 16, 2008 from http://nces.ed.gov/fastfacts/display.asp?id=72

Vogel, E. (2005). Playgrounds swing away from merry-go-rounds to "composite structures." *Las Vegas Review-Journal, July 5, 2005*. Retrieved June 22, 2009 from http://www.reviewjournal.com/lvrj_home/2005/Jul-05-Tue-2005/news/26818011.html

Zhao, Y. (2008). What Knowledge Has the Most Worth? *The School Administrator.* Retrieved June 22, 2009 from http://www.aasa.org/publications/saarticledetail.cfm?ItemNumber=9737

Chapter 8
Games for Peace:
Empirical Investigations with PeaceMaker

Cleotilde Gonzalez
Carnegie Mellon University, USA

Lisa Czlonka
Carnegie Mellon University, USA

ABSTRACT

This chapter presents an investigation on decision-making in a dynamic and complex situation, the solution of international conflict, and the achievement of peace. The authors use an award winning video game to collect behavioral data, in addition to questionnaire surveys given to players. The Israeli-Palestinian conflict is one of the most difficult political problems of our times, and PeaceMaker represents the historical conditions of the conflict and provides players with an opportunity to resolve the conflict. Students in an Arab-Israeli history course played PeaceMaker from the perspectives of the Israeli and Palestinian leaders at the beginning and end of the semester. The authors recorded and analyzed their actions in the game and information on their personality, religious, political affiliation, trust attitude, and number of gaming hours per week. Results indicate the number of actions taken in the game alone cannot distinguish between good and bad performers. Rather, individual identity variables such as religious and political affiliation, personal affiliation to the conflict, and general trust disposition relate to the scores obtained in the game. They discuss the implications for policy and general conflict resolution and present their ideas for future research.

INTRODUCTION

Dynamic decision making (DDM) is a field of research dedicated to the study of how individuals make decisions in dynamic, complex environments involving multiple components: alternatives, events, and outcomes; high uncertainty; and many constraints, including time, workload, and resources (Brehmer, 1992). In the Dynamic Decision Making Laboratory (http://www.cmu.edu/ddmlab/), we have studied human decision making in dynamic tasks across a variety of contexts including military command and control, medical decision-making, and supply chain management, among others.

DOI: 10.4018/978-1-61520-739-8.ch008

Conflict resolution can be conceptualized as a dynamic decision making process, in which the resolution of the problem is obtained by making a series of interdependent decisions in the face of changing realities, interests, and relationships between the conflicting parties (Kelman, 2008). One of the psychological concepts useful in understanding conflict resolution is experience: decision makers often tend to adhere to past decisions and rely on established procedures and technologies as the safest course of action (Kelman, 2008). In fact, in DDM research, it is known that decision makers often make decisions from experience, based on what is learned from past decisions and their consequences (Edwards, 1962; Gonzalez, Lerch, & Lebiere, 2003).

Unfortunately, little is known about the socio-psychological aspects that influence DDM and the use of experience in addressing a conflict. For example, a concept useful in the understanding and amelioration of the violent manifestation of international conflict is identity, or the need for belonging to an ethnic or national group (Shamir & Sagiv-Schifter, 2006). It is well known in several socio-psychological theories that conflict increases people's attachments to their "own" group and generates hate for the "other" group, because people tend to think in terms of social categories or groups. Thus, group membership provides security at many levels, but it is also the source of many conflicts (Gartzke & Glenditsch, 2006). For example, other versus self accountability is an antecedent to anger, and often this emotional reaction is foreseeable when the self moral values and beliefs are jeopardized by the "other" (Cheung-Blunden & Blunden, 2008).

In this research, we investigate the effects of identity variables on the aspects of conflict resolution as they relate to experience. We are interested in determining the effects of these identity variables as an individual becomes more familiar with a conflict. Our research follows an innovative approach to the study of conflict, by executing controlled laboratory experiments using a video game. The video game used in this research is *PeaceMaker*, developed by ImpactGames (Impact Games, 2008). PeaceMaker simulates realistic Israeli-Palestinian interactions, with the player assuming the role of either the Israeli Prime Minister or the President of the Palestinian Authority. In each case, the leaders attempt to make effective policy choices leading to peace, while having to respond to external events like suicide bombings, army raids, and the demands of public opinion. The goal of the game in either role is to establish a stable two-states solution to the conflict.

The research reported here will present the use of PeaceMaker in an example of conducting empirical investigations with video games to build theoretical models of socio-psychological variables that influence DDM. In what follows, we describe PeaceMaker and a laboratory study conducted to analyze participants' decision making strategies and socio-psychological factors that relate to the solution of the conflict in the game. We will present some of the results from this study and discuss their implications for future directions in using video games for behavioral research.

BACKGROUND

The popularity of video games worldwide is undeniable. Specifically, the use of interactive videogames for conducting research and improving learning in the classroom is becoming very common. Psychology has adopted video games as research tools for quite some time (Donchin, 1995), but more recently, other disciplines are adopting simulations and games for teaching and research, including Engineering (Foss & Eikaas, 2006), Business and Management (Zantow, Knowlton, & Sharp, 2005), Medicine (Bradley, 2006; Griffiths, 2002), and Political Science (Kelle, 2008; Malaby, 2007; Mintz, Geva, Redd, & Carnes, 1997).

A key characteristic of DDM research is the use of *interactive* problem solving tools (i.e., Microworlds or Decision Making Games [DMGames]) in laboratory experiments (Gonzalez, Vanyukov, & Martin, 2005). Researchers have used DMGames in the study of decision processes for many years, but they refer to those tools by various names, including "microworlds," "synthetic task environments," "high fidelity simulations," "interactive learning environments," "virtual environments," and "scaled worlds," just to name a few (Brehmer & Dorner, 1993; DiFonzo, Hantula, & Bordia, 1998; Gonzalez et al., 2005; Omodei & Wearing, 1995).

DMGames are valuable for decision-making research because they represent a compromise between experimental control and realism. Traditional behavioral decision-making research typically boasts a higher level of experimental control, but involves tasks with a high degree of artificiality. The most common method is to present a hypothetical scenario, on the basis of which participants must imagine the possible consequences of presented alternatives, and to then have participants self-report their preference for which alternative to choose. With DMGames, decisions are actually made and feedback resulting from the decisions is received from the environment, which is used to make future decisions from experience. Thus, DMGames enable researchers to conduct laboratory experiments with dynamic and more realistic environments in which they can also obtain experimental control.

The use of games to teach international conflict in the classroom dates back to 1966. John Gearon created a simple game to be used in the classroom to introduce ninth graders to anarchy and the concept of *balance of power* that emerges in most situations of conflict (Bueno de Mesquita, 2006). Years later, Mintz et al. (1997) used a "decision board" to collect data in the classroom on what information is acquired during the decision-making process in political scenarios. The tool helped determine how much information was needed during the decision making process. More recently, Kelle (2008) used an arms control simulation in an international relations course. Kelle's analysis of students' reflections concluded that the simulation helped facilitate knowledge of the complexity of international negotiations.

Another example of the use of games in the area of international conflict is that presented by Krolikowska and colleagues (2007). This study presents an example of an environmental conflict emerging from the tensions between nature protection and economic development. A role-playing simulation game was designed and used in a course to give the students direct experience with the challenges involved in solving difficult social-ecological tensions. Data from this study suggested that the biggest barrier to successful problem solving was personal beliefs and viewpoints. This study concluded that conflict resolution is highly dependent on the personal interactions and that adopting others' perspectives is key to finding a lasting solution.

PeaceMaker represents the historical events and the different aspects of a vexing international problem: the Israeli-Palestinian conflict. PeaceMaker was originally developed as a student project at Carnegie Mellon University's Entertainment Technology Center, but it has inspired many to think of the conflict in a different way (ImpactGames, 2008). PeaceMaker has gained popularity and won multiple awards including the University of Southern California's Center on Public Policy on Reinventing Public Diplomacy through Games contest (May 8, 2006) and the Games for Change annual contest (June 27, 2007). Gonzalez, Czlonka, Saner, & Eisenberg (under review) utilized PeaceMaker in a classroom setting to impact students' learning of the historical background of the Israeli-Palestinian conflict and to measure the decision-making strategies that occur with learning. Students in an Arab-Israeli history course played PeaceMaker from the perspectives of the Israeli and Palestinian leaders at the beginning and again at the end of the semester. We recorded and analyzed the students' actions in the game, as well as information on their personality, religious affiliation, political affiliation,

Games for Peace

Figure 1. PeaceMaker Screenshot. © 2007, ImpactGames LLC. Used with permission

and general trust attitude. Results revealed that political and religious background correlate to game performance at the beginning of the semester, but that the correlation disappears by the end of the semester. This study suggests that learning to stand in the others' shoes may reduce the effect that religious and political preconceptions may have on our actions.

The results we present in this book chapter are a follow up analysis to those results reported in Gonzalez et al. (under review). Here we discuss some new patterns and relationships discovered, which present a new perspective of how performance in PeaceMaker is impacted by aspects of individuals' background and identity.

PeaceMaker

The PeaceMaker (see Figure 1) video game was inspired by real events in the Israeli-Palestinian conflict. ImpactGames (2008) developed Peace-Maker with the assistance of U.S., Israeli, and Palestinian Authority advisors to ensure that the game did not just reflect personal world views. Many prominent reviewers (Associated Press, 2006; Musgrove, 2005; Thompson, 2006) consider the PeaceMaker's representation of the Israel-Palestinian conflict to be a fair and accurate reflection of the historical and current events and relations. Peacemaker is now a commercial video game, and it has become popular worldwide and has received a series of accolades and awards in competitions (ImpactGames, 2008).

In PeaceMaker, players can play one of two roles: the Israeli Prime Minister or the Palestinian President. The game can be played on a Macintosh or PC, and it comes in three languages: English, Arabic, and Hebrew. Some generic assumptions of the game are that the player will win the game only if a two-state solution is reached. A two-state solution to the conflict is obtained when Israel and Palestine are able to coexist side by side as independent nations.

Players choose actions to take in the game and accumulate points according to the effects those actions have on the approval ratings of various interest groups. The scores are calculated by a function within the game, and reflect the satisfaction or unhappiness of different nations and political groups, both within the region and around the world (e.g., Israel, Hamas, United Nations, etc.). If the player manages to reach a balanced and highly positive score for both the Israeli and the Palestinian groups, the player wins and the game is over. If a player cannot sufficiently please his/her own people (called ScoreOwn) or the other side's constituents (called ScoreOther), the player loses and the game is over. To win the game, both ScoreOwn and ScoreOther need to be at 100 points each. If either score falls below -50, however, the player loses the game.

There are three difficulty levels in the game: calm, tense, and violent. The levels differ in the frequency and consequences of turbulent events that are beyond the player's control (AI actions); with the frequency highest for the violent level and lowest for the calm level. Using the information from these events and information available by clicking on maps, cities, and polls (other viewed), the user can formulate a strategy and take actions from three main categories: security, political, and construction. Within the three main categories are a variety of subcategories (i.e., security: send police; political: request meeting with foreign leaders; construction: build new housing settlement). Most actions within each main category can be either pro-social (i.e., increase police funding) or antisocial (i.e. decrease police funding). PeaceMaker provides a dynamic environment in which the effects of the actions on the polls (thus, the score) is dependent upon the role the user is playing, the level of violence, the history of the user's past actions, and the current levels of satisfaction at a given time. In addition, every new game will present players with a unique combination of AI actions. Thus, a player cannot win the game using the same strategy for each role or for each difficulty level.

In partnership with Impact Games, the game was modified to collect information about each action taken by a player and the state of the environment at the moment the action was taken. A log file is created for each play of the game. This log file contains information about the role being played; the difficulty level selected; the time it takes for the user to make each decision; the number and type of actions taken in the game (such as AI actions, security, political, or construction); and the resulting score (further broken down by change in points for all satisfaction polls). The output from these log files is our main data source for analysis of performance in-game. That data can also be correlated with data collected by other means such as questionnaires on religion, cultural and political identity, and beliefs.

PEACEMAKER IN LABORATORY RESEARCH

The laboratory study reported in this book chapter is an extension of a laboratory experiment presented in Gonzalez et al. (under review). This study was conducted with undergraduate students enrolled in a course on the history of the Arab-Israeli conflict. Students played the role of the Israeli Prime Minister and the Palestinian President at the beginning and end of the semester, either in the calm or violent difficulty levels. The goal was to understand what aspects of an individuals' background impact performance in the role-playing game. In the next section, we discuss the methods and instruments used to collect information about the identity variables that characterize the participants. Then, we will present new results on the performance in PeaceMaker as it relates to the categories of the participants' background variables.

Methods

Students were asked to complete a survey to gather general information on their backgrounds and identities. Variables measured in the background questionnaire included the Myers-Briggs Type Indicator (MBTI) measure of personality, a self-rated measure of personal involvement with the conflict, religious affiliation, U.S. political affiliation, typical hours spent gaming, and their trusting personality, measured in a scale originally developed by Yukawa (1985).

The MBTI has four dimensions of personality: (extraverts (E) vs. introverts (I); sensors (S) vs. intuitors (N); thinkers (T) vs. feelers (F); and judgers (J) vs. perceivers (P)). Often, thinkers, feelers, judgers, and perceivers are used to identify differences in problem solving and decision-making (Chwif & Barretto, 2003). In the current study, students were asked to visit the Team Technology website (2008) and make selections on the strength of their agreement between pairs of statements. A report was generated that indicated the personality (MBTI) type with which their responses had the highest percentage of agreement. In our sample, 30% were extraverts, 30% were sensors, 17% were thinkers, and 23% were judgers.

We expected that personal involvement with the conflict, religious affiliation, and U.S. political affiliation would be correlated with the participant's performance in the game, given the nature of the conflict. That is, the conflict has its roots in religious beliefs and the long-standing nature of the conflict extends beyond the region to involve many countries and political groups. We asked students to report their personal affiliation to the conflict (either with Israel, Palestine, other Middle East, or none), evaluated as having grown up in the area, having a significant portion of family from the area, etc. Due to the small number of participants that rated personal affiliation with Palestine, the Palestinian and other Middle Eastern affiliation were coded the same: Other Middle East.

We also expected to find that the hours that participants generally spent playing video games would influence the performance in the game. For instance, Lintern and Kennedy (1984) found that playing Atari's *Air Combat Maneuvering* accounted for 48-86% of the variance of performance on a carrier-landing flight simulator. Further, Donchin (1995) has noted that participants with more gaming experience in general had better performance during an experiment involving a video game task.

Also, researchers pointed out that the video games serve as a natural environment for the development of trust (Malaby, 2007). Trust of the "other side" should influence the actions taken in the game and therefore, game performance. We were interested in finding out whether and how much a person's general trusting nature would impact his or her performance in PeaceMaker. Yukawa's (1985) trust scale originally included 60 items related to trust, while Yamagishi (1986) and Yamagishi and Sato (1986) developed and utilized a brief version called the Social Values Orientation (SVO) scale. This version, published by Chaudhuri, Khan, Lakshmiratan, Py, and Shah (2003), was used to study the correlation between trust attitude and likelihood of exhibiting behavior that contributes to the public good. It has also been used in a prisoner's dilemma simulation to correlate trust to cooperation (Parks, Henager, & Scamahorn, 1996; Parks & Hulbert, 1995).

A total of 22 participants played at the calm level (50% female, M=19.9 years, Std. Dev.=0.80) and 20 participants played at the violent level (25% female, M= 20.3 years, Std. Dev.=0.86). Students played the game at the beginning and at the end of the semester and in each session they were asked to play the game twice: in the Israeli and in the Palestinian roles. In both sessions, students were counterbalanced in the order which they played the role of the Israeli Prime minister or the Palestinian president. In the first session, all participants completed the same game tutorial, after which they played the game twice. It took participants approximately two hours to complete the first session at the beginning of the semester. The second session took place two months later at the end of the semester. In the interim, students had no contact with the game, but learned about the history of the Israeli-Palestinian conflict in class. In the second session, students again played the game twice, once in each role.

Figure 2. Percentage of actions in the game taken in successful and unsuccessful games

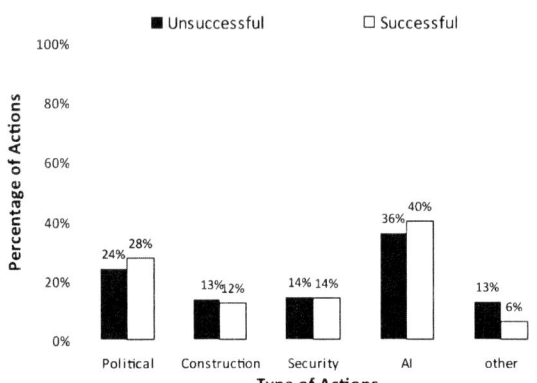

We describe below the results of players' performance and how they relate to each of the background measures we collected.

Results

Our results uncovered a mixture of expected and unexpected relationships. First, as one would expect, game outcomes were such that more players lost in the violent level (15 out of 20 participants, 75%) than the calm level (9 out of 22, 41%). The difference between violent and calm is considerable, but it only indicates that the game was well calibrated in terms of its difficulty. Furthermore, across violence levels, 16 out of 42 participants, 38%, lost all four games; 20 out of 42 participants, 48%, won only one of the four games they played; 5 out of 42, 12%, won two of the four games played, and only 1 out of the 42 participants, 2%, won all four games. Also, more participants lost the games in the beginning of the semester (29 out of 42, 69%) versus the end of the semester (21 out of 42, 50%). This gives a rate of learning of 19%, which is surprisingly low. It was interesting to find that more participants lost when playing as the Israeli Prime Minister (32 out of 42, 76%) than when playing as the Palestinian President (19 out of 42, 45%). More on their behavior at the beginning and end of the semester will be presented next.

To better understand this variability in gaming performance, we initially looked at actions that players took in the games. We expected to find that better players (those with a positive score by the game's end) would have taken different strategies compared to less successful players of the game (those with negative score by the game's end). We expected that their different strategies would be reflected in the numbers and types of actions taken during the games. To evaluate this possibility, we calculated the number of actions taken in the games and compared the more successful games to the less successful games. To classify games into those that were and were not successful, we used the outcomes of the game (Score Own and Score Other). The scores were averaged and the resulting average score was taken as successful if it was greater than zero and not successful it if was less than zero. The percentage of each possible type of action taken was then calculated for both successful and non-successful games.

Figure 2 shows the percentages of each type of action taken by the participants in successful and unsuccessful games. In general, it is observed there is little difference in the number of actions taken in successful and unsuccessful games. For example, in successful games political actions comprised 28% of the total number of actions, while in unsuccessful games they were 24% of the total. Similar, non-significant differences were

Figure 3. Mean performance by dimensions of the MBTI personality type

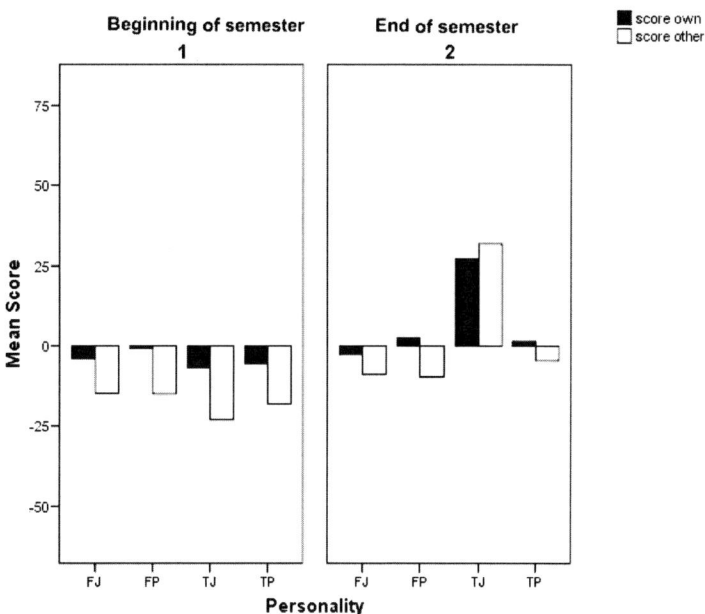

exhibited in other types of actions as well. Thus, these results suggest that what leads to more successful games is not the number of actions and the types of actions taken in the game. Rather, other variables may influence performance in this game.

We expected that the identity variables of the participants would determine their success in this game. We examined the variables measured in the background questionnaires and determined how they were related to scores in the game. Figure 3 shows the mean scores in the game at the beginning and end of the semester sessions for four personality combinations that have been identified as related to decision making performance (Chwif & Barretto, 2003): Feeling and Judging (FJ); Feeling and Perceiving (FP); Thinking and Judging (TJ); and Thinking and Perceiving (TP). The average scores in general were negative at the beginning of the semester, regardless of the personality type of the player, with the lowest scores for the Thinking/Judging personality type. Interestingly, as seen in the figure, by the end of the semester, the Thinking/Judging participants improved their performance in the game considerably, becoming the best players among all personality types.

We also expected that performance would be affected by how closely involved someone was with the conflict. Figure 4 shows the scores at the beginning and end of the semester separated by the participants' reported personal affiliations (i.e., Israel, None, or Other Middle East). Those with Other Middle Eastern affiliation were the ones with the lowest scores among the groups, both at the beginning and at the end of the semester. Those with Israeli and no affiliation had negative scores at the beginning of the semester, but by the end of the semester, they were able to raise their scores to the positives.

As with the personal involvement, we also expected that participants' religious affiliation would influence their performance in-game. Figure 5 presents the scores at the beginning and end of the semester by religious affiliation. At the beginning of the semester, those with religious affiliation

Figure 4. Mean performance by personal affiliation at the beginning and end of the semester

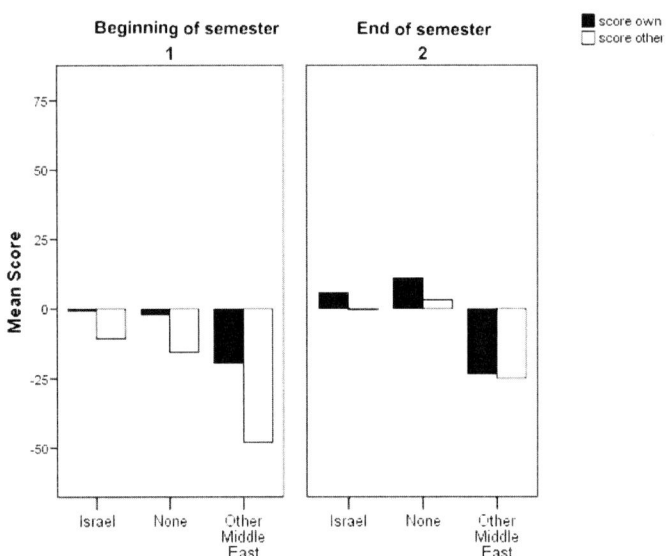

to Christianity, Hinduism, and Judaism had the worse scores among all the groups. By the end of the semester, only those reporting Hinduism continued performing poorly, and a considerable improvement was seen in the Christianity group. We had a separate category for Islam, but no players reported Islam as their religious affiliation.

For the scores broken down by political affiliation, those with Democratic and Republican U.S. political affiliation were poor performers at the beginning of the semester, but they improved by the end of the semester. Figure 6 shows the performance differences at the beginning and end of the semester by U.S. political affiliation.

Figure 5. Mean performance by religious affiliation at the beginning and end of the semester

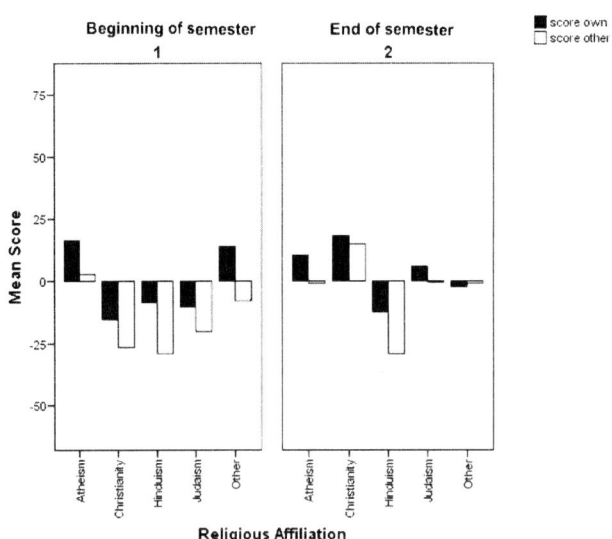

Figure 6. Mean performance by U.S. Political affiliation at the Beginning and end of the semester

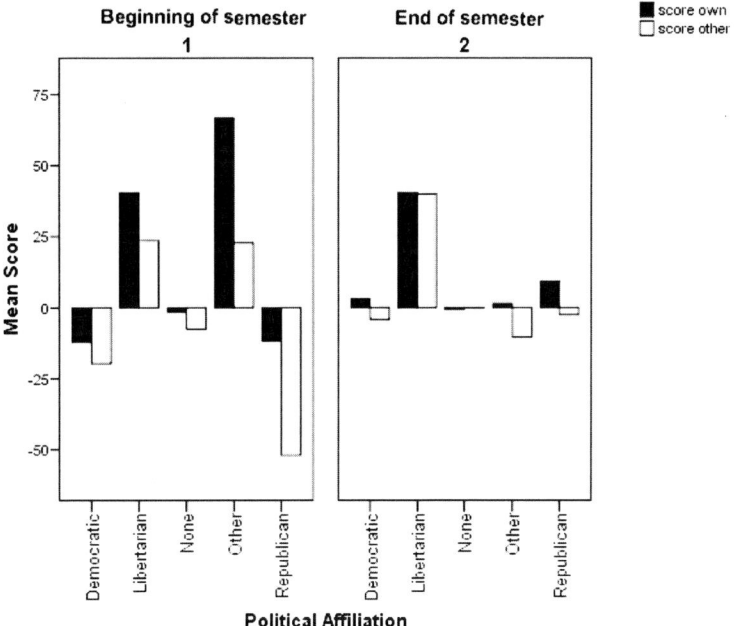

The Libertarians were consistently positive at the beginning and at the end of the semester. Although those with Democratic, Republican, or no affiliation improved performance, they were still negative by the end of the semester. Those with other political affiliation initially had the highest scores and ended up with lower scores by the end of the semester.

We expected that the general trusting tendency of participants would also have an effect on how they would play the game. Figure 7 presents the scores at the beginning and end of the semester, for those participants that were relatively trusting (score on the questionnaire fell below the mean) versus those that were relatively untrusting (score fell above the mean). Figure 7 shows that untrusting participants initially had lower scores than the trusting participants, but at the end of the semester, untrusting individuals scored better than the trusting individuals.

Finally, Figure 8 shows the results broken down by the number of weekly gaming hours participants reported. We separated the scores for those that reported playing video games for less than six hours a week from those that reported playing video games for six or more hours each week. The graph shows that in general, all participants that reported less hours of gaming practice per week tended to improve their performance in the game from the first to the second session, while those with more gaming practice per week tended to have a positive score but don't improve their performance from the first to the second session.

Discussion and Relevance of These Findings

The results above are interesting in several ways. First, we show that it is difficult to determine how a player becomes successful at the game. We found that the actions taken in-game are not good indicators of performance. The concrete, but insensitive, measure we used to come to this conclusion is the

Figure 7. Mean performance by trust scores at the beginning and end of the semester

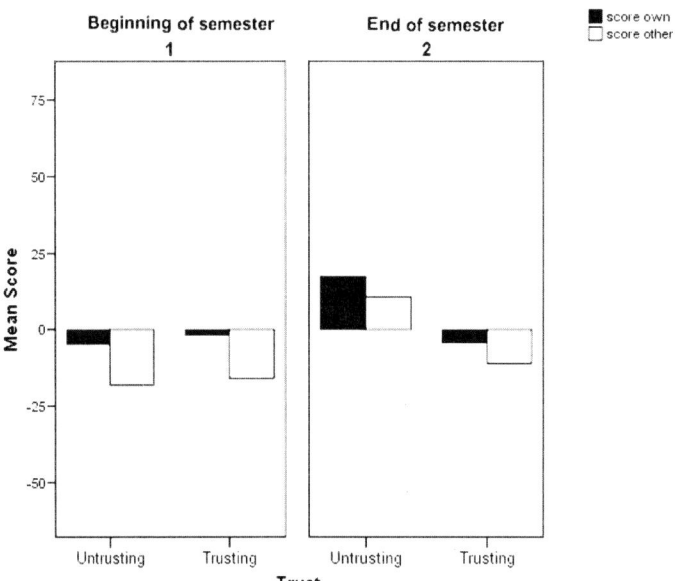

number of actions taken in a game. We believe, however, that more work is needed to determine the strategies used by successful and unsuccessful players. For example, in PeaceMaker, the key to success might be in the *timing* of the decisions and the particular *sequence* of actions taken in the game rather than the number of actions. This idea will need a more detailed type of analysis to examine the actions taken by the players.

Interestingly, performance in the game was clearly influenced by the player's personality, affiliation to the Middle East, religion, political affiliation, and trusting attitudes. For example, participants that were able to improve their performance in the game from the beginning to the end of the semester were those with a Thinking and Judging personality. People with a Thinking personality focus on analytical decision-making, based on a desire for fairness; those with Judging personality prefer control over their lives, seek closure, are organized, and plan accordingly (Filbeck & Smith, 1996). Thus, it is interesting that participants sharing these characteristics were the ones able to improve their performance considerably over the two sessions.

It is also interesting to note that individuals that reported Christianity were able to reverse their performance from negative to positive score from the first to the last session. Those that reported other religion went from a positive to a negative score from the first to the second session. Surprisingly, very little is known about the effect of religion in decision making and conflict resolution, but there are many discussions on the way religion influences judgments in several situations. According to a poll conducted by the Pew Research Center (Lugo, 2007), Americans' religious beliefs influence their views on a range of political issues, including foreign policy. In addition, conventional wisdom suggests that cultural differences matter in the conduct of international affairs, but it is unclear what elements of cultural identity matter most, and it is also unclear what relationships those elements have to political ideologies, to conflict, and to conflict resolution (Gartzke & Gleditsch, 2006). Thus, more research is needed to interpret

Figure 8. Mean performance by gaming hours at the beginning and end of the semester

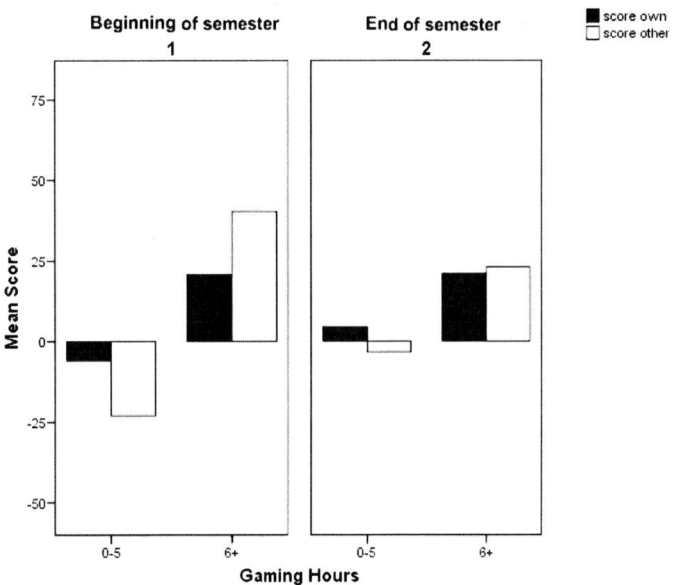

the patterns of religious and political affiliation influences that were found in this study.

Results showed that those with a general tendency to be untrusting scored better by the end of the semester than did those generally trusting of people. This seems to be a counterintuitive result that indeed is opposite to results in a cooperative game (Parks et al., 1996; Parks & Hulbert, 1995). When an action is taken in the game, such as holding negotiations or giving aid to the poor, some constituents in the game would like it and some others would not. Thus, it is possible that those who do not have as much trust for people are skeptical about their actions being accepted by all. Because they are aware that people are going to be critical even of good gestures, they may be more cognitively prepared for these results and to respond better than someone who is more trusting.

Finally, the results on gaming hours confirm the difficulty of the game. PeaceMaker is not a game of skill acquisition in which more gaming hours necessarily translate into more learning. Those with more gaming hours did, in fact, not improve their performance from the beginning to the end of the semester. Thus, even participants that usually spend more than five hours per week playing video games had difficulty winning the game. PeaceMaker does not seem to be a game that a video game player can pick up and conquer quickly.

FUTURE RESEARCH DIRECTIONS

It is clear that this study needs to be extended and expanded to account for a larger diversity of participants in all senses: religions, affiliations, and gaming hours. We have started to do that, and have conducted a data collection session similar to the one described in Doha, Qatar. In addition, we are building partnerships with the Peres Center for Peace in Israel. This is a non-profit organization that is currently conducting workshops in Israeli and Palestinian high schools using PeaceMaker. The future analyses of these data collection efforts are expected to confirm the current relationships, provide stronger relationships, and uncover other aspects of the social, religious, and cultural aspects that influence performance in this game.

We have also continued data collection in the U.S., manipulating other experimental elements under our control. Most recently, we have focused on improving the motivation players are given to win the game. In our first studies in the classroom (Gonzalez et al., under review), students were told they needed to play the game and write a descriptive paper on their experience. There was no specific motivation to perform well in the game which may account for why 38% of the participants lost all games played. Without specific incentives, the detachment naturally inherent when playing a computer game may not motivate players to achieve success. Some of our most recent experiments tie performance in a repeated games design to the amount of credit received, and thus provide clear benefit for the individual. We also plan to manipulate other experimental variables, such as training time and feedback. For instance, Gonzalez, Martin, and Hansberger (2006) in a different video game study were able to control for textual feedback (advice from experts) and showed an improvement in performance in the game after repeated trials.

Given the above proposals, some would argue that the risk of doing experimental studies with a realistic game, like PeaceMaker, is that the game may not reflect the real situation and may limit the degree to which findings can be extrapolated. While research is replete with the challenging tradeoff between external validity and experimental control (Greitzer, Hershman, & Kelly, 1981), we would argue that it is important to be able to manipulate the world in the game. At this point, PeaceMaker does not allow a player (or an experimenter) to create a new scenario or situation to play with. It would be interesting to be able to collect data, as we do in other experiments in the laboratory, in situations in which we can control and play "what if" scenarios where we can change the world directly. It remains to be seen whether PeaceMaker will develop to the point that this becomes possible, but our research suggests that there are still many aspects of the game that can be examined further.

CONCLUSION

Initial laboratory experiments with PeaceMaker highlighted interesting findings regarding individual factors related to dynamic decision making. These findings are evidence of a large opportunity for future research to learn more about decision making through video games like PeaceMaker, by diversifying the population sample and by manipulating variables such as motivation and feedback. An even greater opportunity lies in the potential for researchers and video game developers to collaborate to develop video games that have the capability to manipulate the internal variables of the game: creating experimental tools for researchers. It is important to the advancement of game studies for these relationships to be forged between the academic and gaming world. The study presented here shows an example of how a video game, educational and fun, can be used as an experimental tool to generate and create new behavioral theories.

ACKNOWLEDGMENT

This research was partially supported by the Richard Lounsbery Foundation award to Cleotilde Gonzalez. We thank Eric Brown and Asi Burak for their helpful comments and for their development of the inspiring video game that motivated this research, PeaceMaker.

REFERENCES

Associated Press. (2006). A video game that seeks to make peace, not war; 'PeaceMaker' simulates Israeli-Palestinian conflict in all its complexity. *MSNBC*. Retrieved from http://www.msnbc.msn.com/id/12423759

Bradley, P. (2006). The history of simulation in medical education and possible future directions. *Medical Education*, 40, 254–262. doi:10.1111/j.1365-2929.2006.02394.x

Brehmer, B. (1992). Dynamic decision making: Human control of complex systems. *Acta Psychologica*, 81(3), 211–241. doi:10.1016/0001-6918(92)90019-A

Brehmer, B., & Dörner, D. (1993). Experiments with computer-simulated microworlds: Escaping both the narrow straits of the laboratory and the deep blue sea of the field study. *Computers in Human Behavior*, 9(2-3), 171–184. doi:10.1016/0747-5632(93)90005-D

Bueno de Mesquita, B. (2006). Game theory, political economy, and the evolving study of war and peace. *The American Political Science Review*, 100(4), 637–642. doi:10.1017/S0003055406062526

Chaudhuri, A., Khan, S. A., Lakshmiratan, A., Py, A., & Shah, L. (2003). Trust and trustworthiness in a sequential bargaining game. *Journal of Behavioral Decision Making*, 16, 331–340. doi:10.1002/bdm.449

Cheung-Blunden, V., & Blunden, B. (2008). Paving the road to war with group membership, appraisal antecedents, and anger. *Aggressive Behavior*, 34(3), 175–189. doi:10.1002/ab.20234

Chwif, L., & Barretto, M. R. P. (2003). Simulation models as an aid for teaching and learning process in operations management. In S. Chick, P. J. Sánchez, D. Ferrin, & D. J. Morrice (Eds.), *2003 Winter Simulation Conference* (pp. 1994-2000).

DiFonzo, N., Hantula, D. A., & Bordia, P. (1998). Microworlds for experimental research: Having your (control and collection) cake, and realism too. *Behavior Research Methods, Instruments, & Computers*, 30(2), 278–286.

Donchin, E. (1995). Video games as research tools: The Space Fortress game. *Behavior Research Methods, Instruments, & Computers*, 27(2), 217–223.

Edwards, W. (1962). Dynamic decision theory and probabilistic information processing. *Human Factors*, 4, 59–73.

Filbeck, G., & Smith, L. (1996). Learning styles, teaching strategies, and predictors of success for students in coporate finance. *Financial Practice and Education*, 48, 1299–1300.

Foss, B. A., & Eikaas, T. I. (2006). Game play in engineering education: Concept and experimental results. *International Journal of Engineering Education*, 22(5), 1043–1052.

Gartzke, E., & Gleditsch, K. S. (2006). Identity and conflict: Ties that bind and differences that divide. *European Journal of International Relations*, 12(1), 53–87. doi:10.1177/1354066106061330

Gonzalez, C., Czlonka, L., Saner, L., & Eisenberg, L. (under review). *Learning to stand in the other's shoes: A computer video game experience of the Israeli-Palestinian conflict*. Manuscript under review.

Gonzalez, C., Lerch, J. F., & Lebiere, C. (2003). Instance-based learning in dynamic decision making. *Cognitive Science*, 27(4), 591–635.

Gonzalez, C., Martin, M., & Hansberger, J. T. (2006). Feedforward effects on predictions in a dynamic battle scenario. In *Proceedings of the Human Factors and Ergonomics Society 50th Annual Meeting* (pp. 265-269). Santa Monica, CA: Human Factors and Ergonomics Society.

Gonzalez, C., Vanyukov, P., & Martin, M. K. (2005). The use of microworlds to study dynamic decision making. *Computers in Human Behavior*, 21(2), 273–286. doi:10.1016/j.chb.2004.02.014

Greitzer, F. L., Hershman, R. L., & Kelly, R. T. (1981). The Air Defense Game: A microcomputer program for research in human performance. *Behavior Research Methods and Instrumentation, 13*(1), 57–59.

Griffiths, M. (2002). The educational benefits of videogames. *Education and Health, 20*(3), 47–51.

ImpactGames. (2008). *PeaceMaker [game]*. Retrieved from http://www.peacemakergame.com

Kelle, A. (2008). Experiential learning in an arms control simulation. *Political Science & Politics, 41*, 379–385.

Kelman, H. C. (2008). A social-psychological approach to conflict analysis and resolution. In D. Sandole, S. Byrne, I. Sandole-Staroste, & J. Senehi (Eds.), *Handbook on conflict analysis and resolution* (pp. 170-183). London: Routledge [Taylor & Francis].

Krolikowska, K., Kronenberg, J., Maliszewska, K., Sendzimir, J., Macnuszewski, P., & Dunaksi, A. (2007). Role-playing simulations as a communication tool in community dialogue: Karkonosze Mountains case study. *Simulation & Gaming, 38*(2), 195–210. doi:10.1177/1046878107300661

Lintern, G., & Kennedy, R. S. (1984). Video game as a covariate for carrier landing research. *Perceptual and Motor Skills, 58*(1), 167–172. doi:10.2466/PMS.58.1.167-172

Lugo, L. (2007). International obligations and the morality of war. *Society, 44*(6), 109–112. doi:10.1007/s12115-007-9021-0

Malaby, T. M. (2007). Beyond play: A new approach to games. *Games and Culture, 2*(2), 95–113. doi:10.1177/1555412007299434

Mintz, A., Geva, N., Redd, S. B., & Carnes, A. (1997). The effect of dynamic and static choice sets on political decision making: An analysis using the decision board platform. *The American Political Science Review, 91*(3), 553–566. doi:10.2307/2952074

Musgrove, M. (2005). Video games give world peace a chance. *The Washington Post*. Retrieved from http://www.washingtonpost.com/wp-dyn/content/article/2005/10/15/AR2005101500218.html

Omodei, M. M., & Wearing, A. J. (1995). The Fire Chief microworld generating program: An illustration of computer-simulated microworlds as an experimental paradigm for studying complex decision-making behavior. *Behavior Research Methods, Instruments, & Computers, 27*, 303–316.

Parks, C. D., Henager, R. F., & Scamahorn, D. S. (1996). Trust and reactions to messages of intent in social dilemmas. *The Journal of Conflict Resolution, 40*, 134–151. doi:10.1177/0022002796040001007

Parks, C. D., & Hulbert, L. G. (1995). High and low trusters' responses to fear in a payoff matrix. *The Journal of Conflict Resolution, 39*, 718–730. doi:10.1177/0022002795039004006

Shamir, M., & Sagiv-Schifter, T. (2006). Conflict, identity, and tolerance: Israel in the Al-Aqsa Intifada. *Political Psychology, 27*(4), 569–595. doi:10.1111/j.1467-9221.2006.00523.x

Team Technology. (2008). Free personality test: MMDI™ questionnaire. *Team Technology*. Retrieved from http://www.teamtechnology.co.uk/mmdi-re/mmdi-re.htm

Thompson, C. (2006). Video games: Saving the world, one video at a time. *The New York Times*. Retrieved from http://query.nytimes.com/gst/fullpage.html?res=9901E3DB163FF930A15754C0A9609C8B63&sec=&spon=&pagewanted=1

Yamagishi, T. (1986). The provision of sanctioning as a public good. *Journal of Personality and Social Psychology, 51*, 110–116. doi:10.1037/0022-3514.51.1.110

Yamagishi, T., & Sato, K. (1986). Motivational bases of the public goods problem. *Journal of Personality and Social Psychology, 50*, 67–73. doi:10.1037/0022-3514.50.1.67

Yukawa, M. (1985). *Structural and psychological factors in social dilemmas.* Sapporo, Japan: Hokkaido University.

Zantow, K., Knowlton, D. S., & Sharp, D. C. (2005). More than fun and games: Reconsidering the virtues of strategic management simulations. *Academy of Management Learning & Education, 4*(4), 451–458.

Chapter 9
Play's the Thing:
A Wager on Healthy Aging

Mihai Nadin
antÉ – Institute for Research in Anticipatory Systems
University of Texas at Dallas, USA

ABSTRACT

This study highlights the findings of researchers who, since the early 1980s, recognized the potential of engaging seniors in game interactions as an alternative to passive activities. Against this background, the perspective of anticipatory processes for evaluating specific gaming needs of the aging and providing games with anticipatory features is introduced. The hypothesis informing this work is that aging results in diminished adaptive abilities, resulting from decreased anticipatory performance. To mitigate the consequences of reduced anticipatory performance, we address brain plasticity through playing. Since anticipation is expressed in action, the games conceived, designed, and produced for triggering brain plasticity need to engage the sensory, cognitive, and motoric. The AnticipationScope, i.e., integration of motion-capture data and physiological sensors, is the platform for identifying individual characteristics and for validating the results of game participation. The output is the Anticipatory Profile. Implementations inspired by this original scientific framework are presented.

INTRODUCTION

The year was 1958. William Higinbotham, working at Brookhaven National Laboratory, conceived of a tennis game consisting of a Systron analog computer, handheld commands, some relays, and an oscilloscope. Simulated on a screen was a side view of a tennis court. The two players could rotate a knob that changed the angle of the "tennis ball," or press a button to send the ball toward the opposite side of the virtual court. If the ball hit the net, it rebounded at an unexpected angle. If the ball went over the net, but was not returned, it would hit the court floor and bounce again at a natural angle. Since this was an analog computer-based game, the ball sometimes disappeared from the screen. A *Reset* button could be activated, causing the ball to reappear and remain stationary until the *Hit* button was pressed (Brookhaven National Laboratory, 1981; Ahl, 1983).

CATCHING UP WITH SENIOR ADULTS

Today, games for health, and in particular games addressing the rapidly growing aging population, represent a large segment—evaluated at 26%—of the entire effort to conceive, design, produce, and market games of all kind. One of the games used in this effort—Surprise! Surprise!—is tennis. And one of the most important observations made so far is that playing tennis with the computer is interesting at the beginning. However, playing with someone else, for instance, over the Internet, is what users want, regardless of whether they are beginners or those handicapped seniors who once upon a time used to play real tennis every day. The social aspect of playing is actually more important for the aging than for any other demographic group: "Games will entice the aging to remain fit and mentally active, *to connect with others*" (Montet, 2006).

During the 50 years that have passed since Higinbotham conceived his game, many more observations have been added to the one regarding the social nature of human interactions through games. In our days, there are many attempts to make games, usually associated with entertainment and therefore compared to the success of Hollywood. In the meanwhile, games became part of the awareness of the medical community (after being successfully adopted by the military, by education, and even by politics). There are competitions—in the USA alone, over 30 posted on the Internet in 2008—funding opportunities (associated with the NIH, NSF, AASHA/CAST and several private foundations), classes offered in various settings—from training sessions to Ph.D. studies. As far as medical applications are concerned, there are many research activities, ranging from addressing obesity in the young (and not only), to PTSD, autism, cancer treatment, etc. The body of scientific contributions in peer-reviewed journals has increased spectacularly over the last five years. Many journals—from *Nature* and *Neuroscience* to *Science*—report on progress in understanding the various aspects of a game-oriented culture, and in particular of game adoption by healthcare professionals. The number of experiments dedicated to the extremely important issue of game impact validation doubled in the last three years. Indeed, to have a good time with a game is not really dependent on what various experts measure: How fast do you get bored? How long until you abandon the game? What gets your "juices" running? and so on. Experts continue to measure such characteristics anyway—because this is how funding can be secured, and because the game industry wants numbers for marketing purposes. Since the game industry is extremely competitive, but also very impatient, even a little knowledge (as dangerous as it can be) is considered better than none. But once you project expectations related to health, the name of the game (pun intended!) changes. As a matter of fact, a game for health—for young or old—will be judged almost like a physician is. Forget liabilities for a moment—which have already attracted the attention of lawyers and public advocates. The promise implicit in any treatment entails the necessity to produce the data validating the promise. Moreover, from casual games to massively multi-player online role playing game (MMORPG), what drives the effort in the first place is art and design, combined with good computer knowledge, extending into artificial intelligence and virtual reality. Once a certain game platform is made available, technological considerations are translated into game opportunities. Technological progress—faster processors, new forms of interaction, better interfaces, better I/O devices, etc.—have led to new game platforms. And this translates into opportunities. This is true also in respect to games for health. Consider only the impact of Nintendo's Wii platform on new game developments taking advantage of more intuitive forms of interaction. What in the first place drives the effort in the area of games for health is medical knowledge. This knowledge tends to be specialized: brain science, physiology, cognitive science, gerontology, etc. Not surprisingly, game companies tend to secure ownership rights over such knowledge as a way to maintain competitive advantage. Published science might be in the public domain, but when science needs to be translated into game specifications, proprietary solutions are usually the result (for example, Posit Science).

At the meeting point of high scientific competence, innovative game design, and technology development—often requiring new means and methods for game interaction, art and design, and feedback monitoring—not only engaging games, but also means for evaluating their usefulness and attractiveness are expected by the public. The major difference between any other type of games—including those designed for training—and the category lumped under the label "games for health" is that without scientifically based validation, such games compete with traditional medical care, but without offering to people in need of alternatives any proof for their effectiveness. Validation is also important in the competition among the many products claiming their share of the market. So far, games remain the equivalent of over-the-counter products—some more convincing than others, and the majority not qualifying for medical insurance coverage. The aging, in particular, are likely to fall prey to the latest crop of snake-oil peddlers operating in the gray zone of hope against resignation.

With 1958 as a reference point for games themselves, let us look at some of the earliest attempts to associate games with health and the many aspects of addressing the particular needs of the aging. Way back in 1976, efforts were made to involve the relatively small population segment of the aging in computers (Jaycox & Hicks, 1976). Interestingly enough, it was the interaction with students, i.e., young people, which proved most stimulating for the elderly, not their exposure to computers. So far, nothing unexpected—games or not, interaction with the young always helped people facing the many debilitating effects of aging. Slowly, in the course of events, the subject changed and the focus shifted to self-learning and self-sufficiency (Hoot & Hayship, 1983). In this context, Weisman (1983) reported on modified computer games made available to a group of 50 nursing home residents. The study was conducted with the help of the Georgetown University School of Medicine and involved games such as *Little Brick Out*, *Ribbit*, *Country Driver*, and *Hangman*. Evidently, one should not even try to compare the research possibilities of the 1980s (the last century, imagine that!) to the technology of our time or the challenges we face in addressing the various roles games can play in the lives of a rapidly growing aging population. Still, it is encouraging to see how far ahead the thinking went in respect to the possible impact of games. Even game addiction was brought up at that time.

Now fast forward to this century—almost ten years old—and you will notice that one question lingers: "Video games for the elderly: an answer to dementia or a marketing tool?" *The Guardian* (McCurry, 2006) and *The Examiner* (Barker & Bernhard, 2008) quote my own work on games and the aging ("…the best ones …are those that require players to be active and to socialize with other players"), together with the work of other researchers on the subject. There is no reason to compare *Little Brick Out* or *Ribbit* (mentioned here for those few who might still remember them) to *Guild Wars*, played Sandra Newton (age 62 at the time) or to *Final Fantasy*, played by Barbara St. Hilaire (age 70, when mentioned), both interviewed on National Public Radio (Cohen, 2006). In the years that Whitcomb (University of South Carolina) presented his paper at the *Symposium on Computers and the Quality of Life* (1990), focus was on how the elderly (as they were called in those days of less than politically correct expression) can "keep up with recent developments ...," and on how to improve "the range and quality of public services, of which the elderly are heavy consumers" (Whitcomb, 1990, p. 112; he also cited Elton, 1988). Yes, the elderly of those years, not unlike the seniors or aging of our own time, got better in playing the more they played (some even became addicted to games). And they enjoyed games as a more active form of entertainment than watching TV (never mind staring at an empty wall in isolation). What changed from those early beginnings is the understanding that games are not only for entertaining senior adults, but that the process of aging itself can be influenced. In some cases (e.g., epilepsy), conditions can be aggravated by the wrong types of games. This improved understanding was not reached over night, and, frankly, is not yet as conclusive as we would sometimes like to believe—or are prodded to believe by the slick marketers of their wares.

UNDERSTANDING THE GAME EXPECTATIONS OF SENIOR ADULTS

For reasons of methodological clarity, let us distinguish between three different perspectives:

1. *The demographic of the commercial games on the market*, that is, what Ijsselsteijn *et al.* (2007, p. 17) describe as "a significant potential niche market for game developers." The more seniors can be enticed to get involved with games, whether for health or entertainment, the larger the market.
2. *Motivations of seniors to engage in gaming*— to learn, interact, have a good time, fill available time, benefit from gaming (earn money, gain recognition, improve health).
3. *Conceiving, designing, producing, deploying, and evaluating games* with therapeutic aims.

The three aspects are not independent of each other. Young and old, male and female, educated or not, poor or rich, regardless of race and cultural identity, enjoy good games, i.e., games with an attractive story line, rich interactions, good design, and ability to engage. Some people— those who can afford the games—benefit from gaming; others, aware of their limited means, seek community-based game-playing opportunities. The motivation factor in respect to seniors is quite different from that of their children and grandchildren. Seniors will rather use stationary games than those available on cellular phones. And if there is anything that can make seniors lessen the burdens of aging (aches due to various causes, diminished performance, dependence on others, insecurity, etc.), they will use their resources (time, financial), or they will gladly partake in community-supported activities.

Numbers don't tell the whole story, but they are part of the story. According to recent statistics (ESA, 2008), the fastest growth in the adoption of games is in the senior population (double digits for the last five years). Gamers under age 18 (25% of all gamers) used to dominate the market. Now gamers 50 years old and older make up a slightly larger segment (26%), which is increasing very fast. At almost half the market, gamers 18-49 years of age exercise, through their expectations, the most powerful influence on the industry. They want excitement and push for everything that gets their adrenaline high, and that fills the last available second (after text messaging, games are the biggest application in the mobile phone industry). The most telling change is from playing alone to playing with others: the percentage more than doubled from 29% in 2004 to almost 60% today. For the aging—who also play more and more with others, of the same age or younger—this translates into better social interaction.

Action games and entertainment games are still the most popular among seniors; role-playing games (RPG) also rank high. However, the dynamics is such that games for health—still a recent category—are increasingly acquired by those in the middle years and by seniors. While *Halo 3* (for the Xbox 360) and *Call of Duty: Modern Warfare, Super Mario Galaxy* dominate the market, *Brain Age* has made inroads; the Wii play version with remote rated as second among the 20 top selling games in 2007. In 2008, many newcomers have tried to penetrate the market. Among the types of online games played most often, strategy and role-play figure quite high; so do puzzle and board games. Competitive gaming, preferred by younger gamers, is yet another trend in the fast growing senior involvement with games. As for therapeutic games, *exergames* of all kind—from the ubiquitous DDR used for weight control in younger populations (Carson, 2006; 2008) to mobile phone platforms for working out—satisfy the needs of youngsters, while a wide variety of mind-and-body games—ranging from Nintendo's brain training to Xavix's golfing and fishing—are geared towards the aging. Therapeutic games, including Second Life developments, make up almost 65% of the entire games for health market.

It is clear to any observer of the game market that the industry is assiduously data mining these and many other numbers. Nintendo's strategy is living proof of their understanding how to effect a shift of focus from 18-year-olds and young adults to affluent middle-agers and seniors. Other companies, such as Xavix, and even Sony, offering products at least as good, have not been able to penetrate the market with a similar rate of success. What is not clear is how the industry is meeting its obligations and promises to aging players without having developed its own research agenda for games for health, or without having funded such research. In view of such questions, the next part of this study will focus on the attempt to make therapeutic games available to the rapidly growing senior population, and especially on providing means for validating the results of involvement with such games.

FROM MENTAL EXERCISE TO INCREASED ADAPTIVE SKILLS

Again, a short look back: Clark, et al. (1987, p. 83) noticed that games show a promising potential "to improve selection" abilities. Ten years later, Goldstein et al. (1997, p. 346) "studied the effects of playing video games (*Super Tetris*) on the reaction time, cognitive perceptual adaptability, and emotional well-being of 22 non-institutionalized elderly people aged 69-90." The result: "significant improvement in ... reaction time, ... relative increase in self-reported well-being." Important note: The strongest effects were recorded on measures of reaction time, and the weakest on cognitive performance. When Nintendo reached the conclusion that a neurologist, Ryuta Kawashima (2003), author of a popular book on brain training, could inform their own effort to expand into the market of senior gamers, the company was aware of articles such as the ones mentioned above. But science does not simply translate into effective and engaging games. The neurologist Kawashima—who eventually became a game character because the company built on his credibility as a scientist—translated his observations on regular mental exercises into clean competitive actions: counting the number of syllables in phrases, memorizing words, performing simple arithmetic operations. This is how *Brain Age: Train Your Brain in Minutes a Day*, played on Nintendo's DS hand-held system, came into existence. Holding the dual-screen device open like a book, players can use the built-in microphone or the touch-sensitive screen. These are easy, very intuitive ways to interact with the game. And the output is much more directly related to mental health than is the outcome of playing with *Super Tetris*. In the game, a so-called "brain age" is defined, and the goal is to improve it over time, that is, to get a "younger" mind by playing, and playing, and playing. Fourteen basic activities and additional side games are made available. Once the number puzzle game *SuDoKu* became popular in the USA, it was integrated in *Brain Age*. There's nothing like success!

At the time the game was released, the connection between engaging the mind and improvements in cognition was not convincingly documented. The gamers' improvement in the game itself was quite impressive; but the supporting data regarding how this translated into real-life situations is still not available. Michael Merzenich (2008)—popular through the sponsoring of a fundraiser for PBS, where he (and others) explained aging, and how his own game could delay it—built on neuroscience expertise that reaches back to brain plasticity (Merzenich, 2008; Ellison, 2007). The *Brain Fitness Program* (marketed by Posit, a company in which the scientist has a minority interest) starts with simple fine-tuning exercises targeting the auditory system. It is different from crossword puzzles in the sense that it engages more sensory information and animation. With some other game-based activities, Merzenich focuses on treating individuals affected by schizophrenia. Although some improvement was documented in clinical trials, the results are not yet conclusive. Posit sponsors evaluations by independent practitioners in the healthcare sector, and this is encouraging. But as games on their own merit, Posit products are at best rudimentary. This applies to the scripts, to design, to interactions, and to user interfaces. Even before we have conclusive data regarding their therapeutic effectiveness, what needs to be acknowledged is the path: address the fact that brain plasticity can be affected by engaging individuals in interactions beneficial to their cognitive performance.

WE ARE MORE THAN OUR COGNITION

George Bernard Shaw, the famous playwright, elegantly summed up the problem of aging: "We don't stop playing because we grow old, we grow old because we stop playing (Shaw, G.B., n.d.)."

This highly descriptive but rather unscientific hypothesis predates video games and computer-based games. It was formulated at a time when playing was more than activating buttons on a console. Play included chess and backgammon, poker and bridge; but it also referred to games in which the entire body and the mind were engaged. A 90-year-old would probably avoid hide-and-go-seek, not to say tennis, golf, and bowling, which involve a physical component that could expose the player to the risk of falling, bruising, losing balance, or injury. But is there a way to be wholly involved, mind and body, without inviting the risk of getting hurt? If such ways exist, growing old would no longer entail giving up play because it is dangerous, or because the element of competition involved implies physical resources no longer available. Interesting enough, from the pen of the same writer came the statement, "I dream things that never were, and I say, Why not?" This might have made little sense when it was printed (well before word processing, mind you), but today, in the age of games, it sounds almost prophetic.

As suggestive as they are, Shaw's ideas cannot guide the effort of conceiving games for the health of the elderly more than any other generality can. Let's be clear: only a good scientific foundation, representative of the wealth of knowledge accumulated in respect to aging can be accepted as a premise for such games. When it comes to health, to life, to individuals who earned the right to be treated with maximum consideration, there is an obligation to work under the requirement of zero error. This translates into the obligation to consider the end of life not as a stage of resignation and defeat, but of taking advantage of opportunities—especially the opportunity to continue an independent life in which so much more can be learned, enjoyed, appreciated, and contributed to the lives of others. But games are not, and should never be, presented as yet another miracle cure for what might ail women and men willing to enjoy the rewards of their toils.

Nintendo's successful attempt at addressing cognitive characteristics and the attempt of Posit's Brain Fitness program are testimony only to an attitude of responsibility. Their limitation is the reductionist idea that if we train the mind, we will be able to arrest further inevitable mental deterioration. This is not a goal to be underestimated. Millions of aging gamers invest time and trust in such games on a daily basis. These games were, naturally, copied immediately; and it is almost impossible to say if the imitations are drastically different from the originals.

An alternative approach, also building on the hope connected to brain plasticity, is to consider the individual as the unity of cognitive, sensory, and motoric characteristics and capabilities. Here the focus changes from maintaining cognition to enhancing the expression of the unity between cognitive, sensory, and motoric activities in actions. Indeed, the goal is to maintain and even augment adaptive capabilities of aging individuals. Project *Seneludens*, initiated by the antÉ – Institute for Research in Anticipatory Systems, was begun in the following context:

1. Aging, along with the associated costs (not just in dollars), is a major health problem (see the *United Nations Programme on Ageing* Website). Many resources are utilized for fighting aging and its effects at its core (genetic, molecular, neuronal), but few for attenuating the consequences of the basic aging process. The fact that the post-World War II baby boom generation will be reaching elderly status in the coming years adds urgency to the need to address foreseeable problems before they occur. This situation is tantamount to a Sputnik for research in aging, similar to what became the impetus for the American space program. Let us again be as blunt as possible: If the meltdown of financial markets (since April 2007 and ongoing) is an indication, not addressing the consequences of aging in due time could result in an even less manageable situation. Not only money is involved, but also the fabric of society.

2. In respect to the research on aging and anticipation, which is the focus of this study, an exceptional situation was created by the start of a new program—Arts and Technology, ATEC for short, at the University of Texas at Dallas—focused on game design and development. This development stimulated ways to integrate results from the study of anticipation—in the broadest sense—and new developments in game design. Consequently, the following rationale for the project was articulated almost four years ago: Aware and respectful of the research in progress, especially in the area defined as brain plasticity, or rewiring, we attempt to integrate its results with a new approach: combine the will of the aging to remake themselves, to live in a dignified manner, to enjoy quality of life, with available means other than medication for maintaining characteristics that make life worth living; in particular, with means that engage the individual, such as interactive games—understood in a very broad sense—that stimulate an active path of maintaining anticipatory characteristics. Games are open-ended, competitive, nonlinear stories with an embedded wager (reward mechanism). Involvement in games can be physical, cognitive, emotional, and/ or social. It always entails learning, which is why games are adopted for all kinds of training. Documented through research is the impact of game involvement in active recall of information; coordination skills, especially eye-hand coordination; reaction time; predictive thinking; decision making; and collision avoidance. These are all the result of the anticipatory characteristics of human beings. More to the point: maintaining these characteristics can become concrete objectives in elaborating new games aimed at benefiting senior adults.

Figure 1. Mapping from the brain to game design

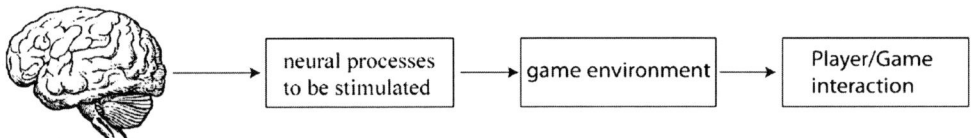

Why is the emphasis placed on anticipatory characteristics? To start at the end: Senescence is the stage at which anticipation degrades to such an extent that the body is practically reduced to its physical-chemical reality (cf. Rosen, 1985; Nadin, 1991). Anticipatory characteristics of the living—human beings, as well as other life forms—are fundamental to the preservation of life, moreover, to the performance of rewarding activities (learning, sports, creative acts in general, orientation in one's immediate environment and beyond, etc.). Anticipation is understood as "the sense of context" (Nadin, 2003, p. 27); better yet, as the expression of the human being's adaptive performance. This translates into the ability to successfully operate in a world in which knowledge of the order and repetitiveness of physical phenomena is insufficient for coping with life's unexpected aspects, in particular, change. The anticipation of danger, as well as a successful course of action, in a competitive or non-competitive context, is part and parcel of our existence. We negotiate steep slopes of downhill skiing not through reaction only, but especially through anticipation. We learn how to use tools (including the new "virtual" machines) with the aid of our anticipatory characteristics. We define goals and methods for small tasks or for an entire life in anticipation of ever faster changing contexts.

Games for Health, whether for seniors or adolescents, can take many forms, and can be guided by a variety of premises. As already pointed out, mental health—which some understand as memory performance, the ability to think and reason, stimulus recognition, and the like—is addressed by games that engage the mind. Ideally, playing such games would result in higher confidence in handling tasks associated with daily life. Activation of mental abilities corresponds to the recruitment of various areas of the brain. Therefore, the game designer for such particular games has to be able, under the guidance of medical knowledge and in permanent interaction with healthcare professionals, to map from the desired goal of stimulating specific neural processes to the means to be used. As obvious as the reasons for closely working with competent healthcare professionals are, they are worth repeating. People who develop games for the military, for instance, or for training firemen, would not play armchair expert; they would involve experts, from the initial concept to testing and evaluation, and to periodic follow-up and adjustments. This is the reason why students of game design were taught in the spirit of interfacing with medical experts (neuroscientists, physiologists, therapists, etc.).

In order to identify significant neural processes, we worked closely with experts in the field. Furthermore, we came up with new ideas for game environments (such as new I/O devices). The interaction between senior adults willing to play and the game became our focus. Furthermore, we cannot make any progress in this area if we do not realize that human beings do not age the same way. Therefore, accounting for differences is by far more important than assuming some averages. All this became the focus of graduate seminars and of research work that involved gamers, designers, story-tellers, computer scientists, virtual reality experts, and medical personnel. Students

Figure 2. Physical activity is linked to cognitive performance. The need to integrate cognitive and motoric capabilities is illustrated by brain activity

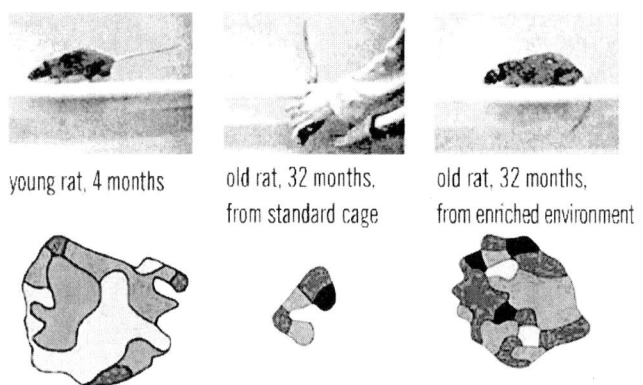

were also guided towards understanding the need to develop validation procedures and to test the effectiveness of their games (Nadin, 2006).

VALIDATION AND FORMULATION OF HYPOTHESES

Games for Health, in particular for senior adults, became subject to evaluation in various settings, including brain imaging. As we performed our own work, we learned that the data linking mental training to improved cognitive performance was not necessarily evincing more than improved performance on the tested tasks. In other words, regardless of how appropriate a game was, the more the seniors played it, the better they performed in the game. Therefore, the path we chose was a more complex behavioral therapy that integrates games supporting coordinated physical and mental engagement mediated through sensory input. A variety of studies make it clear that physical exercise and cognitive stimulation go hand in hand. In particular, we interacted with the Institute for Neuroinformatics (Ruhr University-Bochum, Germany, headed by Dr. H. Dinse), known for it studies on animals that can easily (and plausibly) be generalized to humans. Animals (rats, in Dinse's case studies) placed in enriched surroundings show improved dendritic branching and neurogenesis. Physical activity proves to be linked to higher performance on cognitive tasks. The exercise-triggered neurotrophic factors and the better flow of blood to the brain (so-called *vascularization*) speak in favor of extrapolating such results to humans.

Figure 2 illustrates how the enriched environment affects brain activity. It is worth pointing out that the enriched environment includes motoric activities. This is even more relevant for human beings who in general already benefit from a level of comfort higher than in the past, but which usually results in diminished physical activity (sedentary patterns).

Consequently, a working hypothesis was formulated, integrating knowledge of anticipation and studies such as those mentioned: In order to address the fact that an individual's adaptive capabilities decrease with aging, we need to stimulate brain plasticity through games that simultaneously address cognitive, sensory, and motoric capabilities. The metrics for assessing anticipatory characteristics and how game playing impacts them corresponds to characterizing the

Figure 3. The goal of project seneludens

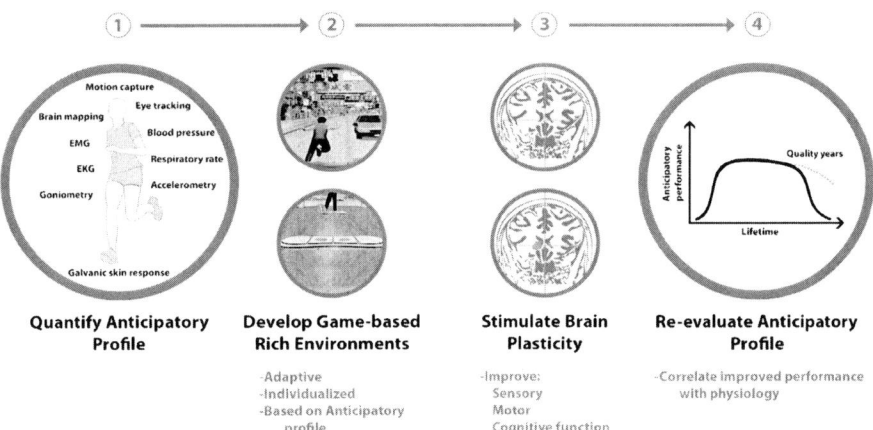

performance of an individual in action. Therefore, the project's goal becomes the characterization of human actions and the focused attempt to improve performance through game playing.

Figure 3 divulges that the entire effort presented in this study starts with defining the *Anticipatory Profile*™ of each individual (to be discussed in some detail shortly) and leads to individualized involvement with games that integrate sensory, cognitive, and motoric aspects. Validation, in this case via brain imaging (although other methods are acceptable), returns information on how successful the game behavioral therapy has been. The hope and goal are that decline in adaptive performance can be postponed, i.e., that independent living for senior adults can be extended.

Does all this, which some game designers might consider totally beyond their interest and focus, mean anything to those who want to engage in games for health, in particular to senior adults? The question is not rhetorical. Gamers do not, and should not, care about the theory behind the game. They should never come up with their own theories. But the credibility of games as therapeutic means cannot be disconnected from the foundations of the effort. To these foundations belong, of course, conception of games, design, development, production, and testing, and their improvement as predicated by experience gained in using them. It is obvious that game companies will not—and actually should not, even for the sake of a very attractive market share—begin the kind of animal research mentioned above, not to say research on human subjects. But it is also obvious that ethical considerations, as well as risk mitigating factors, speak in favor of interacting with the scientific community in order to identify possibilities, as well as limitations. The same holds true for interacting with game designers, artists, engineers, psychologists, computer scientists, etc.

In what follows, I will describe the very promising path, informed by the science of anticipatory systems, that we pursued. It is not the only conceivable path, and so far others have followed it rather marginally (Meyer, et al., 1997). The advantage of the methodology to be described is that it offers practical answers to game-specific questions. It also provides validation mechanisms that rely on the same procedures through which we generate characters for games. As already stated, no one who is seriously committed to Games for Health can avoid working within a context of validation, and without subjecting the work to the same.

BEHAVIORAL THERAPY –
PROJECT SENELUDENS

In the development of anticipatory characteristics, games of all kinds—from simple interaction (hide-and-go-seek, tennis, bowling, golf) to elaborate machine-supported performance (e.g., interactive games)—play an important role. Each game is a physical entity—described by the physics of the respective game (different in basketball, golf, football, chess, etc.)—and an anticipatory process: the outcome (future) defines the current state of those playing (tennis, pinball machines, cards, or some other game). Aging affects not only high performance (in athletics, chess, monopoly, backgammon, etc.) and learning, but even the willingness to play and to continue acquiring knowledge. Limitations brought on by aging—deteriorating vision and hearing, limited tactility, reduced agility and strength, reduced sense of balance, impaired cognition, and a sense of alienation or even exclusion—make games difficult to handle and even to accept (the "What for?" syndrome). Thus games, and play in general, are progressively eliminated. Physical and cognitive challenges, as well as social attitudes encountered, affect the abilities of the majority of the older people. To overcome such obstacles, we need to engage healthcare professionals, insurance companies, families, and the media. And we need to offer *attractive* video or computer games. Until recently, games and play were eliminated from the lives of the elderly when they actually need them most. To reverse this situation is not easy, but the process already started. Video games and computer-based games are making a difference.

Appropriately conceived games could fill the void of increasingly isolated men and women by maintaining physical and cognitive characteristics and social interconnectedness. Project *Seneludens* (www.seneludens.utdallas.edu) is focused on designing games and other therapeutic behavioral environments with the specific aim of maintaining anticipatory characteristics during the aging process. (The title is derived from two words: *senescence,* from the Latin *sene*, "old age", referring to changes that take place in a living organism as time advances; and *ludens*, as in *homo ludens*, playful human, playfulness being a characteristic of humankind.) It is known, and amply documented in scientific literature, that play, as an active form of involvement in a rewarding activity, fulfills necessary functions. It is practice for real-life situations. Play develops physical, mental, and social skills, just as it relaxes the mind and body, without the deleterious effects of extended televiewing and other forms of passive distraction.

Based on the cognitive and psycho-physiological mapping of human characteristics at the stage of life when aging is noticeable, we defined the structure and nature of a new category of games, and other play-conducive environments, pertinent to this phase. Based on results from current advanced research in the cognitive neurosciences (accumulated in clinical observations and other experiments, especially in brain research), we initiated the design of games and interactive behavioral therapy environments that support the *active* maintenance of anticipatory characteristics of the aging. Such games have an important learning (and reward) component.

The metaphor of mirror-neurons (Rizzolati, et al., 2000; Gallese, 2001) i.e., neuronal configurations generated as people attempt to imitate an action (mimetic procedure), was repeatedly confirmed in game situations. Indeed, watching a football game or a tennis match engages the viewer in ways reminiscent of their own playing. More challenging situations (jumping, jogging, driving) experienced in a game make the player emulate—even if only partially—the condition. Neuronal configurations and muscle engagement testify to this.

The image in Figure 4 is suggestive of something we all experience: showing someone else how to perform an action (let's say, putting in golf). Once the person we train starts perform-

Figure 4. Neuronal configurations and mind-body engagement triggered by virtual actions

Fig. 1: Some of the games developed will require coordination skills and "mind control."

ing, we kind of partake in the action, almost like "pushing" with our mind in sync with our apprentice. This is the simplest experience of mirror neurons at work. Watching a game (tennis, baseball, boxing) often engages viewers almost as though they were playing. Sure, the physical effort is not duplicated, but the mental experience is real (and rewarding).

Many neuroscientists believed that the capacity of brains to reorganize was restricted to early development. Only later was it recognized that both young and mature brains can be modified by experience. Research in progress, at UT-Dallas (the Center for BrainHealth™) and at many other centers in the USA and abroad, focusing on cognition and the brain resulted in evidence (some only in regard to animals, others including the human being) that reveals the following:

- Environmental enrichment and behavioral training increase brain responses.
- In order for plasticity to occur, the triggered sensory experiences should be behaviorally relevant (i.e., the nature of the task is important).
- Task difficulty plays an important role in inducing brain plasticity.

Only in a limited way are these findings sufficient for specifying the type of activities conceived for the aging that should be made available through games. They constitute a methodological premise in the attempt to define the individual nature of the behavioral therapy made possible through the medium called *game*. These findings also fully confirm the anticipatory perspective, i.e., the proactive nature of the wellness model that project *Seneludens* pursues. Virtual activities (fishing, boating, swimming, climbing, bowling, etc.) are immediate candidates for games built on the premise of mirror neurons. The degree of realism, which new technology supports better and better, is reflected in their effectiveness. Games and behavioral therapy are not a cure. They facilitate improvements, but not a return to full performance. Actually, a more responsible description is that they are rather methods of prevention. Their precise objectives are maintenance of the human being's anticipatory characteristics, and delay of the onset of cognitive, sensory, and physical decline. Even for these objectives, games have to be engaging and stimulating; they have to offer a challenging learning curve.

Essential Game Characteristics

Very important is the creativity intrinsic in the project's goal. To conceive of games, i.e., play structures, that translate into behavioral therapy is to invent a category of artifacts that can be individualized, are engaging, display adaptive properties (i.e. reflect the dynamics of aging), and support brain plasticity. For example: a tennis game, with simulated forces (the serve, returning a fast ball, the serve) will have to be adjustable in terms of speed and physical effort. These games will have to offer multi-sensory environments in order to stimulate the interplay between sensory driven actions and mind-initiated actions. Since sensory performance is affected by age, we have to provide ways to compensate for lower performance by providing cognitive cues. Some examples: an image can suggest the sound no longer perceived due to deafness; movement, as animation or realistic rendition, can carry haptic and temporal information. This is a challenge, but also an opportunity. Moreover, the games need to be designed in a manner that supports the customization effort.

Whether the game's object is space exploration (with time somehow left our or of marginal significance) or the experience of time, what is essential is to associate the playing aspect with the desired impact on the cognitive and motoric performance. Aspects as simple as: Can the player pause the game? Does the game allow for synchronous actions? Who controls the game? (the so-called internal control aspect) translate into engaging brain plasticity in a desired direction. To alter the game environment (i.e., in a game like Lemmings, this is possible), or even to add content (such as new features in second Life) are not to be understood only as attraction features, but as means for stimulating the player's active contribution. Even game decisions such as whether the player should see the whole game area (such as in *Badminton*), or become familiar with it as the game unfolds (*BioFighters* is a good example), need to be addressed from the very precise perspective of cognitive performance. Each game that is focused on the health of senior adults has to provide options for customization.

Taking a Cue from Memory Research

In studying various processes accounting for human memory, scientists have advanced the complementary model of an action and drug-based cure (Tully, 2005) (see Figure 4). The goals of *Seneludens* are farther-reaching than memory maintenance (although memory is an implicit theme). They include replacing experience as a generic term through the variety of interactions that games and play afford; associations, mind-driven actions (Nicolelis, 2005); mirror neuron generation; and heuristics (the discovery of new possibilities). Neural circuitry becomes neuronal configuration, and expressions become performance level in the game. We do not exclude the possibility of using appropriate medication (Dinse, 2005), although this aspect is not a specific goal of the *Seneludens* project.

For people involved in developing games that address the well-being of adult seniors, it is advisable that they consider non-game attempts (such as drug-based remedies and even classic therapies) and try to duplicate their structure. This is not an easy path, and it always requires the effort of medical experts in addition to game designers. But it is a good model, at least because medical practitioners are no less inclined than patients themselves to look for no-drug/no-surgery alternatives for healing. The aging themselves prefer to entertain non-invasive methods instead of medication for maintaining their viability.

The current science and technology that will eventually be embodied in new products to aid the aging were not conceived from the anticipatory perspective. As we know from the games developed to help train the military, they are rather the expression of the science of the physics, but not of the art of living. The new generation of creative designers and artists, who also acquired expertise in the computational, will have to express their creativity in a class of products that defies the industrial model of one (product, in this case, game) for all. The alternative is the customized, individual product that specifically targets the cognitive dysfunction of the individual to be helped. The metaphor of mirror neurons (see Figure 4) suggests that we can conceive of games that can "teach" their players how to improve their motoric performance as well. We have the science and technology for achieving this. But we do not yet have the knowledge and art necessary for creating variable products with adaptive characteristics.

Figure 5. From drug-based memory enhancement to game-based behavioral therapy

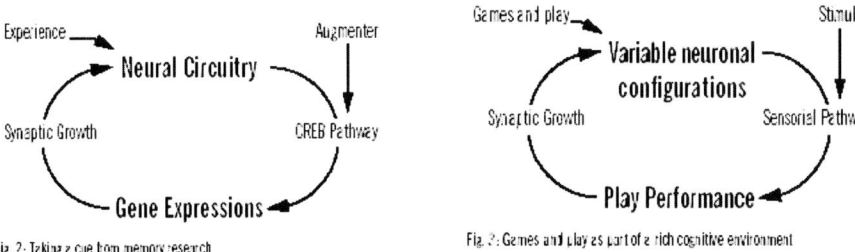

HOW DO WE GENERATE DATA FOR THE NEW GAMES?

In view of the unavoidable requirement of working with medical professionals, as project *Seneludens* was defined, I contacted major health professionals in Dallas. From the many contacts established for this purpose, one dialog with the Medical Director for Research at Presbyterian Hospital (Dr. Mark Feldman, MD, FACP) remains vivid in my memory. I explained the context of anticipation, as well as the opportunities to conceive games intended to help maintain adaptive performance in aging men and women. Dr. Feldman had only one question: How are you going to measure anticipation? He added, "We healthcare practitioners follow very precise research protocols. Our focus is always on documenting changes associated with therapy. And for doing so, we measure." Since that conversation, I have always credited Dr. Feldman for starting me on the path of quantifying anticipation.

This was the beginning of the *Anticipation-Scope™*. Anticipation is expressed in action—avoidance (of danger, such as slipping, tripping, or being struck); successful performance (catching an object we did not even see coming); preparation (for changing position, such as raising arms, an example to which I shall return). Therefore, the measuring process has to reference the action. In other words, if we want to see how raising one's arms is preceded—anticipation is something that happens before—by cognitive and motoric performance, we need to juxtapose a description of the action and the cognitive and motoric parameters to be measured.

Leonardo da Vinci (1498) observed "...when a man stands motionless upon his feet, if he extends his arms in front of his chest, he must move backwards a natural weight equal to that both natural and accidental which he moves towards the front." That is, a physical entity modeled after a human being (in wood, stone, or metal) would not keep balance if we raised its arms because its center of gravity changes. Figure 6 makes the point clear.

Muscles Region

Compensation slightly precedes the beginning of the arm's motion. In short, compensation occurs in anticipation of the action, not in reaction to it (as Libet's, 1980-84 work on the brain's readiness potential confirmed).

The *AnticipationScope* quantifies the process by capturing the motion, not on film or video, but in the mathematical description (as a matrix) corresponding to high-resolution motion capture technology. This description allows us to focus on single points on the body and to pick up even the slightest tremor or motion characteristics (of joints, body segments, muscles, etc.). Known sensors,

Figure 6. Before arms are raised, muscles in the back of the legs tighten

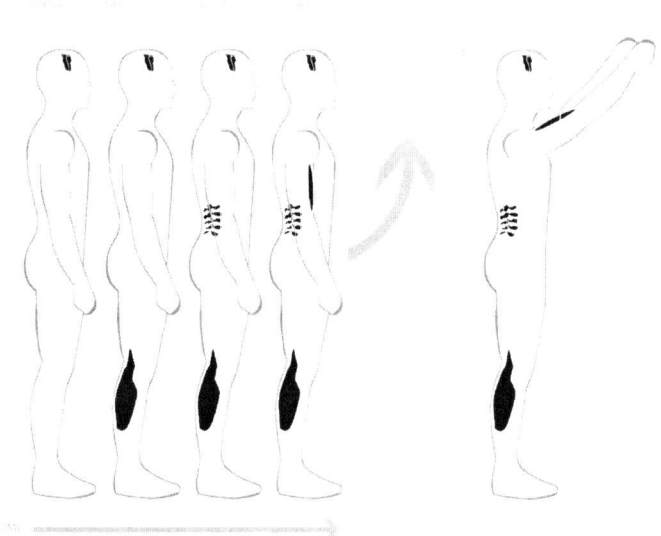

such as EMG, goniometry, accelometry, blood pressure, EEG, among others, and new sensors, developed as significant parameters are identified, are applied to different parts of the body. These sensors capture the various preparatory processes, as well as the reactive components of maintaining balance.

The Game Program at ATEC/UT-Dallas has benefited from a motion capture lab early in its inception. Based on this felicitous situation, I suggested that instead of filming a subject raising his/her arms, we should motion capture the action. In motion capture, the description results in a matrix corresponding to high-resolution "scanning" of the body in action. Such a quantitative description allows for focusing on single points on the body and for picking up even the most evasive tremor. However, in order to fully capture the expression of anticipation, we need specific values of physiological parameters affected in the action. Therefore, I designed an integrated system of motion capture and sensors attached to specific body parts. The first objective was to create a measuring environment that does not affect the free movement of the subjects of quantification. The second objective was to integrate motion capture data and sensor data, which are of different formats and different timescales. The third objective was to design a data processing program that allows for examination of the multitude of information collected, and for eventual aid in identifying the expression of anticipation, moreover, in describing it quantitatively.

This architecture further informed the information processing model. It needs to be emphasized that even working with medical professionals does not automatically mean that all questions pertinent to the procedure have a well-defined answer. For example: Which physiological parameters should be monitored? In medicine, attempts were made to describe some functions (e.g., cardiac profile, which means all the known elements that affect heart functioning). Still, even medical practitioners could not name all the physiological parameters that need to be considered. We knew one thing: as we try to quantify anticipation, we are focused on the totality of the human being. Therefore, we look at descriptors that few will associate

Figure 7. Architecture of integrated AnticipationScope

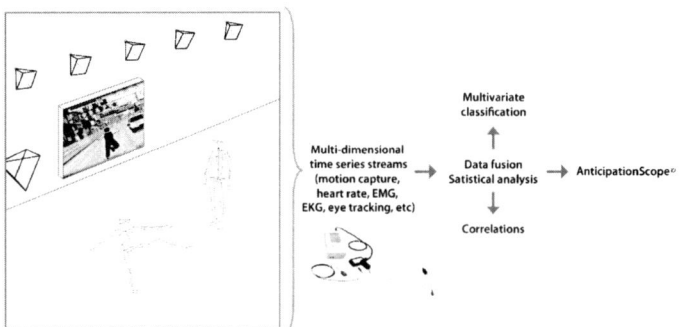

with how well we adapt. Example: tympanic temperature. Dr. Michael Devous (University of Texas-Southwestern Medical Center) suggested this parameter as we tried to find out how we can capture the brain activity component when the subject jumps, catches a ball, or goes up and down stairs. At the current stage of technology, brain imaging or other brain monitoring techniques cannot be considered. It turns out that tympanic temperature is representative of the engagement of the brain in certain actions. Therefore, it provides some information regarding brain activity.

Together with Dr. B. Prabhakaran (School of Engineering and Computer Science, UT-Dallas) and Gaurav Pradhan (Ph.D. candidate in Computer Science), we designed and implemented the data acquisition and data processing procedure. From this effort, a number of research goals were accomplished (Nadin et al., 2007a, 2007 b, 2008). They are the result of my concept for an *AnticipationScope* and of efforts to engage a large sample of subjects (over 75 in 3 years) in a variety of experiments relevant to the final goal, i.e., the development of games that can contribute to the well-being of senior adults.

It should be made clear that this preliminary research was not geared towards producing medical data or developing algorithms relevant to integrating a variety of data types. My goal was, and continues to be, to see how the quantified anticipation eventually translates into specifications for games and how such games can be used effectively for compensating, through brain plasticity, for lost anticipatory characteristics. Therefore, parallel to elaborating the *AnticipationScope* and defining the *Anticipatory Profile*, I worked with a post-doctoral research assistant, Dr. Navzer Engineer (Ph.D. in Neuroscience) in defining the typology of experiments, and in developing meaningful experiments for our ultimate purpose. If indeed we want to create game-based behavioral therapy, which is also "individualizable," we need to find ways to translate the *Anticipatory Profile* into those cognitive and motoric actions that make up a game capable of stimulating brain plasticity.

Our efforts did not take place in isolation from the broader game economy. While ideas originating from our research were adopted by others (with or without appropriate credit given), it became clear that we developed not only a good premise for game development, but also a very valuable evaluation platform. Figure 10 suggests the variety of approaches relevant to the premise formulated so far:

The reader will easily notice that the *AnticipationScope*, as conceived and implemented, serves as a development platform—where we obtain the data that will inform the concrete game development. But it also serves as an evaluation platform. Indeed, as game involvement therapy

Figure 8. Information processing model

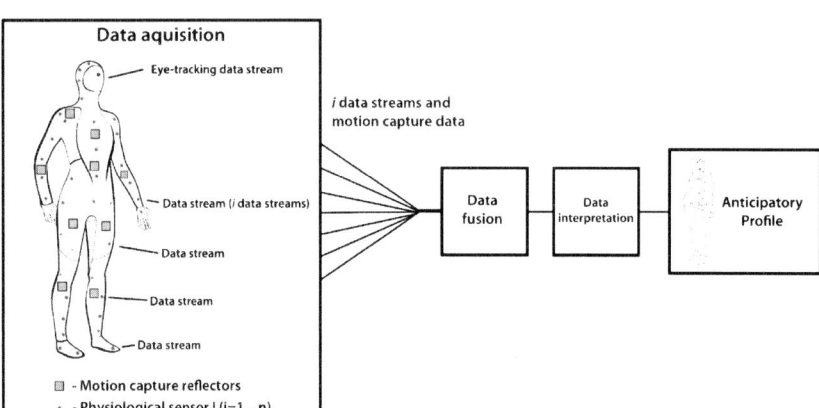

is practiced—under the strictures of any therapy—we expect that the aging will benefit from improved adaptive capabilities. To document the improvement, the *AnticipationScope* is used to see how the *Anticipatory Profile* changes after active involvement with the game and successive "fine tuning."

Anticipation is carried out on the neuronal level and precedes action by milliseconds. These milliseconds can make the difference between protection ("knowing" how to fall, for example) and injury. Anticipation declines as people age, making them susceptible to injury (damage resulting from not "knowing" how to fall), which is complicated by advanced age. Furthermore, anticipation is a major factor in successful game playing, as well as in sports and high-performance activities. By their nature, games and sports (e.g., card games, chess, tennis, hockey, downhill skiing) imply an anticipatory element in that the player who can best anticipate is successful. This project relies on data from experiments focused on brain plasticity. Success entices a player to play better, hence our aim to develop games that can be targeted. Success entails positive emotions and attitudes; it is self-motivating.

A *Mind Gym* as Test Bed for the Scientific Hypothesis Pursued

The following examples of games is indicative of the direction followed, i.e., create games with a focus on stimulating anticipatory action and evaluate their effectiveness. With this aim in mind, we have worked with XaviX, Inc. since 2005 and accumulated experience that can be generalized. (XaviX is also providing equipment to UC-San Diego researchers at the School of Medicine who are studying exergames as an option for motivating and sustaining the physical activity of adolescents.) In continued collaboration with XaviX, Inc., project *Seneludens* is evaluating the effectiveness of several types of games now on the market in maintaining and/or improving cognitive-motoric abilities. The C.C. Young Retirement Community (Dallas) has installed the XaviX *MindGym* for use by their residents. Deploying the *AnticipationScope*, we measure changes in the adaptive capabilities of the persons involved (ranging in age from 55-90) and refine theory-based principles. The unique contribution lies in the quantifying anticipatory capabilities ("anticipation") of the persons participating in the data collection. Although the researchers are using games provided by XaviX, Inc., along with Nintendo Wii games, in the future, they might use proprietary prototype games

Figure 9. The anticipatory profile

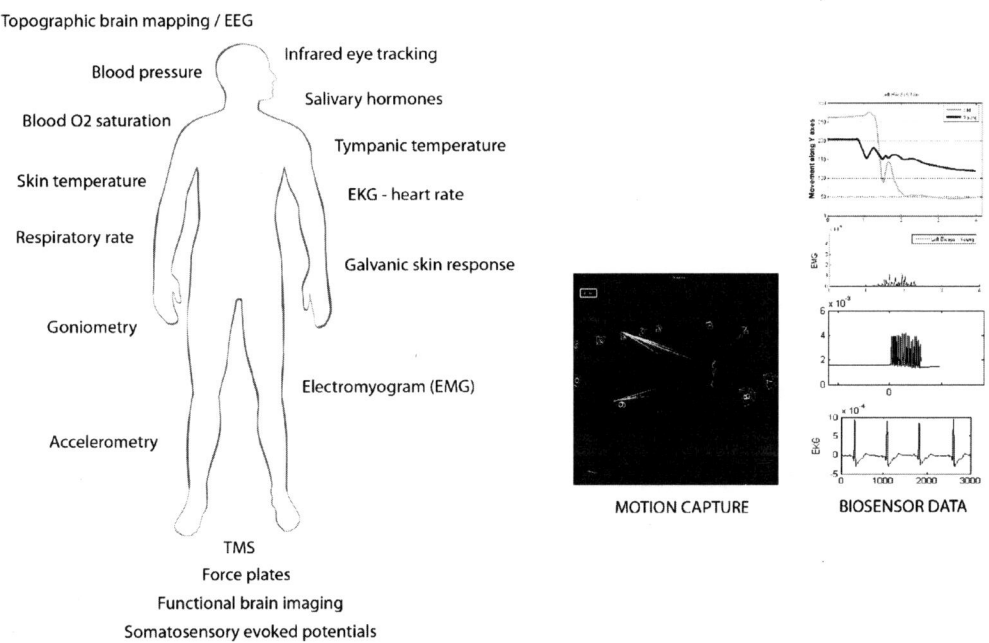

Figure 10. Various forms of individualized game-based behavior therapy (illustrative selection)

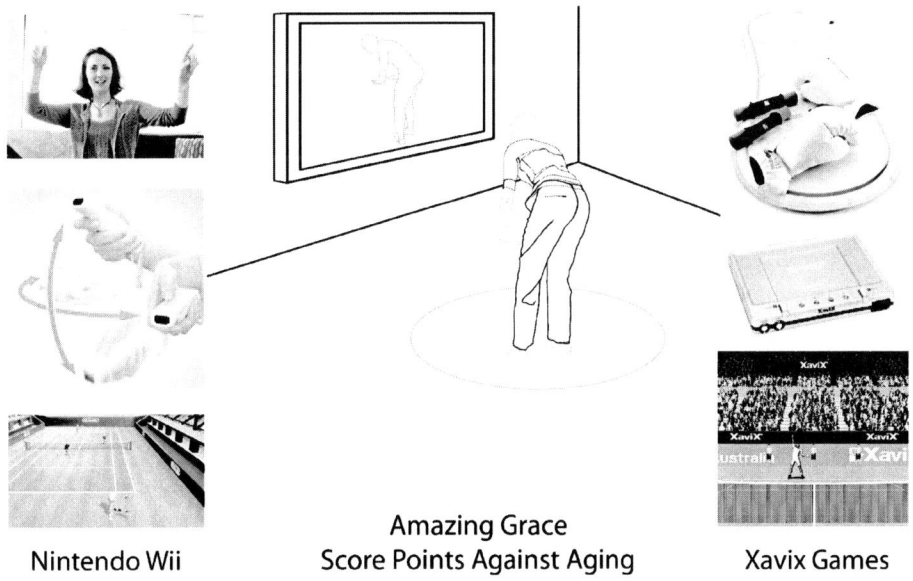

developed by the ATEC faculty and students for the US Army (a project begun in 2006). This also corresponds to our final aim of elaborating a new type of interactive games and gaming environments for targeted therapy.

XaviX EYEHAND, for example, is a multi-task training interactive game that stimulates the integration of cognitive and motoric abilities. These are expressed as decisions, reflexes, and dexterity. On-screen instructions inform the player about actions performed using *Glove Sensors* attached to the hands. The EYEHAND training system uses a CIS (cmos image sensor) camera mounted on the system cartridge, together with motion sensor technology, that accurately recognizes and captures the movement of the gamer's hands. The players are requested to move their hands in sync with the two orbs animated in a fixed pattern. This mode stimulates the performance of multiple tasks simultaneously. Other tasks involve touching the panel of the specified color that appears on screen. As the player scores points for each correct action, the variety of goals increases. In this particular mode, the purpose is to develop discrimination capabilities and improve reaction time. Other scoring possibilities involve improving the speed of actions ("catch butterflies with both hands"), ordering abilities (grab numbers in ascending or descending order), memory performance (memorize patterns and positions of cards). This game benefited from the input of a cardiologist (Dr. Yutaka Kimura, professor at the Kansai Medical University).

Another game of similar anticipatory focus is the CardioFitness workout routine (which can also be called "Run and Have Fun with Jackie Chan"—who is the animated character selected for the purpose of engaging players). Walking, running, jumping, performing squats, and avoiding obstacles are part of this personalized set of ten gradually more challenging routines. Evaluation of the improved anticipation of gamers of all ages provided us with valuable data regarding the integration of cognitive abilities and motoric performance. Actually, even a handicapped senior adult can fully enjoy the games using leg(s) and body movements.

The way in which knowledge pertinent to anticipatory processes can inform the design of such games remains the major focus of project *Seneludens*. Quantification of the results of game involvement of the aging is by now the subject of a Ph.D. thesis (Melinda Andrews is the Ph.D. candidate) in progress that will benefit from the interaction of the antÉ Institute, the C.C. Young Retirement Community, XaviX, the School for Behavioral and Brain Sciences, and the School of engineering and Computer Science (both at UT-Dallas).

AmazingGrace™

This application is based on documented evidence of the many benefits associated with African dance movements aiding villagers in Senegal and Benin (as well as in other countries in Africa) develop and maintain physical and mental well-being. In particular, these movements, in which anticipation is expressed in a variety of actions meant to maintain the dancers' sense of balance, rhythm, and coordination. They were introduced to us by Germaine Acogny, one of the most important dancers in the world today. Expression of rhythm, movement, and coordination in a simple dance could help in addressing problems of the aging within populations that are becoming more sedentary. In their wisdom, African villagers understood that mind and body reach their harmony as they stimulate each other. Acogny, in turn, understood that the dances that inspired her art are also an expression of knowledge. When I explained to her that we could actually show how someone who practiced her selection of movements would be positively affected by them, she spontaneously offered her cooperation. One year later (in April 2007, after our first meeting in May, 2006), she and her drummer came to the *AnticipationScope*. During five days of sessions, we were able to fully

Figure 11. Motion capture and rendering of germaine acogny's dance selections (Jeff Senita, a PhD candidate active in the ATEC program, provided the modeling and animation)

motion capture her selections, to look at details, and to see how these translate into stimulating brain plasticity. Based on this, we aimed at developing customized families of dance games that allow for individual and collective sessions. *AmazingGrace* was conceived as a video game in a manner that does not require technical skill for using it.

Acogny herself paid attention to the therapeutic aspects of the dances from which she extracted her "exercises." As far as I know, some of this work concerning the role of rituals and dancing in supporting the health of senior adults was validated by researchers in Europe and the USA. Let me mention the work of Heather Lundy (2002), especially her attempt to incorporate dance and other therapy techniques. Mirror neuron research, which our Institute considers as a very powerful technique for documenting how dance affects young and old, also produced some important data validating the claim. The interested reader can also refer to Hanna (1978, 1995).

Like all our work, this game was devised in a context in which we can also quantify the benefits of involvement with it. That is, we can see how effective daily dance exercise is. There is no standard dance to be performed, there are selections, and there are procedures for customization. Each selection corresponds to what the individual can afford from a cognitive-sensory-motoric viewpoint. For the aging, the level of individualized selection is determined by their particular abilities and the goal of the action (focused stimulation of brain plasticity

Convincing the aging to participate in such a game is not a barrier, as some are inclined to believe. So far, each time senior adults were addressed, they showed interest and determination to follow through. One can get involved with *AmazingGrace* while in a wheelchair. There are limited movements, which can be performed even by someone bedridden. This is what distinguishes the game from other exercise routines. Brain plasticity can be stimulated without putting anyone in a dangerous situation. What needs to be further addressed is individualization, that is, customization to the person. This applies to the

Figure 12. Comparing patterns of movement

The Anticipation Scope makes it possible to compare an individual's movement pattern with the basic Acogny movement reference base:

Acogny Pattern   Individual Pattern

movements and rhythms, as it applies to coordination when two or more persons are involved in the dance-game.

Germaine Acogny generously offered time, knowledge, and experience. During her visit to the Institute, researchers performed high-resolution full motion capture of her dance movements. Using the *AnticipationScope*, we also evaluated the significance and effect of each movement. Based on this data, we are now working on further generating the animation for each sequence. The animated character/characters drive the interactive game. Our challenge is to integrate the animation with the person/persons, young and old, who will eventually use the game. We need to develop a mapping from the movement to the performing individuals so that characteristics such as cursivity, expressiveness, synchronicity, balance, etc. can be monitored by the game. To dance means to learn, and this learning addresses cognitive and motoric characteristics. To dance also means to enjoy movement and rhythm in the company of others.

The transfer of knowledge, from the dance captured to the aging affected by sedentary life, remains a major objective. In concrete terms, we proceeded from the living dancer to the motion captured images to modeling and animation.

The image associates the experience of dancing in the Senegalese village (Toubab Dialaw) where Acogny works to a routine she developed with the aim of helping individuals in need of improving their adaptive capabilities. Also documented is the sequence from motion capture to the creation of the avatar that interacts with the users of the game. Since the gamer is engaged in the dance, we had to conceive of procedures for comparing the details (Acogny pattern vs. individual pattern).

Based on these detailed comparisons, the goal of matching the original movement and the movements of the senior adults involved in the game also became a scoring component. Indeed, each game is driven by a "wager," the reward mechanism being yet another example of human anticipatory characteristics at work.

One of the challenges is to keep the interactive platform to a minimum. (In order to fully evaluate the movement of a gamer and compare it to that of the game avatar, we would need a full motion capture setting, with at least three cameras, which at this juncture in the development of technology is not yet a practical solution.) Therefore, we developed minimal tracking algorithms (very similar to triangulation) and a feedback mechanism with a reduced set of control possibilities.

Figure 13. Scoring based on the degree to which the senior adult players match the original movements

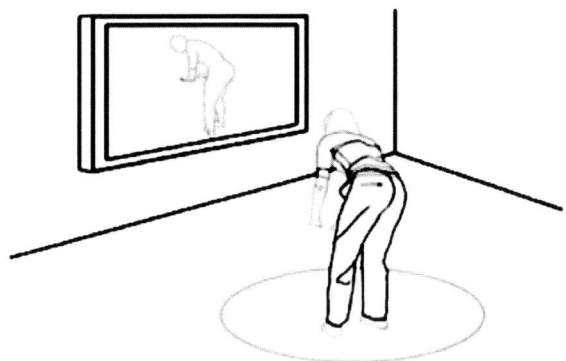

Figure 14. Testing the game implementation

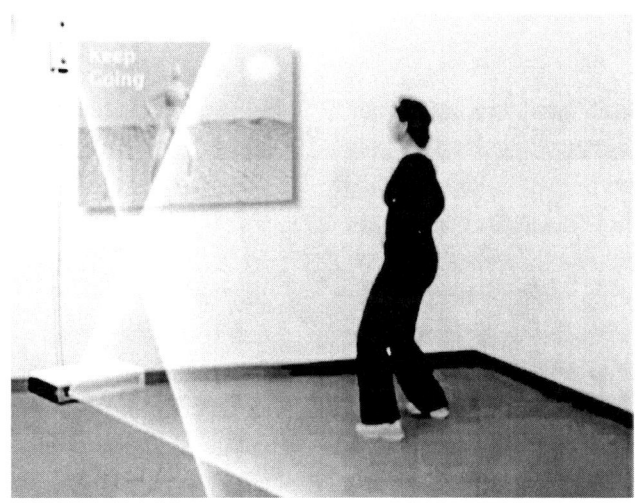

FURTHER RESEARCH AND APPLICATIONS

Based on knowledge accumulated from the *AnticipationScope* and from project *Seneludens*, a number of associated game projects were derived, and chances are good that the development platform we conceived and tested so far will be used for a wide variety of games with a focus other than the aging, but always relevant to the health of senior adults. We ourselves are pursuing games addressing the broad category of spectrum disorders (Parkinson's disease, Alzheimer's, PTSD, autism, etc.), as well as games pertinent to extreme situations. One of such games was prompted by the tragic lack of preparation for Hurricane Katrina, in which many adult seniors lost their lives (see Figure 15).

The major hypothesis for the real-time-game-based alert system is the following: The action of anticipation cannot be distinguished from the person's perception of the anticipated (Nadin, 2005). This is especially important when dealing with extreme events in nature—earthquakes,

Figure 15. From real life to a networked game receiving real-time data

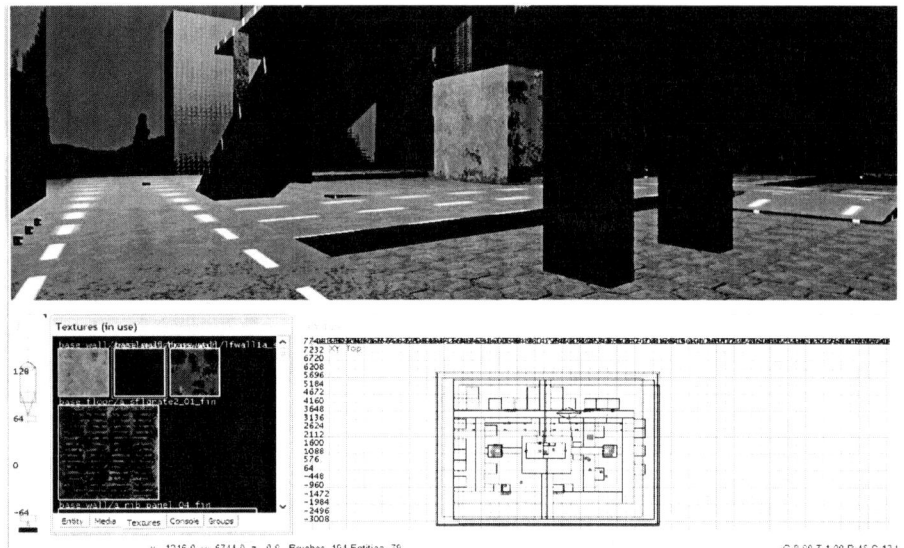

hurricanes, tornados, etc.—and the associated actions that people initiate: for example, how they act when they become part of the extreme event. The game as a story line in the tradition of action and adventure games can be applied here. It has a learning monitoring function because the player is supposed to learn features pertinent to the environment (street configurations, access to help facilities, and difficult to handle natural features, such as brooks that might suddenly swell with water, etc.). Instead of the traditional Japanese approach of training for earthquakes, the game makes players feel familiar with their environment, but also aware of possible dangers (for example: What happens if the levees break? What if a tree falls? What if the roof blows off?). The game is connected to the Internet and receives real-time meteorological data, as well as input from the players themselves (this is a distributed, multiplayer game). As conditions require, the game becomes an alert system: "Go immediately to the tornado shelter." For the aging, the game provides connections to community support groups that will assist the individual or deploy help, as the situation requires.

Play, in such cases, is preparation for real situations, i.e., life under extreme event conditions. This particular development aims to integrate real-time data of an extreme event in a game environment. "Playing" an ongoing event tests and improves a person's anticipation capabilities before they become a matter of life and death. Performance that relies on reaction only was shown to fail under extreme conditions. Therefore, in maintaining anticipatory characteristics, the game increases survival chances, especially for those senior adults who otherwise end up confused, or even ignored in the context of drastic events. The passing of time can put pressure on senior adults. Their game proficiency can be severely tested. But if in games of racing cars or shooting everything that moves only the score is affected, in our game survival is at stake. Therefore we had to develop a metaphor that connects the time in the game to the time in the world outside. Richard Gray, aka Levelord, the well-known level design artist, helped the students involved in the project in addressing this concrete objective.

Related to this development is the use of games for unobtrusive monitoring of the health condition

Figure 16. From real life to a game of emulating traumatic circumstances

Figure 17. How a dance is taught: A proposed framework for digi-dance

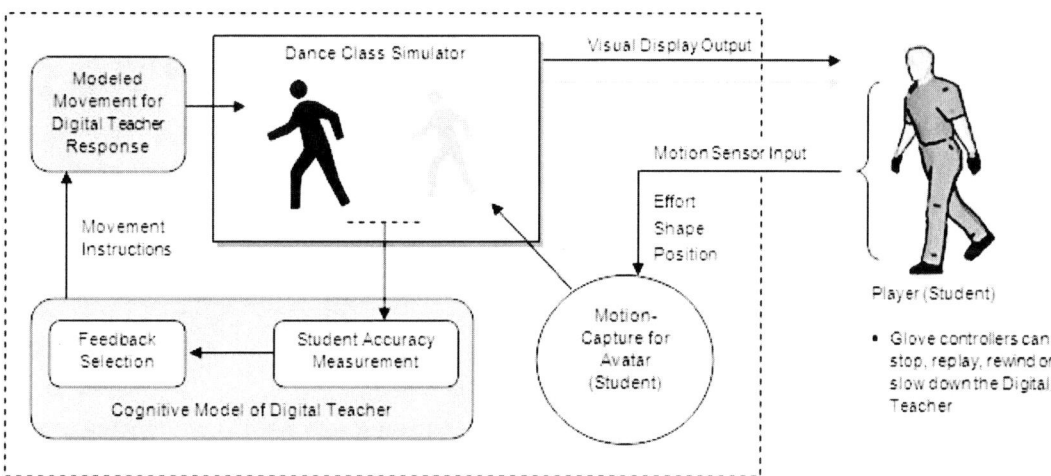

of senior adults (living independently or even in retirement communities). Patterns of interaction between a player and a networked game are very telling in respect to cognitive performance and motoric skills. When the "game" detects deterioration in performance, it can alert medical centers for further evaluation and eventual remedy.

Finally, from the games for health focused on aging, a new approach developed in respect to post-traumatic stress disorder (PTSD). PTSD has been fully documented in young and old, in male and female, in all races. The major observation guiding the research is that PTSD entails diminished anticipation. Anatomical changes (Brenner et al, 1995) point to reduced hippocampal volume, and thus explain the decline in anticipation. In addition, processes pertinent to preparation of movement or anticipation of upcoming stimuli are severely inhibited. Consequently, performance is either reduced or rendered impossible. We build on expertise acquired in defining the *Anticipatory Profile*, and from this we derived an integrated PTSD profile. It represents the multiple correlations among sensory data and motion descriptions focused on the specific symptoms of post-traumatic experiences. But what is relevant to game development is the ability to map from the PTSD profile to games that could assist individuals afflicted by traumatic experiences in recovering some of their adaptive capabilities. One particular project focused on aspects of PTSD associated with the events of 9/11. In short, Carolina Dabbah pursued the path of a virtual reconstruction of the traumatic event. Exposing subjects affected by PTSD to a virtual reconstruction—the game of "What if?"—(obviously under supervision, so as to avoid further trauma) resulted in progressive increase in movement preparation in anticipation of stimuli associated with explosions, fire, building collapse, etc.

The findings are still in a preliminary phase. Such games, like all others conceived in the framework of project *Seneludens*, require a high degree of individualization. The traumatic event needs to be reconstructed for each subject.

As a final note: Currently, Sheri Segovia, a Ph.D. candidate in the Humanities, and Catherine Turocy are in the process of generalizing from the experience spelled out in this study to a new category of games in which learning to dance is associated with the benefits of dancing. The two authors define *Digi-Dance* as an interactive video game involving full-body, real-time motion capture and the creation of digital agents as dance teachers, and even as dance partners. It reinvents the role of social dance in everyday life by joining player-avatars in a Wi-Fi-enabled environment sharing a library of selected dances (from the 16th century to the present).

Through the ages, dance can be viewed as the original social network. The aging (but not only) face the progressive loss of such a network. *Digi-Dance* offers the possibility to build new social networks. Therefore, it can contribute as well to the maintenance of adaptive performance of senior adults. With light-weight, attachable motion sensing controls, players learn dance steps and patterns demonstrated by a digital teacher, while seeing their own movement on screen in avatar form. A dance can be enjoyed solo (as in the Shim Sham or a Clog), with a digital partner (as in a minuet or a tango), or with a Wi-Fi-enabled group (as in a Texas line dance or a Cotillion). Besides simply spreading the joy of dancing, *Digi-Dance* inspires creativity and fitness by providing physical and mental challenges, and a somatic experience that can address the physiological effects of depression. It also prolongs mental sharpness. Casual dance involves brain activity in rapid decision-making, as well as memorization of sequence.

Digi-Dance is accessible to all ages, independent of language, and is culturally inclusive. With its breadth of dance genres, it will likely encourage cultural and social interaction. Adults struggling with depression and/or the physiological effects of aging would have a self-directed activity in their own home that is physically and mentally beneficial. In addition, interaction with children and grandchildren is facilitated through a gaming experience that can be rewarding to all.

FORESEEABLE BENEFITS

Last but not least, games that "turn back the clock" of aging through maintenance and possible enhancement of anticipatory characteristics will add a new field of broad social and commercial significance to the economy. Based on market estimates (cf. Kaiser Permanente Foundation), by 2025 the share of the population age 65 and above will reach 22%, up from 12% in 2000. This translates into a dollar amount requested for addressing the needs of the aging population in the range of 8-10 trillion dollars (included here are Social Security, Medicare, additional medical insurance costs, drugs, various forms of therapy, medical care, homes for the aged, etc.).

Individuals—old and young—and society at large should benefit from the research presented herein. For each lifespan that an individual gains, his/her younger family members will be able to provide affectionate support, unburdened by expensive and time-consuming care and physical assistance. Society would save the money currently spent on addressing the needs of the aging (drugs, therapy, assisted living). As we know too well, some of the aged end up abandoned by family members. Others, for reasons beyond their control, have no recourse to family support. Society would also benefit from the behavioral therapy implicit in game-supported maintenance of skills and the learning it entails. The knowledge accumulated through games can only add to the lives of individuals who are rich in experience, but getting poorer, as they get older, in respect to their physical and cognitive faculties. This late-in-life knowledge will in many ways reward a society that makes life-long learning a priority across generations.

REFERENCES

Ahl, D. H. (1983). Editorial. *Video & Arcade Games, 1*(1), 4. (See also http://www.atarimagazines.com/cva/v1n1/editorial.php) Barker, T., & Bernhard, B. (n.d.). Wii leagues mix virtual exercise with real results. *Examiner*. Retrieved on November 3, 2008 from http://www.examiner.com/a-1669072~Wii_leagues_mix_virtual_exercise_with_real_results.html

An integrated mobile wireless system for capturing physiological data streams during a cognitive-motor task: Applications for aging. (2007, November 11-12). In *IEEE Engineering in Medicine and Biology Workshop, IEEE Dallas University of Texas at Dallas*, (pp. 67-70).

Analyzing Motoric and Physiological Data in Describing Upper Extremity Movement in the Aged (with G.N. Pradhan, N. Engineer, B. Prabhakaran). (2008, July 16-19). In *International Conference on Pervasive Technologies Related to Assistive Environments (PETRA 2008), ACM International Conference Proceeding Series,* Athens, Greece, (Vol. 282, pp.8).

Bremner, J., Randall, R., Scott, T., Bronen, R., Seibyl, J., & Southwick, S. (1995). MRI-based measurement of hippocampal volume in patients with combat-related posttraumatic stress disorder. *The American Journal of Psychiatry, 152,* 973–981.

Brookhaven National Laboratory. (1981). *Video Games – Did They Begin at Brookhaven?* Retrieved November 3, 2008 from http://osti.gov/accomplishments/videogame.html

Carson, L. (2006). *WV Games for Health*. Retrieved on November 17, 2008 from statewide-ddr-wv.ppt

Carson, L. (2008, November 3). Video games promoted in fight against childhood obesity. *Canada.com*. Retrieved on November 17, 2008 from http://www.canada.com/topics/news/story.html?id=3dfaf40b-0c53-4a89-9b31-306deecbcf5e

Clark, J. E., Lanphear, A. K., & Riddick, C. C. (1987). The effects of video game playing on the response selection of elderly adults. *Journal of Gerontology, 42*(1), 82–85.

Cohen, A. (2006). Video-Game Makers Discover New, Older, Market. *NPR*. Retrieved on November 17, 2008 from http://www.npr.org/templates/story/story.php?storyId=6589941.

Da Vinci, L. (1498). *Trattato della Pittura*. Milan, Italy.

Dinse, H. R. (2001). The Aging Brain. In N. Elsner & G.W. Kreutzberg (Eds.), *The Neurosciences at the turn of the century: Vol. 1. Proceedings of the 4th Meeting of the German Neuroscience Society,* (pp. 356-363). Thieme: 28th Göttingen Neurobiology Conference.

Dinse, H. R. (2005). Treating the aging brain: cortical reorganization and behavior. *Acta Neurochirurgica, 93*(Suppl.), 79–84. doi:10.1007/3-211-27577-0_12

Ellison, K. (2007, May 21). Video Games vs. the Aging Brain. *Discover Magazine*. Retrieved on November 5, 2008 from http://discovermagazine.com/2007/may/the-elastic-brain/article_print

Elton, M. When will the information explosion reach older Americans? *The American Behavioral Scientist, 31*(5), 564–575. doi:10.1177/000276488031005005

Entertainment Software Association. (2008 August). *Statistic of the Month: Essential Facts—Sales, Demographic, and Usage Data*. Retrieved November 4, 2008, from http://www.theesa.com/newsroom/esa_newsletter/august2008/index.html

Gallese, V. (2001). The "Shared Manifold" Hypothesis. From Mirror Neurons to Empathy. *Journal of Consciousness Studies*, 5(7), 33–50.

Goldstein, J., Cajko, L., Oosterbroek, M., Michielsen, M., van Houten, O., & Salverda, F. (1997). Video games and the elderly. *Social Behavior and Personality*, 25(4), 345–352. doi:10.2224/sbp.1997.25.4.345

Hanna, J. L. (1978). African dance: Some implications for dance therapy. *American Journal of Dance Therapy*, 2(1), 3–15. doi:10.1007/BF02579589

Hanna, J. L. (1995). African dance: some implications for dance therapy. *Journal of Alternative and Complementary Medicine (New York, N.Y.)*, 1(4), 323–331. doi:10.1089/acm.1995.1.323

Hoot, J. L., & Hayship, B. (1983). Microcomputers and the Elderly: New directions for self-sufficiency and life-long learning. *Educational Gerontology*, 9, 493–499. doi:10.1080/0380127830090513

Integration of Motion Capture and EMG data for Classifying the Human Motions (with Balakrishnan Prabhakaran, Gaurav Pradhan, Navzer Engineer). (2007, April 17-20). In *Proceedings of the IEEE 23rd International Conference on Data Engineering Workshops* (pp. 56-63).

Isselsteijn, W., Nap, H. H., deKort, Y., & Poels, K. (2007). Digital game design for elderly users. In *Proceedings of the 2007 Conference on Future Play*, Toronto, Canada (pp. 17-22). New York: ACM.

Jaycox, K., & Hicks, K. (1976). *Elders, Students, and Computer—A New Team* (ISEAC No. 7). Urbana, IL: University of Illinois, Illinois Series on Educational Applications of Computers.

Kawashima, R. (2003). *Train Your Brain*. Teaneck, NJ: Kumon Publishing Co.

Lundy, H. (2002). *Using dance/movement therapy techniques to augment the effectiveness of therapeutic holding with children*. Master of Arts thesis, College of Nursing and Health Professions, MCP Hahnemann University. Archived at http://hdl.handle.net/1860/1100

McCurry, J. (2006). Video games for the elderly: an answer to dementia or a marketing tool? *The Guardian*, March 7. Retrieved on November 3, 2008 from http://www.guardian.co.uk/technology/2006/mar/07/nintendods.games

Meyer, C., Ganascia, J.-G., & Zucker, J.-D. (1997). Learning Strategies in Games by Anticipation. *International Conference on Artificial Intelligence*, Nagoya, Japan. Merzenich, M. (n.d.). *On the Brain*. Retrieved on November 3, 2008 from http://merzenich.positscience.com/. See also Posit Science: Brain Fitness Program http://www.positscience.com/products/brain_fitness_program/

Montet, V. (2006). Video games aim to spice up old people's lives *Yahoo News*, October 1. Originally appearing at http://fullcoverage.yahoo.com/s/afp/afplifestyleusitgameshealthelderly. Retrieved on November 3, 2008 from http://www.nadin.ws/archives/477

Nadin, M. (1991). *Mind—Anticipation and Chaos*. Stuttgart, Germany: Belser Presse.

Nadin, M. (2003). *Anticipation—The end is where we start from*. Basel, Switzerland: Lars Mueller Verlag.

Nadin, M. (2004). *Project Seneludens*. Richardson, TX: University of Texas at Dallas, antÉ – Institute for Research in Anticipatory Systems.

Nadin, M. (2005). Anticipating Extreme Events – the need for faster-than-real-time models. In *Extreme Events in Nature and Society*, (Frontiers Collection). New York: Springer Verlag.

Nadin, M. (2006). *Anticipation, Games, and Brain Plasticity*. Syllabus for the graduate course, Special Topics in Art & Technology. Retrieved on November 19, 2008 from http://www.utdallas.edu/bbs/news_events/2005/ATEC7390.001.html and http://ante.utdallas.edu/agbp/

Nicolelis, M. (2005). *Neural Control of Artificial Limb Movement*. Paper presented at the symposium: Reprogramming the Human Brain, April 2005, Dallas, TX.

Rizzolati, G., Fadiga, L., Fogassi, L., & Gallese, V. (1999). Resonance behaviors and mirror neurons. *Archives Italiennes de Biologie, 137*, 83–99.

Rosen, R. (1985). *Anticipatory Systems. Philosophical, Mathematical and Methodological Foundations*. Oxford, UK: Pergamon Press.

Shaw, G. B. (n.d.). Retrieved on November 3, 2008 from http://thinkexist.com/quotes/george_bernard_shaw/

Tully, T. (2005). *From Genes to Drugs for Cognitive Dysfunction*. Paper presented at the symposium: Reprogramming the Human Brain, April 2005, Dallas, TX.

United Nations Programme on Ageing. (2002). Retrieved on November 5, 2008, from http://www.un.org/ageing/.

Weisman, S. (1983). Computer Games for the Frail Elderly. *The Gerontologist, 23*(4), 361–363.

Whitcomb, G. R. (1990). Computer Games for the Elderly. Symposium on Computers and the Quality of Life. In *Proceedings of the Conference on Computers and the Quality of Life,* George Washington University, Washington, DC, (pp. 112-115). New York: ACM SIGCAS.

Chapter 10
Re-Purposing a Recreational Video Game as a Serious Game for Second Language Acquisition

Yolanda A. Rankin
IBM Almaden Research Center, USA

Marcus W. Shute
Clark Atlanta University, USA

ABSTRACT

Serious games designed for educational purposes promote acquisition of knowledge and skills that are valued in the both the virtual realm and the real world. However, the million dollar question is how do we design serious games that produce positive learning outcomes without sacrificing the element of fun? The authors' answer is simple but no less profound. Don't recreate the wheel; instead use it to create new technology! Using this premise, they re-purpose the recreational Massively Multiplayer Online Role Playing Game (MMORPG) EverQuest® II as a serious game, leveraging the entertainment value and readily accessible developer tools to promote learning in the context of Second Language Acquisition (SLA). They outline the process of transformation, first identifying the affordances attributed to MMORPGs and then evaluating the impact of gameplay experiences on SLA. Promising results from experimental studies reveal that in-game social interactions in the target language between native speakers and non native speakers provide a higher degree of engagement and significantly increase second language vocabulary acquisition and reading comprehension skills compared to traditional classroom instruction. They conclude with the design of two game modules that promote vocabulary acquisition, reading comprehension and conversational fluency.

DOI: 10.4018/978-1-61520-739-8.ch010

INTRODUCTION

Interactive digital media such as video games serve primarily as a source of entertainment, surpassing both the movie and music industries as the number one form of entertainment in America (Electronic Arts, 2007; Mainelli, 2001). Additionally, video games attract thousands of players who spend numerous hours mastering game objectives (Roberts et al., 2005). However, video games are often criticized for being mindless entertainment, lacking in educational value. Researchers argue that video games embody an underutilized, ideal learning environment with clearly defined goals, resources for completion of game tasks, adaptability to players' skills, immediate feedback and rewards, and natural progression of increased difficulty that contributes to increased engagement (Aldrich, 2005; Begg et al., 2005; Gee, 2003; Koster, 2005; Prensky, 2001). Video games provide an active, personal experience that is difficult to duplicate for each student in the traditional classroom environment. We perceive video games to be models of virtual learning environments where the player represents a user model of the active learner. Recreational video games create a "flow" or high level of enjoyment that correlates learning with the obstacles encountered during gameplay (Csikszentmihalyi, 1990; DeBold, 2002). Therefore, gameplay activities which support learning in the virtual realm can transfer to positive learning outcomes recognized in the real world.

Researchers have investigated the plausibility of video games as effective pedagogical tools in both traditional and informal learning environments. Pillay et al. (1999) posit that video games engage players in complex cognitive processes such as problem-solving tasks associated with scientific reasoning. Because video games create gameplay experiences that closely emulate the learning process, game designers have created a new genre of video games known as *serious games,* games that do more than just entertain (Aldrich, 2005; Chatham, 2007). Unlike recreational video games, serious games create opportunities for players to acquire and develop knowledge or life skills that are valued in both the virtual and real worlds (Aldrich, 2005; Chatham, 2007; Mayo, 2007). Successful integration of traditional learning objectives with the elements of entertainment, play, and fun becomes the goal for developing serious games. As the serious games movement gathers momentum, the game industry and researchers are asking challenging questions: What is the educational value of a particular video game as it relates to a specific domain? What do players learn as a result of gameplay? How can we realize the benefits of video games in the context of educational assessment? How do game designers merge the positive aspects (e.g. flow, high level of engagement, etc.) of gameplay with domain-specific objectives? More importantly, how can game designers create serious games that produce positive learning outcomes without sacrificing entertainment value or accumulating expensive production costs? One solution lies in developing game modules (game modding) based upon experimental studies of players' interactions during gameplay that transform existing commercial video games into serious games with positive outcomes (Rankin et al., 2008).

This chapter outlines the process for re-purposing recreational video games as serious games. First, we examine the potential of one particular genre of video games, Massive Multiplayer Online Role Playing Games (MMORPGs), as unorthodox language learning tools, correlating gameplay experiences to second language pedagogy. Secondly, we discuss the results of experimental studies of English as Second Language (ESL) students playing a commercial video game to determine the feasibility of MMORPGs as Second Language Acquisition (SLA) tools, specifically English proficiency. Finally, results from experimental studies inform the design of two game modules, an in-game dictionary that facilitates vocabulary acquisition and reading comprehension, and a chat prompt module that promotes conversational fluency.

COMPUTER ASSISTED LANGUAGE LEARNING

Standard American English remains the dominant language in the United States, and yet 14% of American students live in homes where English is not the primary language (Watts-Taffe & Truscott, 2000). As English proficiency becomes an increasing important issue, educators face the challenge of creating inclusive learning environments for students who possess different linguistic capabilities, emphasizing the need for second language pedagogical supports in both mainstream classrooms and informal learning environments. Foreign language instructors utilize Computer Assisted Language Learning (CALL) tools to mediate communication between native and non native speakers in the target language (Chapelle, 2001; Warschauer, 1996). CALL tools (e.g. emails, self-paced tutorials, chat rooms, etc.) reinforce development of reading and writing skills in the target language, creating opportunities for native and non native speakers to interact with one another as non native speakers develop proficiency in the target language (Backer, 1999; Beauvois, 1992; Beauvois & Eledge, 1996; Chapelle, 2001; Hudson & Bruckman, 2002).

VIDEO GAMES AS LANGUAGE LEARNING TOOLS

If we examine how young people spend their time at home, we find that more than 84% of youth between the ages of 8 and 17 years old play video games for recreational purposes (Roberts et al., 2005). A tremendous number of young people hailing from all walks of life and with variable linguistic capabilities currently spend hours each day playing video games (Eck, 2006). Hubbard (1991) understands that video games offer an appealing alternative to traditional CALL tools, primarily because video games possess an element of fun that causes players to repeatedly participate in game tasks. Video games include social interactions as part of the gameplay experience and become public social places for people of different ethnicities, culture and languages to meet and communicate with one another (Hudson & Bruckman, 2002; Steinkuehler, 2005). In contrast to self-pace tutorials that focus on reading, writing, listening and speaking skills, video games require players to follow rules and process information to accomplish game tasks where language is just a means to an end and not the end itself. Rather than relying solely on text as the means for providing critical information about the virtual environment, computer games leverage sophisticated graphics, sounds, gestures, and intelligent Non Player Characters (NPCs) to afford opportunities for foreign language students to develop proficiency in the target language. Interpretation of visual information becomes a critical part of developing communicative competence in the target language as players understand information and respond accordingly. Hubbard (1991) understands that video games offer an appealing alternative to traditional CALL tools, primarily because video games possess an element of fun that causes players to repeatedly participate in game tasks. Such activities overcome the lack of motivation attributed to text-based Multi-Object Oriented virtual environments (Backer 1999; Falsetti & Schweitzer 1995).

Recently, video games have begun to break new ground as unorthodox language learning tools, representing an alternative CALL tool. Johnson et al. (2004) developed the Tactical Language Training System (TLTS) for the purpose of training military soldiers to practice their Arabic listening and speaking skills as they interact with Non-Player Characters in culturally relevant scenarios (Johnson et al., 2003). However, TLTS was purposely designed to assist with SLA. In comparison, deHaan and Diamond (2007) developed a video game that required players to match rhythmic foot patterns as they were exposed to Japanese vocabulary words during gameplay. While this approach investigated the impact of physical interaction during gameplay on second language acquisition, results indicated that the rhythmic foot patterns interfered with players' ability to recall Japanese vocabulary words (deHaan & Diamond, 2007). Thus, this study produced negative learning outcomes. In another study, deHaan (2005) evaluated an intermediate Japanese foreign student's ability to increase his Japanese vocabulary as a result of playing a baseball game in Japanese. Though the student demonstrated positive learning gains due to the constrained context of language use and repetitive game tasks, the genre of sports video games may alienate a larger and diverse group of second language students, possibly producing lower levels of engagement and negative learning outcomes. Furthermore, the small sample size (N = 1) does not maximize social interactions as a pedagogical tool that allows non native speakers to practice their communication skills in the target language.

We can maximize the appeal of video games to provide motivation for development of SLA. Furthermore, we can design interactive interfaces that utilize audio, graphical images, displayed text and social interactions with both Player Characters (PCs) and Non Player Characters (NPCs) characteristic of video games and extend these characteristics to language learning.

AFFORDANCES OF GAMEPLAY: MMORPGS FOR LANGUAGE LEARNING

By their very nature, MMORPGs represent a genre of computer-based games that allow players to portray a fictional character in a fantasy world, requiring players' to invest in virtual character development. This creates an ideal language learning environment that encompasses investment and involvement while closing the distance between foreign language students and interaction with native speakers. For these reasons, we posit that MMORPGs create the ideal learning environment for second language acquisition in addition to the social interaction among players that supports meta-thinking abilities for semantic, syntactic, and contextual knowledge of a foreign language.

Because MMORPGs attract players of various backgrounds, video game interfaces need to support both native and non native speakers of the target language. Game developers of *serious games* that support language learning face the challenge of designing video games that integrate learning objectives into gameplay without sacrificing the entertainment value. We propose that playing MMORPGs affords the following benefits for SLA:

- Active learners
- Immersive learning environment as the context for Second Language Acquisition
- Online social interactions during gameplay

Active Learners

Second language students become active learners who continually make decisions about various game tasks during gameplay. The structure of MMORPGs supports character evolution as a means of game progression and so the first task of gameplay is the selection of a virtual character. See Figure 1. Active language learners assume the role of the virtual characters they have created and consciously commit to the advancement of these characters in the virtual world (Gee, 2003). The avatar masks the true identity of foreign language students and creates less threatening social interactions with native speakers in the virtual world (Steinkuehler & Williams, 2006).

Additionally, video games supply authentic environments for learning, complete with suffi-

Figure 1. Overview of the froglok character in the game of EverQuest® II

cient opportunities for foreign language students to practice, develop, and test their emergent communicative abilities. Foreign language students develop reading comprehension skills in the target language as they read Non Player Characters' (NPCs) dialogue displayed on the screen. Though NPCs share pertinent information regarding game activities, MMORPGs are designed to foster partnerships with other Player Characters (PCs). Players establish relationships via chat messages displayed on the screen. Text-based chat windows support reflective thinking as second language students compose messages read by PCs. This requires second language students to practice their conversational skills and understand cultural nuances (e.g., inventory of items) attributed to the game world, increasing communicative performance skills.

Immersive Environments as the Context for Second Language Acquisition

Acknowledging that learning does not occur in a vacuum, context plays a crucial role in the learning process. Role-playing fantasy games motivate players, creating a virtual world as the context for foreign language students to concentrate on accurate and coherent use of the target language to communicate intent and to assist with completing game tasks. Video games emulate the experiential approach of SLA by providing an immersive learning experience. Game-informed practices influence the design of interactive digital interfaces embedded in video games, leveraging context in the virtual world via animations, textual information, and sound to assist students with proficiency in the target language. Text is displayed on the screen, giving visual cues to determine context of

Figure 2. Non player character's dialogue in French displayed on the screen

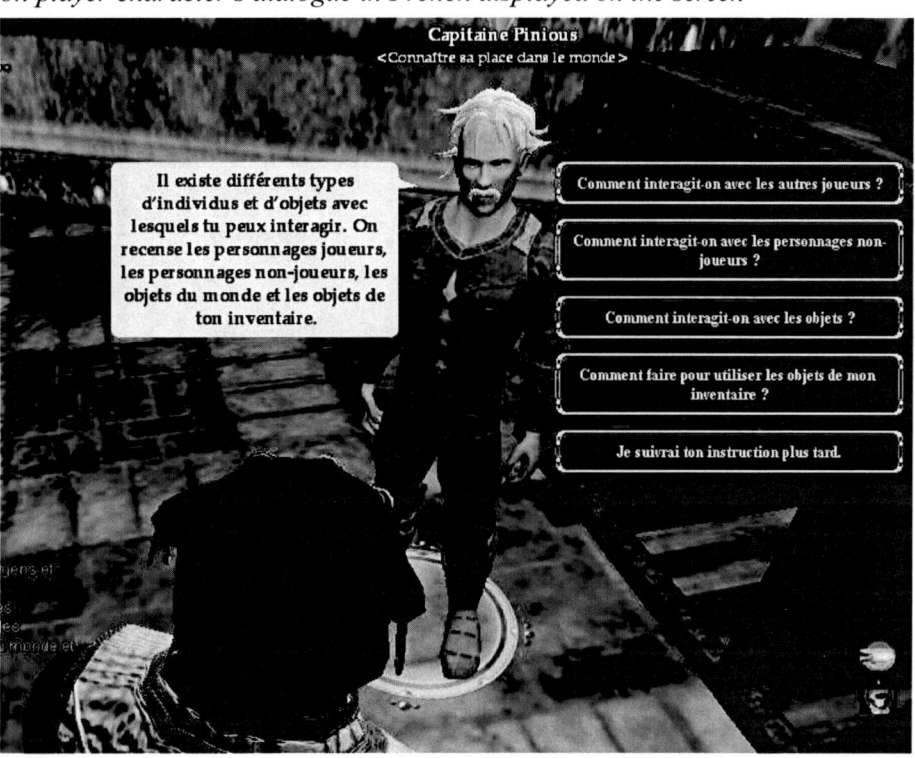

meaning of second language vocabulary. Three dimensional animations provide additional context for vocabulary words and cultural nuances present in the virtual world. MMOPRGs display dialogue between NPCs and PCs to assist foreign language students with syntactical structure of the target language. The combination of text and graphical images in the virtual world provides shared context of meaning, enabling players of various linguistic capability to derive the appropriate meaning of potential vocabulary words. The target language is spoken by virtual characters, providing foreign language students the opportunity to hear the accents and intonation specific to that language. Players select a response to NPC questions which feature potential second language vocabulary and information relevant to quests in the game (Rankin et al., 2006a; Rankin et al., 2006b). See Figure 2. Interactions between PCs facilitate communicative performance as foreign language students develop an understanding of what constitutes an appropriate response while engaging in conversations with native speakers via the chat window. The practice of producing language that is evaluated for meaning by other role-playing characters constitutes authentic dialogue between native and non-native language speakers. Video games create immersive virtual worlds for developing proficiency in a foreign language. Thus, MMORPGs supply an immersive environment and opportunities to participate in social interactions with native speakers in the target language, creating an effective digital learning environment for SLA. As a result, language becomes a necessary component of successful gameplay.

Online Social Interactions

MMORPGs supply a social infrastructure that permits like-minded players to form groups as

Figure 3. Two player characters work together to complete a quest

evident in multiple user dungeons (MUDs) and other role-playing games (Brown & Thomas, 2006; Nardi & Harris, 2006; Steinkuehler & Williams, 2006). Consequently, MMORPGs are designed to create and support social networks of gamers. Experienced gamers realize successful progression through the virtual world that depends on well-formed affiliations that lead to avatars that possess advanced skills needed to complete increasingly difficult game tasks. Powerful alliances play a key factor in gamers' abilities to defeat enemies and accomplish tasks that are virtually impossible to perform alone. MMORPGs sustain social interaction between players and serve as the catalyst for fostering students' communicative performance (e.g., conversation skills) as students chat in a foreign language while playing the game. See Figure 3. Quests provide common ground that fosters social interactions between players of diverse background (Girgensohn & Lee, 2002; Lee et al., 2001). Social interaction serves as a prerequisite to students' language proficiency (Hadley, 2001). Without social interaction, students lack motivation, opportunities for practicing target language skills, and immediate feedback; all three components are crucial if students desire to increase their communicative abilities in the target language (Hadley, 2001; Krashen, 1991). The social practices that exist in MMORPGs model cultural norms that are emphasized in gameplaying activities and define the community of players. In a similar manner, second language teaching methodology encourages foreign language students to participate in cultural practices associated with the target language (Hadley, 2001; Krashen, 1991). As a result, students develop proficiency and communicative performance in the target language as they communicate with native speakers. Hence, MMORPGs are transformed into CALL tools and serious games for successful SLA for novice, intermediate and advance language students (Rankin et al., 2006a; Rankin et al., 2006b; Rankin et al., 2008).

TRANSFORMATION FROM RECREATIONAL TO SERIOUS GAME

Initially, we considered designing an original, serious game to facilitate SLA but realized the disadvantages of this approach. First, the challenge of designing an educational game containing enticing storylines or plots and the sophisticated graphics typically associated with commercial games is not easily accomplished. While we acknowledge that there exist several commercial educational games on the market, how many times have we heard the complaint "that game was no fun?" Effectively embedding learning tasks in meaningful gameplay experiences is quite a challenge. Educational games have been critiqued for 'dressing up' learning objectives in fancy graphics and offering boring gameplay experiences that students prefer not to repeat (Elliott & Bruckman, 2002). Pouring chocolate over green peas does not necessarily cause you to enjoy eating them. Similarly, placing educational content in the form of a video game does not guarantee a fun learning experience (Elliott & Bruckman, 2002). If the educational content supersedes creation of an enjoyable gameplay experience, then the students lose the benefits of gaming (e.g., immediate feedback and awards, virtual collaboration, active learners, and exploratory virtual worlds). In contrast, we believe that both the learning objectives and an enjoyable game experience should receive equal attention throughout the game design process which is not trivial or easily achieved. Unlike the negative connotations of learning associated with traditional classroom settings, Barab et al. (2005) argue that video games provide a meaningful context for learning where players have fun as they learn concepts. This creates a higher level of engagement as players help one another complete with game quests. It is this higher level of engagement that becomes the goal for designing game modules that offer communication tools for second language students as they develop proficiency in the target language.

Secondly, commercial game development teams consists of hundreds of people, including user interface developers, game engine programmers, modelers and animators of avatars and in-game creatures (Fullerton et al., 2004). The product development process for a "big budget" commercial game takes, on average, 2 to 3 years to complete (Fullerton et al., 2004). Independent game developers have smaller budgets and fewer resources available which extends the game development cycle and impacts the quality of the game. While we applaud the efforts of small game development teams who get the job done, it is an unrealistic expectation and daunting task at best to expect an individual to spend 2 to 3 years designing a quality serious game from scratch.

EVALUATION OF MMORPG FOR SECOND LANGUAGE ACQUISITION

To determine feasibility of MMORPGs as language learning tools, we formulate the hypothesis that MMORPGs can facilitate SLA and evaluate Sony Online Entertainment's EverQuest® II (EQ2), a Massive Multiplayer Online Role Playing Game (MMORPG) not intended for SLA. To test our hypothesis, we conduct a between-subjects experimental design to answer the following questions: How do MMORPGs increase second language (L2) vocabulary acquisition? How does EQ2 support other SLA skills? The assumption is that EQ2 compared to traditional classroom instruction will provide adequate support for L2 vocabulary acquisition.

Methods

Eighteen Advanced English as Second Language (ESL) Chinese students enrolled at a southern liberal arts college were randomly assigned to three conditions: 1. Six ESL students who attended three hours of class instruction; 2. Six ESL students who played EQ2 for four hours; 3. Six ESL students were grouped with Native English Speakers (NES) in a different physical locale played EQ2.

To accommodate the learning curve associated with understanding the game objective and maneuvering the game controls, ESL students spent an extra hour becoming familiar with EQ2. All eighteen Advanced ESL students were enrolled in an Intensive English Program at a southern liberal arts college. Prior to participation in each condition, participants took an assessment that measured their prior knowledge of L2 vocabulary words that were modeled in NPC speech during gameplay. The assessment required ESL participants to use the potential L2 vocabulary in a sentence demonstrating prior knowledge. We purposefully selected L2 vocabulary that was not specific to the game and represented college level academic words (e.g., coagulated, coalesce, fervent, revive). ESL students who attended class participated in drill and rote exercises (e.g., define the L2 vocabulary word and use it in a sentence) while the ESL students who played EQ2 were given the tasks of completing quests 1–8. Once both groups had completed the designated hours for the study, each participants took three assessments. The first post-test assessment asked students to use L2 vocabulary in sentences demonstrating understanding of the word. Sentences were evaluated for appropriate use of L2 vocabulary and not grammatical correctness. The second post-test assessment used a recognition task based on gameplay scenarios where ESL participants selected the correct meaning from multiple choice options of L2 vocabulary words. The third assessment was a rational cloze assessment which measured ESL participants' semantic knowledge of L2 vocabulary words outside the context of gameplay in addition to their ability to select L2 vocabulary words based on contextual clues located in the clause, the sentence, and in the text. Finally, ESL students completed a post-game questionnaire which reported their perceptions of the gameplay experience.

Data Analysis

A one-way Analysis of Variance (ANOVA) of post-test scores for sentence usage revealed a significant difference in post-test scores for ($F[2, 15] = 9.65$ for $p < 0.05$). See Figure 4. ESL students who participated in traditional classroom instruction had an average post-test score of 54.78 out of 100 compared to the average score 16.16 for ESL students who played EQ2 and an average score of 13.10 for the ESL students grouped with NES. Refer to Figure 4. Significantly higher post-test scores on the second language (L2) sentence usage assessment of ESL students who participated in traditional classroom instruction revealed the near transfer task of drill and rote practices to sentence usage of vocabulary words. However, a one-way Analysis of Variance of post-test scores for L2 vocabulary in the context of game tasks revealed a significant difference in post-test scores ($F[2, 14] = 4.162$ for $p < 0.05$) of ESL students grouped with NES players. See Figure 5. This suggests that the social interactions between native and non-native speakers played a pivotal role in L2 vocabulary acquisition. It is important to note that the performance of those ESL students who played EQ2 alone was similar to those ESL students who participated in traditional classroom instruction. Therefore, these results suggest that MMORPGs show great promise as a second language pedagogical tool, provided game designers leverage the benefits of MMORPGs for SLA.

DISCUSSION OF RESULTS

The drill and rote practices used for L2 pedagogy in the classroom instruction condition proved beneficial for post-test assessment, especially since it required ESL students to define and use the vocabulary words in sentences. We can conclude that the sentence usage post-test resembled a typical classroom learning exercise, a near task of transfer for this group of participants. Though both groups of ESL students who played EQ2 did not improve significantly, the majority of participants who played EQ2 did increase their understanding of vocabulary words, acquiring ap-

Figure 4. Average post-test score for second language vocabulary sentence usage

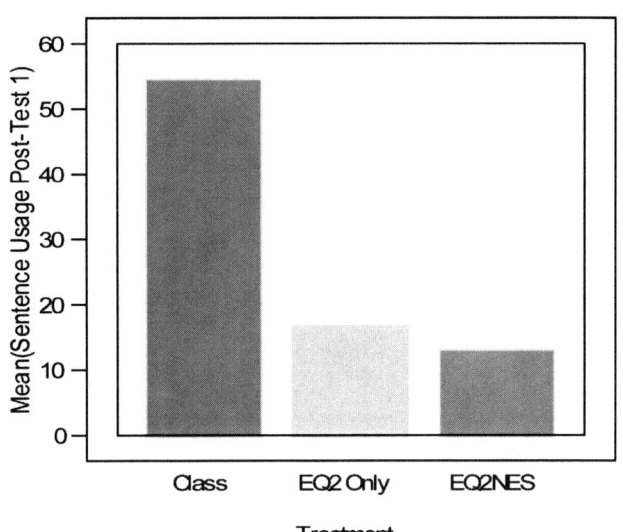

proximately 2 to 3.5 words. See Figures 4 and 5. Results suggest that gameplay experiences did not provide the same kind of structured pedagogical support for L2 vocabulary acquisition as traditional classroom instruction. Unlike the drill and rote practices associated with classroom instruction, the ESL students struggled with learning more than just the twelve L2 vocabulary words on the post-test. L2 vocabulary words were present in NPC speech, displayed on the screen as feedback (e.g., "Parry!"), and used as labels for fictional beings (e.g. goblins). Of course, this implies that the ESL students who played EQ2 did not give equal attention to all the potential L2 vocabulary words, which is to be expected due to exposure of a substantial number of L2 vocabulary words. Despite the overflow of information, 82% of the ESL students who played EQ2 did show minimal increases (i.e. acquisition of two additional vocabulary words) in English proficiency. Thus, the expectation is not that MMORPGs would outperform traditional classroom instructions but that MMORPGs accommodate SLA, supplementing traditional classroom instruction.

Results for the One-way ANOVA of post-test scores for L2 vocabulary in the context of gameplay indicate that the ESL students who were grouped with NES performed significantly better than both the traditional classroom instruction condition and the ESL students who played EQ2 alone condition. This suggests that the social interactions between native and non-native speakers played a pivotal role in L2 vocabulary acquisition. We attribute the positive learning outcomes to the conversations centered on completion of the eight game quests. Because the majority of the L2 vocabulary words were used to provide details about the quests, ESL students often asked their NES teammates for assistance in completing the quests, providing a context for understanding the meaning of L2 vocabulary words. The ESL students who participated in traditional classroom instruction were unable to select the appropriate meaning of L2 vocabulary words in the context of gameplay scenarios, despite the surrounding words used in the sentence. Therefore, the conversations concerning game strategy between native and non-native speakers represented a form of

Figure 5. One-way analysis of variance of ESL students' post-test scores in the context of gameplay activities

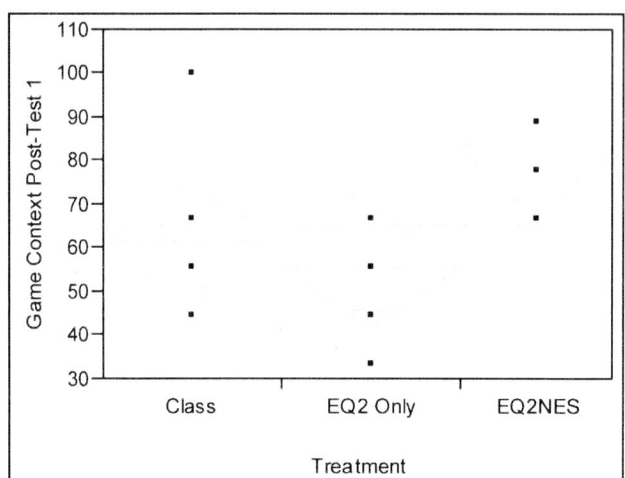

language socialization, a crucial component of SLA. Interestingly enough, the performance of those ESL students who played EQ2 alone was similar to those ESL students who participated in traditional classroom instruction, indicating that once again, gameplay experiences do promote SLA. Therefore, these results suggest that MMORPGs show great promise as a second language pedagogical tool and give insight as to how game designers can leverage the benefits of MMORPGs for SLA.

Post-gameplay survey results demonstrated that the majority of ESL students interpreted their overall gameplay experiences to be positive. Participants expressed their enjoyment of playing EQ2 and often played the game longer than the time allotted for the game sessions. This behavior signaled a desirable level of engagement due to the various game tasks and social interactions inherent to the infrastructure of EQ2 that would be difficult to duplicate due to limited resources for an independently developed serious game. We consider enjoyment to be one of the key affordances of gameplay that should not be compromised. The remainder of the chapter discusses the process of transforming a recreational game into a serious game for SLA. We discuss the design specifications and implementation of two game modules that correlate to conversational fluency and L2 vocabulary acquisition.

DESIGN & IMPLEMENTATION OF GAME MODULES FOR SLA

Post-test scores of ESL students who were grouped with NES participants were significantly higher than both the ESL students who participated in traditional classroom instruction and those ESL students who played EQ2 alone. Based upon post-game questionnaire results, ESL students indicated that the avatar provides a heightened level of engagement for foreign language students and forms the basis for social interactions between players of various cultural backgrounds in MMORPGs. Therefore, interactions with other PCs leads to a higher level of engagement demonstrated by the generation of more chat messages than other ESL students. The interactions between ESL PCs and NES PCs provide sufficient opportunities for ESL participants to socialize and acquire knowledge of L2 vocabulary associated with completion of quests. Consequently, these social interactions foster emergent cooperative relationships with PCs during gameplay. The design goal is to provide in-game pedagogical support that facilitates ESL students' ability to imitate similar behaviors as they interact with other players.

Results from the second phase of playtesting revealed that ESL students who played EQ2 alone did not chat with other PCs. However, those ESL students who were grouped with NES participants generated more chat messages and communication patterns similar to NES. Consequently, the social interactions between ESL students and NES participants played a crucial role in increased L2 vocabulary acquisition and reading comprehension. These results specify the design requirements for a game mod that encourages ESL students to communicate with NES PCs during gameplay. The requirement is to design a game mod that encourages ESL students to chat with PCs thereby developing conversational fluency in the target language, one aspect of communicative performance. The game module provides examples of chat messages that correspond to in-game social interactions, including meeting PCs, working together and scheduling future game sessions. The implementation details of the two game interfaces designed to facilitate conversational fluency in English are discussed below.

Development Tools for EverQuest® II Customized Modules

Sony Online Entertainment (SOE) offers subscribers of EQ2 access to the User Interface (UI) Builder development tool, giving players the ability to create customized game modules visible during gameplay. The SOE UI development team also uses the UI Builder to develop interfaces (e.g., Map of Norrath, list of avatar's skills, chat window, etc.) for EQ2. The interactive graphical tool allows players with no prior knowledge of XML to rapidly create customized user interfaces featuring a library of objects and their properties including pages, style definitions, and widgets (e.g. dialog boxes, text boxes, drop down menus, toolbars, etc.). Developers of game mods (modders) can write XML code from scratch to create new interfaces or choose to drag and drop interfaces visible in the UI Builder workspace to modify existing game interfaces. Interface designers can test drive game mods using UI Builder before integrating them in the actual game. Details of each game module are outlined below.

In-Game Dictionary Module

While the implementation of a dictionary may appear to be an obvious solution to supplementing vocabulary acquisition, the design goal was to first understand the learning needs of ESL students and let those needs motivate the design and implementation of the game modules. Post-game questionnaire results revealed that ESL students requested a dictionary to assist with understanding the meaning of new vocabulary words encountered during gameplay. Furthermore, the ESL instructor advised us that a dictionary is a standard pedagogical tool for vocabulary acquisition and should be an optional resource available during gameplay. In response to this request, we implemented an in-game dictionary module using the UI Builder. The in-game dictionary is a *Dictionary Page object* implemented as a tree data structure. The in-game dictionary represents forty potential English vocabulary words corresponding to academic or advanced English vocabulary used in NPC speech to convey information about the first eight quests on the Isle of Refuge. The Dictionary root object expands into twelve branches that represent the alphabetical index of vocabulary words. Each branch represents a tabbed window pane labeled with two to three letters (*A B* or *X Y Z*) and aligned the top edge of each tabbed window pane. See Figure 6. Players navigate the tabs to search for vocabulary words in the dictionary. Each tab corresponds to page of no more than seven vocabulary words; each entry contains a

Figure 6. Tabbed window pane of dictionary of English vocabulary words

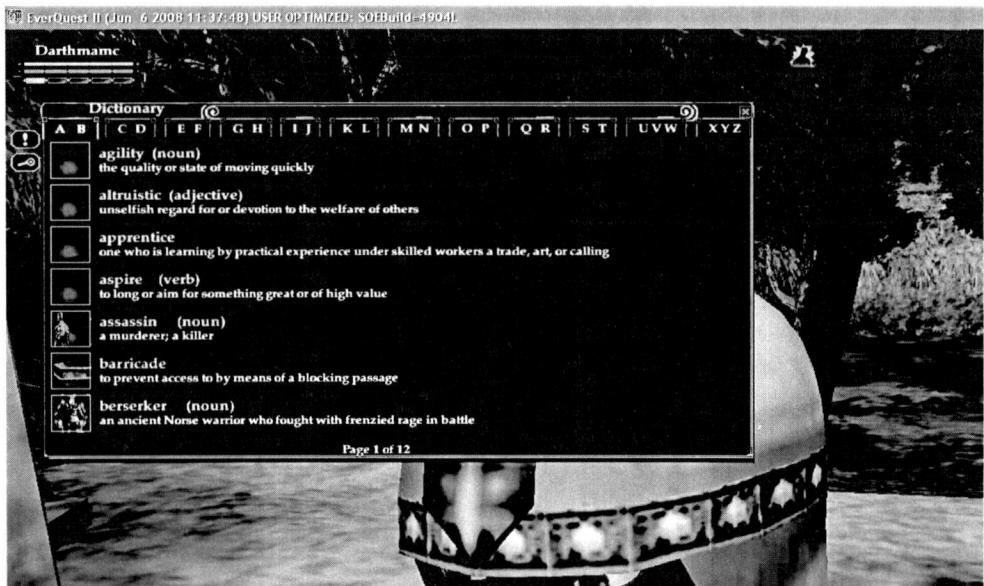

definition and at least one synonym that functions as a "quick and dirty" definition while introducing another second language vocabulary word (e.g., **agility** – *the ability to be nimble; quick*). Definitions were taken from Webster's Online dictionary (http://merriam-webster.com) and the Dictionary website (http://dictionary.com).

Conversational Prompts Module

Data analysis of post-test results indicated that interactions with native English speakers were beneficial to ESL students' vocabulary acquisition whereas those ESL students who played EQ2 independently did not chat with others during gameplay. Thus, we choose to design a game mod that facilitates conversations via chat with other PCs. The physical layout of the game module referred to as the conversational prompts module actually represents a collaborative model of in-game social interactions. Subsequently, each tab represents one of the three categories of social interactions (*Make Friends, Work Together, Meet Again*) attributed to emergent cooperative relationships in virtual spaces and scaffolds ESL students' ability to engage native English speakers in conversation and work together to complete quests.

Remembering the ESL students who independently played EQ2, we recognized that their refusal to chat with other players was caused by two factors: language anxiety or uncertainty about what to say and lack of experience interacting with strangers online. Therefore, we sought solutions that would help ESL students feel more comfortable interacting with strangers in virtual worlds. The first tab, labeled *Make Friends,* offers conversational prompts that encourage the player to introduce his/her self to other PCs during gameplay. The *Make Friends* page consists of common greetings (e.g. "Hi. How are you?"), followed by questions in an attempt to meet PCs and engage them in conversation. The second tab, labeled *Work Together* emphasizes the importance of learning to work with others. The list of conversational prompts can be divided into logical groupings of collaborative social dialogue. The first category, *request for assistance*, encourages PCs to reach out to others in an attempt to solicit

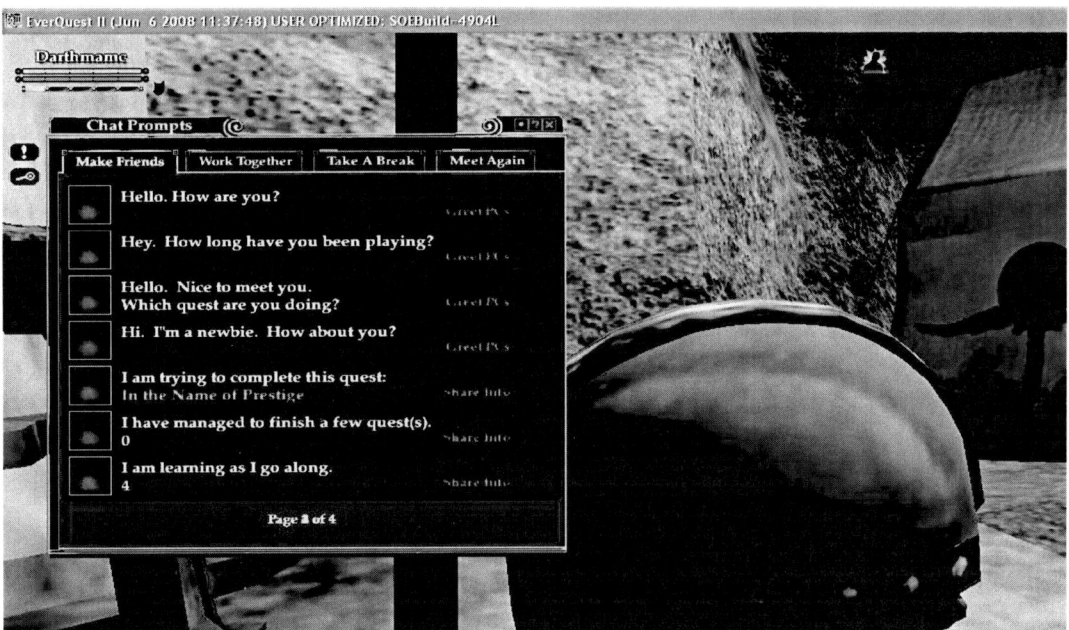

Figure 7. Conversational prompts to facilitate collaboration with other PCs during gameplay

help with a quest (*"I am stuck. Do you know what to do?"*). ESL students need to understand that it is common practice for players to ask for help and should be comfortable doing so. The second category represents *offers of assistance* (*"You look like you need some help. What can I do?"*). This social behavior demonstrates the PC's willingness to help and facilitates an attitude of team work. The third category refers to the PC's ability to influence the future actions of others (*"Let's go kill some stuff"*). The use of collective pronouns like *"we"* or *"us"* suggests a spirit of teamwork and helps establish cooperative relationships in both the virtual and real world. The last category refers to the PC's ability to solicit input from others (*"What should we do next?"*) while empowering others to lead. Good leaders understand the importance of getting buy-in from team members and sharing the responsibility of group decisions with others.

The challenge in virtual environments is maintaining relationships that may not extend beyond the virtual world of Norrath and yet exists in an ever-changing virtual environment of newbies, challenges, and adventures. How do PCs establish connections with other PCs that lead to continuous, collaborative relationships? The same way people do in face-to-face interactions; PCs make commitments to meet online at a designated time to play the game together. Each gameplay session represents an opportunity for L2 students to engage in-game social interactions that foster long-term relationships. The conversational prompts for *Meet Again* assist ESL students with establishing online relationships that give them the opportunity to negotiate future meetings with other PCs to play EQ2. Negotiation is a desired skill set of communicative performance and teamwork. The fourth tabbed page suggests a list of chat prompts that assists with scheduling future play dates. Each conversational prompt indicates that the ESL student will be logging out soon but would like to meet again and play the game. The goal is to solicit agreement from the PC about a future time to collaboratively play EQ2.

The last tabbed page lists conversational prompts for the category of non-social interactions, *Take a Break*. You may ask why the need for such a category. We recalled an incident in which an ESL student temporarily ceased gameplay but did not communicate intentions; the ESL student went to the restroom but the avatar remained poised for action in the game world. Unaware that the ESL student was no longer controlling the avatar and was no longer in front of the computer, the NES participant continued typing in the chat window. After noticing the avatar's lack of movement, the NES speaker repeatedly asked the ESL student for status ("What r u doing?"). Though the ESL student had every right to use the restroom, the ESL student did not realize the common practice of typing "BRB" ("Be right back") in the chat window to indicate temporary absence. Such an example illustrates the ESL student's lack of communicative competence, knowing what to say at the right time, to indicate the student's temporary absence from the computer. While this may appear to be a minor issue, it is important that PCs do not interpret a lack of response as an indication of being ignored as this could hamper collaboration. ESL students need to be cognizant of communicating pauses in gameplay to other PCs. Consequently, we added one more tabbed window pane that lists conversational prompts to indicate temporary absence during gameplay activities, creating the social interaction category of *Take a Break*.

SUMMARY

In conclusion, we demonstrate that recreational video games can be transformed into viable learning tools when given access to development tools that leverage the existing game platform and enable development of customized game modules that promote learning. We discuss the process of re-purposing a commercial MMORPG as a serious game that promotes SLA. Our hypothesis questions the validity of MMORPGs as unorthodox language learning tools. Results from experimental studies indicate that ESL students who interact with native speakers during gameplay produce significant learning outcomes when compared with traditional second language pedagogical methods. Consequently, we developed two customized game modules, an in-game dictionary module that provides graphical images and definitions of potential English vocabulary words, and a conversational prompts module that fosters collaboration behaviors (e.g. requesting assistance with completing game quests) with other PCs. Future work includes evaluation of game modules that facilitate English proficiency in terms of conversational fluency, reading comprehension, and vocabulary acquisition.

REFERENCES

Aldrich, C. (2005). *Learning by doing: A comprehensive guide to simulations, computer games and pedagogy in e-Learning and other educational experiences*. San Francisco, CA: Pfeiffer Books.

Backer, J. A. (1999). *Multi-user domain object oriented as a high school procedure for foreign language acquisition*. Unpublished Ph.D. thesis, Nova Southeastern University.

Barab, S., Thomas, M., Dodge, T., Carteaux, R., & Tuzun, H. (2005). Quest Atlantis: A game without guns. *Educational Technology Research and Development*, *50*(1), 86–107. doi:10.1007/BF02504859

Beauvois, M. H. (1992). Computer-assisted classroom discussion in the foreign language classroom: conversations in slow motion. *Foreign Language Annals*, *25*(1), 455–464. doi:10.1111/j.1944-9720.1992.tb01128.x

Beauvois, M. H., & Eledge, J. (1996). Personality types and megabytes: Student attitudes toward computer-mediated communication (CMC) in the language classroom. *CALICO Journal*, *13*(2), 27–45.

Begg, M., Dewhurst, D., & Macleod, H. (2005). Game-Informed Learning: Applying computer game processes to higher education. *Innovate: Journal of Online Education, 1*(6). Retrieved from http://innovateonline.info/?view=article&id=176

Brown, J. S., & Thomas, D. (2006). You Play World of Warcraft? You're Hired! Why multiplayer games may be the best kind of job training. *Wired (San Francisco, Calif.), 14*(4).

Chapelle, C. A. (1996). *Computer applications in applied linguistics*. Cambridge, UK: Cambridge University Press.

Chatham, R. E. (2007). Games for training. *Communications of the ACM, 50*(7), 36–43. doi:10.1145/1272516.1272537

Chenova, J. (2004). Flow in games and everything else. *Communications of the ACM, 50*(4), 31–34.

deHaan, J., & Diamond, J. (2007, August 04-05). The experience of telepresence with a foreign language video game and video. *Sandbox Symposium 2007*, San Diego, CA, (pp. 39–46).

Eck, R. V. (2006). Digital Game-Based Learning: It's not just the digital natives who are restless. *EDUCAUSE Review, 41*(2), 16–30.

Entertainment Software Association. (2007). *The 2007 essential facts about computer and video games industry*. Washington, DC: Entertainment Software Association.

Falsetti, J., & Schweitzer, E. (1995). SchMOOze University: A MOO for ESL/EFL students. In M. Warschauer (Ed.), *Virtual connections: On-line activities and projects for networking language learners* (pp. 231-232). Honolulu, HI: University of Hawai'i, Second Language Teaching and Curriculum Center.

Gee, J. P. (2003). *What video games have to teach us about learning and literacy*. New York: Palgrave Macmillan.

Gee, J. P. (2004). *Situated language and learning: a critique of traditional schooling*. New York: Routledge.

Girgensohn, A., & Lee, A. (2002, November). Making websites be social places for social interaction. In *Proceedings of Computer Supported Collaborative Work '02,* New Orleans, LA, 136–145.

Hadley, O. (2001). *Teaching Language in Context*. Florence, KY: Heinle and Heinle.

Halverson, R. (2005). What can K-12 school leaders learn from video games and gaming. *Innovate: Journal of Online Education, 1*(6).

Hubbard, P. (1991). Evaluating computer games as language learning tools. *Simulation & Gaming, 22*(2), 220–223. doi:10.1177/1046878191222006

Hudson, J. M., & Bruckman, A. (2002). IRC Francais: The creation of an internet-based SLA community. *Computer Assisted Language Learning, 15*(2), 109–134. doi:10.1076/call.15.2.109.8197

Kelly, H., Howell, K., Glinert, E., Holding, L., Swain, C., & Burrowbridge, A. (2007). How to build serious games. *Communications of the ACM, 50*(7), 44–50. doi:10.1145/1272516.1272538

Krashen, S. D., & Terrell, T. D. (1983). *The natural approach: Language acquisition in the classroom*. London: Prentice Hall Europe.

Lee, A., Danis, C., Miller, T., & Jung, Y. (2001). Fostering Social Interactions in Online Spaces. In *Proceedings of INTERACT 2001: IFIP TC.13 International Conference on Human-Computer Interaction* (pp. 59–66). Amsterdam, The Netherlands: IOS Press.

Mayo, M. J. (2007). Games for science and engineering education. *Communications of the ACM, 50*(7), 30–35. doi:10.1145/1272516.1272536

Pillay, H., Brownlee, J., & Wilss, L. (1999). Cognition and recreational computer games: Implications for educational technology. *Journal on Research in Computing Education, 32*(1), 203–216.

Prensky, M. (2001). *Digital Game-Based Learning*. Chicago: Donnelley and Sons Company.

Rankin, Y., Gold, R., & Gooch, B. (2006a). Evaluating interactive gaming as a language learning tool. In *Conference Proceedings of SIGGRAPH 2006,* Boston.

Rankin, Y., Gold, R., & Gooch, B. (2006b). 3D role-playing games as language learning tools. In *Conference Proceedings of EuroGraphics 2006, 25,* Vienna, Austria.

Steinkuehler, C. (2005). *Cognition and learning in massively multiplayer online games: A critical approach.* Unpublished dissertation, University of Wisconsin-Madison.

Steinkuehler, C., & Williams, D. (2006). Where everybody knows your (screen) name: Online games as "third places." *Journal of Computer-Mediated Communication, 11*(4), Article 1.

Warshauer, M., & Kern, R. G. (2000). Theory and practice of networked-based language teaching. In M. Warshauer & R.G. Kern (Eds.), *Networked-based Language Teaching: Concepts and practice* (pp. 1–19). Port Chester, NY: Cambridge University Press.

1 Watts-Taffe, S., & Truscott, D. M. (2000). Using what we know about language and literacy development for ESL students in the mainstream classroom. *Language Arts, 77*(3), 258–265.

Section 3
Games in Healthcare

Chapter 11
Application of Behavioral Theory in Computer Game Design for Health Behavior Change

Ross Shegog
UT-School of Public Health, USA

ABSTRACT

Serious games are gaining profile as a novel strategy to impact health behavior change in the service of national health objectives. Research has indicated that many evidence-based programs are effective because they are grounded in behavioral and motivational theories and models such as the PRECEDE model, the Health Belief Model, Social Cognitive Theory, the Theory of Reasoned Action, the Transtheoretical Model, Attribution Theory, and the ARCS model. Such theories assist in understanding health behavior problems, developing salient interventions, and evaluating their effectiveness. It follows, therefore, that serious games can be made optimally effective in changing health behavior if they are also informed by these theories. A successful intervention development framework (Intervention Mapping) provides a means to enable game developers to use theory to inform the design of effective games for health. This chapter describes useful theories and models for health game design, introduces the intervention mapping process, and describes a case study of a theory- and empirically-based serious health game intervention that has used these approaches and has been rigorously evaluated.

SERIOUS GAMING AS A STRATEGY FOR HEALTH BEHAVIOR CHANGE

The application of gaming as a method to change health behavior carries with it the burden to impact the consequences of morbidity and mortality. Nowhere, then, is the contention of games having a "serious" purpose more relevant than in the domain of serious games for health. Serious games are gaining profile as a potential strategy to educate the public about health in new and novel ways. Computer games represent an emerging approach to the continued research and development of health education and health promotion programs in the service of national health objectives. Research has indicated that many evidence-based health educa-

DOI: 10.4018/978-1-61520-739-8.ch011

tion and health promotion programs are effective because they are grounded in behavioral theory. It follows, therefore, that serious games might be made optimally effective in changing health behavior if they are also informed by behavioral and motivational theory.

Health Behavior

The Health Objectives for the Nation (US Dept of Health and Human Services, 1980) provided an initial blueprint for a federal strategy for public health education and for monitoring population-wide behavior patterns. Successors of this document include, most recently, Healthy People, 2010, which outlines 28 focus areas and 467 measurable national disease prevention and health promotion objectives. Such strategic plans are particularly relevant foundations for public health education efforts. A critical consideration is that modifiable behavioral risk factors are leading causes of mortality in the United States. Leading causes of death in 2000 were tobacco (18.1% of total US deaths), poor diet and physical activity (16.6%), and alcohol consumption (3.5%), all related to behavioral exigencies. Aside from environmental challenges of microbial and toxic agents, other actual causes of death were also heavily accented to human behavioral causes: motor vehicle crashes, incidents involving firearms, sexual behaviors, and illicit drug use (Mokdad, 2004).

Health behaviors not only include overt actions for health maintenance, restoration, or improvement but also encompass the cognitions and emotions (beliefs, expectations, motives, values, perceptions, affect) that can be measured (Gochman, 1982, 1997). Health behaviors can be categorized in the major categories of risky behavior, preventive health behavior, and sick role behavior. *Risky behaviors* include those behaviors undertaken by individuals despite associated health risks. Tobacco smoking, poor diet and physical activity, alcohol consumption, and unprotected sex have already been indicated as major targets for national campaigns. *Preventive health behaviors* include those that are undertaken for the purpose of preventing or detecting illness in an asymptomatic state and maintaining health. This includes an array of health promoting self-management behaviors that are the antithesis of risk behaviors. These include eating low fat foods or using condoms and also include screening and early detection behavior (e.g. breast, testicular, and skin self-examination). *Sick-role behaviors* include behaviors that are undertaken by those who consider themselves ill with the intention of getting well. In the clinical domain this includes adherence to treatment (e.g. taking a full course of antibiotics) or management plans to control disease (e.g. balanced diet, energy expenditure, and timing of taking insulin; monitoring asthma symptoms to detect early changes) (Kasl & Cobb, 1966a; 1966b]

Health Education and Health Promotion

Changing human behavior is hard to do. It is inextricably linked within the milieu of our lives and it is complex. Health behavior is mediated by cognitions where people's knowledge and thinking precedes their actions. While knowledge is a necessary antecedent of behavior it is not sufficient to produce most behavior changes. An array of behavioral determinants, including perceptions, motivations, skills, and the social environment are key influences on behavior. The expectation that serious games should be conducive to impact behaviors, then, might appear somewhat wishful. Arguably, it is an insufficient expectation that serious games could seriously impact behavior unless they are able to impact an array of behavioral determinants in excess of achievement (knowledge) alone. Health education and health promotion program developers have wrestled with this challenge and have provided guidance in the application of theories and models that can prove invaluable to the armamentarium of the serious health game developer in how to understand the complexity of behavioral change. *Health education* has been broadly defined to include an array of possible learning experiences to facilitate an individual's voluntary adoption of healthier behavior (Green & Kreuter, 1980). *Health promotion* is a more recent term that describes the science and art of assisting people to adopt healthier lifestyles (Glanz, Rimer, and Viswanath, 2008). Efforts to achieve this can include any combination of strategies to raise awareness, change behavior, and alter environments to encourage healthy practices. This addresses, albeit broadly, the interplay of factors that are considered necessary to be accounted for to change health behavior. Health education and health promotion research has laid a solid foundation, though, on how to understand and intervene on health behavior. Methods and strategies have been drawn from theoretical perspectives, research, and practice from the social and health science disciplines including psychology, sociology, anthropology, communications, nursing, economics, and marketing as well as being supported by public health disciplines of epidemiology, statistics and medicine. (Glanz, Rimer, and Viswanath, 2008). The foundations of successes in health education and health promotion can inform, in turn, the science and art of serious games for health.

Application of Behavioral Theory in Computer Game Design for Health behavior change

Computer-based technology has been emerging in the service of health care and health promotion over the past decades, in concert with the evolution of computer science and infrastructure, and is becoming increasingly integral throughout the health continuum in the service of facilitating health-related behavior (Figure 1). Technology-based interventions for health education and health promotion have been collectively termed interactive health communication (IHC) (USD-HHS, 1999). IHC applications have been used in varied settings (e.g., homes, clinics, supermarkets, shopping malls, and schools) and in varied health domains to monitor and change behavior (e.g., cancer, HIV/AIDS, diabetes, substance abuse, smoking prevention and cessation, asthma, violence, and diet). (Strecher, 1995, 1999a. 1999b; Watkins, 1994; Sechrest, 1996; Duncan, 2000; Reis, 2000; Bartholomew, 2000; Barry, 1995; Carrol, 1996; Walters, 2006; Shegog, 2005; Gerbert, 2003) Their value in inducing specific and measurable behavioral change has been documented.(Revere, 2001; Block, 2000; Prochaska, 2000; Hornung, 2000; Kahn, 1993; Deardorff, 1986). Further, reviews of evaluation studies of IHC interventions have reported positive impact of patient behavior, indicating the effectiveness of such interventions (Revere, 2001, Krishna, 1997). The Science Panel on Interactive Communication and Health (USD-HHS, 1999) highlighted the potential of IHC and its use of new media to simultaneously improve health outcomes, decrease healthcare costs, and increase consumer satisfaction.

The advantages of IHC, cited by Owen et. al. (2002) could easily be applied to those of serious games and include core capacities of interactivity, appeal, and engagement (Owen, 2002). Interactivity includes tailoring of program components to user responses and instantaneous feedback that can be personal, normative, or ipsative. Appeal relates to convenience of access across place and time, channel preference where certain users prefer to use an IHC medium, and flexibility for users to choose when material is received and how often. Engagement relates to credible simulations where users can participate in role plays and explore "virtual" environments without risk; openness of communication where the IHC medium provides confidentiality to answer and explore sensitive issues, and multimedia interfaces where still, video, and animated graphics can elicit learning and reduce literacy requirements for material. It is the case that these capacities are as applicable to serious games as to other IHC. Arguably, serious

Figure 1. Current technology-based applications in the continuum of health promotion, disease prevention, and disease management

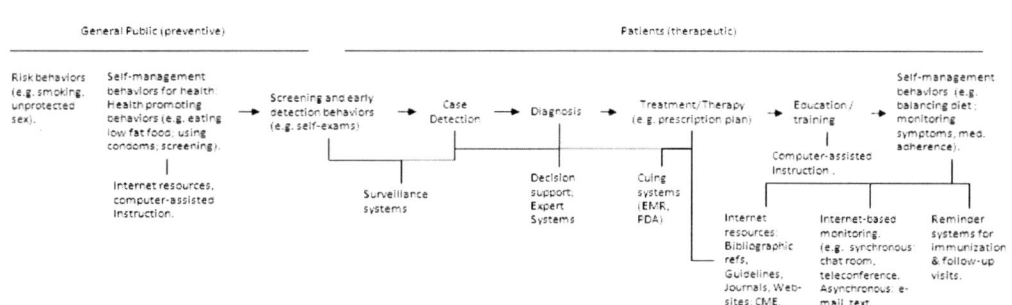

games offer additional advantages so it is not surprising that they are making inroads as a strategy in the health continuum particularly in relation to education and training, and self-management for health and for disease.

Emergence of Serious Games for Health Behavior Change

Gaming has long been considered a core educational strategy for computer applications (Alessi and Trollip, 1991). The intuitive appeal of games has led to a burgeoning interest in harnessing the educational and motivational potential of video game technology to target health behavior change, particularly in child and adolescent learners. Serious games provide the advantages of IHC programs and have some essential advantages over more traditional approaches including their potential to motivate learners and focus their attention on the goal of the game (Schild, 1968). An offshoot of this is the potential to increase the efficiency of learning because learners are likely less distracted. A further bonus, from the perspective of mentoring, tutoring, or social support, is that in discussing increased task performance teachers, health care providers, and parents can be perceived as an ally rather than an authority figure. Serious games help people learn, in part, because they activate prior learning (using previously learned information to move to higher levels of play), provide goals, provide immediate feedback in scoring and in visual and auditory stimulus (allowing immediate behavior modification), enhance skills transfer to real life application, and enhancement of motivation amongst a majority of learners. Of course, a preeminent advantage is that games, inherently, are fun. Authors have grappled with what constitutes fun within the gaming pedagogy. Factors include novelty and powerfulness, appealing presentation, interactivity, challenge, providing a sense of control, and being rewarding (Hsu, 2005). Games also cater to fantasy through imaginary characters and virtual worlds.

Within a taxonomy of serious games Sawyer and Smith (2008) describe 'games for health' as one of seven major domains that also include advergames, production games, and games for training, education, science and research, and work (Sawyer and Smith, 2008). Cross-referenced in this conceptual framework are settings that include Government and NGO, defense, healthcare, marketing and communications, education, corporations, and industry.

Serious games vary in nature and function as much as their purely recreational counterparts. This includes user role (first person, third person), game dynamics (cognitive vs. exergames; puzzle; strategy), platform (proprietary platform vs. generic; 2D vs. 3D; single vs. multi-player),

Table 1. Game types applicable to serious health game application

Game Type	Definition
Action / Arcade	Using the vehicle of an arcade game to teach concepts. Can include shooting games, platform games (where players move between onscreen platforms), or gambling games (where there is generally a large element of chance or money as a motivator).
Activity promoting / Exergames	Games that promote physical activity within the context of play.
Adventure	Player assumed the role of a character in a situation about which little is known. The player must use existing information and resources to solve the problems posed for the character by these situations. Includes action adventure (first person and third person). Teaches problem-solving skills, deductive reasoning, or hypothesis testing. Can include simulation.
Board	Computerized versions of existing games (e.g. chess, checkers). Can be similar to adventure games or logic games.
Fighting / Combat	Games that use combat or violent competition as a primary motivator.
Logic	Player requires the use of logical problem solving to succeed.
Psychomotor	Combine intellectual and motor skill.
Role-playing	Player assumes the guise of a character and acts out that role, thus learning about the character's environment and problems, and how to solve these problems. Similar to simulations. Includes first person and third person.
Sports	Simulating sporting scenarios though not an exergame.
Strategy	Commanding armies within recreations of historical battles.
TV quiz	Instructional games taking the form of an ordinary quiz game.
Word	Teach about words or use words as the basis for the game.

Adapted from Allesi and Trollip, 1991; Herz, 1997; Thompson et. al., 2007; and Baranowski et. al., 2008.

realism; and genre (role play; adventure; combat; simulation). Often a game will provide several roles simultaneously or comprise embedded games within a broader gaming structure (e.g. an adventure game that requires players to complete puzzles, quizzes, and brief psychomotor games). The myriad of game types can lend themselves to the application of serious games (Table 1). The choice of game format is subject to the game objectives, target population, and contextual exigencies. In it's simplest structural form a game consists of introduction, body, and conclusion. Embedded are the devilish details and decision points that developers must grapple with. In the context of serious games for health, developers must extend their focus beyond game dynamics and fun to include how the strategies employed within the game will serve the greater good of influencing behavioral change.

A summit in games for learning by the Federation of American Scientists yielded recommendations for the inclusion of ten attributes to learning for effective educational video games. These include clear learning goals, learner challenge and reinforcement, monitoring of progress and adjustment of content to learner mastery, encouragement of inquiry, contextual bridging between what is learned and its use, time on task, motivation and strong goal orientation, scaffolding for learning, personalization and tailoring, and infinite patience (Federation of American Scientists, 2006). While such summits demonstrate a burgeoning interest and recognition in the use of gaming for learning, reach, and motivation relatively few studies have been conducted demonstrating effectiveness in changing health behavior. Despite books devoted to the subject of serious games and the emergence of peer-reviewed literature on this topic (Lieberman, 1997; Bogost, 2007; Gee, 2007) there is only a single peer-reviewed article on the subject of serious games and health behavior change (Baranowski et. al., 2008). The review compared twenty five articles on video-games that addressed preventive health behavior including diet alone, physical activity alone, physical activity among the physically challenged, diet and physical activity combined, and sick role behaviors including asthma and diabetes self-management (Baranowski et al, 2008). Inclusion criteria were simply that the software was described as a game and that it was targeting lifestyle behavior change. Games that were not video games were excluded. This sample of studies focused on health behavioral change outcomes as distinct from more proximal impact or knowledge change. There was substantial variability across studies in their design, the targets of change, and the characteristics (and reporting of characteristics) of the games. A commonality of these effective games was that they were grounded in, and informed by, behavioral theory.

UTILITY OF THEORY TO UNDERSTAND DETERMINANTS OF HEALTH BEHAVIOR AND HOW TO IMPACT THEM

The truism that most health-related programs produced are done so with the best of intentions belies the reality that not all such programs are equally effective. The current emphasis on evidence-based interventions in medicine, public health, and behavioral medicine has necessitated that programs are founded on sound theoretic principles and empirical evidence.

Behavioral science offers an armamentarium to assist the intervention practitioner in meeting the challenges that are represented by human behavior. Theory helps explain the dynamics of health behaviors, including processes for changing them, and the influences of the many forces that affect health behaviors, including social and physical environments. It provides a road map for studying problems, developing appropriate interventions, and evaluating their successes. Theory and models can help to answer important why, what, and how questions: Why ? ... people do or do not engage in certain behaviors; What ? ... factors should be monitored and measured during program evaluation; How? ... to devise programs that reach the target audience and have an impact (Glanz and Rimer, 2005). The characteristics of a useful theory are that it makes assumptions about a behavior, health problem, target population, or environment that are logical, consistent with everyday observations, similar to those used in previous successful programs, and supported by past research in the same or related areas.

It is an imperative that the games produced for the enhancement of health are done so with a eye to optimizing their effectiveness. This implies that an educational game is not only attractive and fun to gain the attention of players, but that it is effective in motivating behavioral change, an idea aligned to the current emphasis on the dissemination of evidence-based interventions. For the developer of health-related serious games there is great facility for reviewing theory in the context of the targeted behavior to gain immediate insight into the nature of the health problem, the nature of the target population, the nature of the game's objectives, the type of methods the game might employ, and the type of evaluation protocol required. As the use of serious games for health increases so must our sophistication in development, implementation, and evaluation. Systematic reviews have indicated that theory-based interventions can lead to more powerful effects than interventions developed without theory (Ammerman, et al, 2002; Legler, 2002). They allow us to describe the important determinants or mediators of the target behavior. These are constructs such as confidence, intentions, and skills that are antecedents of behavior. Effective serious games that teach self-care do not target the behavior directly. Rather, they target determinants

Figure 2. Seminal theories and models used to inform health education and health promotion program development and evaluation

Figure 2a. The PRECEDE Planning Model

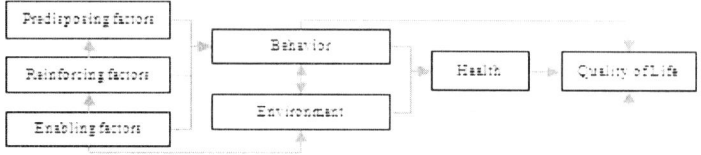

Figure 2b. The Social Cognitive Theory

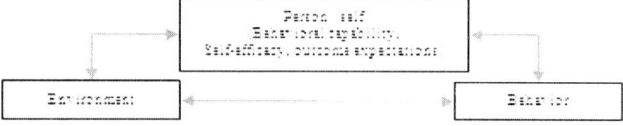

Figure 2c. The Theory of Reasoned Action & Theory of Planned Behavior

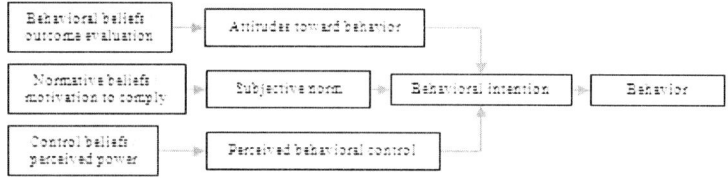

Figure 2d. The Health Belief Model

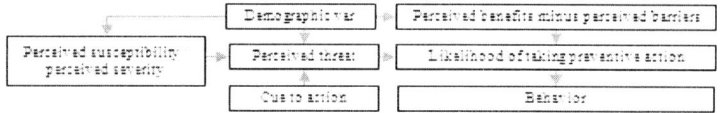

Figure 2e. The Transtheoretical Model

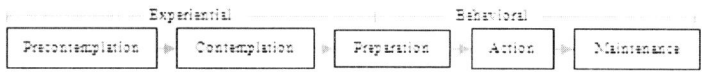

Figure 2f. Attribution Theory

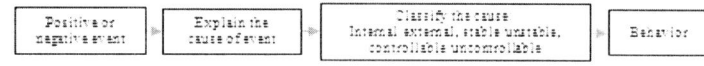

Source: Adapted from Skinner and Kreuter, 1997 (© 1997; Lawrence Erlbaum Associates; Used with permission); and Glanz and Rimer, 2005.

of the behavior. Theory helps the developer identify likely salient determinants and offers guidance on how to impact these to elicit behavior change. Seminal theories and models in health promotion and disease prevention include: Social Cognitive Theory, Theory of Reasoned Action & Theory of Planned Behavior, the Transtheoretical Model, the Health Belief Model, the PRECEDE model, and Attribution Theory (Figure 2).

PRECEDE Model: Understanding the health behavior in the broader context (Figure 2a)

The PRECEDE Model is a widely used planning model that can provide the game developer with a comprehensive understanding of the health problem and its related factors, founding the program on a firm theoretical and empirical footing (Green and Kreuter, 2005). It is very useful for conducting upfront needs assessment for a game development project. The health problem is defined in terms of quality of life (social diagnosis), health parameters (epidemiologic diagnosis), behavior and environment (behavioral & environmental diagnosis), and predisposing, reinforcing, and enabling factors (educational and ecological assessment). The model is useful for health game development because it encourages data gathering from the population at-risk, empirical evidence, and theory. The resultant framework provides an encapsulation of the health problem, an understanding of critical health behaviors related to the problem, and determinants of the health behavior that might be targeted by the game.

Social Cognitive Theory: Understanding Health Behavior in Relation to Cognition, Learning, and Environment (Figure 2b)

Social Cognitive Theory provides a comprehensive view of human behavior and learning. (Bandura, 1986) The theory and constructs from this theory are ubiquitous in health education research and it is currently the predominant guiding theory in games for health research (Baranowski et. al., 2008). For game developers the concepts of triadic reciprocity, self-regulation, and observational learning are particularly salient. *Triadic reciprocity*: Social Cognitive Theory holds that behavior is determined by the interaction of personal, environmental, and behavioral influences, a paradigm termed triadic reciprocity. Personal factors include cognitions that increase or decrease the likelihood of engaging in a particular behavior (e.g., personal values, beliefs, skills, outcome expectations, perceived self-efficacy). Environmental factors include any aspect of the environment (social or physical) that supports or discourages a particular behavior (e.g., influential role models, social or normative support). Behavior is viewed as dynamic, depending on the interaction of the person and environment in which the behavior is performed. Triadic reciprocity is operationalized in a game when the player can experience the interplay of environmental influences, their behavior within the environment (including social support of other players), and their own attitudes and beliefs (including self-efficacy and outcomes expectations described below). This lends itself particularly to consideration of how to simulate these components in the game space. *Self-regulation*: Self-regulation has been cited as particularly useful for targeting behavior change related to the management of chronic health disorders though it provides useful guidance in targeting preventive behaviors also (Clark, 1989; Clark and Zimmerman, 1990). Bandura describes self-regulation as comprising three principle subfunctions: behavior self-monitoring, its determinants, and its effects; judgment of one's behavior in relation to personal standards and environmental circumstances; and affective self-reaction (Bandura, 1991). Self-regulation is operationalized when the player can simulate the problem solving and decision making associated with health behavior management. *Vicarious modeling and observational learning*: Social Cognitive Theory is founded on research in human learning where observers who view a behavior and the reinforcement of the behavior are likely to perform the behavior themselves. In other words, they cognize events and learn not only the viewed behavior but also develop expectations regarding the value of performing the behavior. The modeled behaviors can be in the form of symbolic representations portrayed through graphics as well as live performance.

Modeling that includes vicarious reinforcement is more powerful than modeling alone and can convey which behaviors are appropriate in which settings. In order for modeling to be effective, the player must identify with the model. Further, coping models, who struggle to surmount a problem situation, are usually more effective than models who seem to easily master a difficult situation that may be beyond the competence of the learner. Observational learning influences self-efficacy and outcome expectations, two constructs vital to the performance of any behavior (Bandura, 1991). *Self-efficacy* relates to the confidence an individual has to perform a given behavior in new and novel circumstances. This construct has been adopted by other theories. Skills development through vicarious modeling and through direct guided practice and feedback provide increased behavioral capability and performance accomplishments which, in turn, are a source of increased self-efficacy (Bandura, 1986). Reinforcement can be external (from rewards provided contingent on certain behavior), vicarious (through observing rewards provided to a model contingent on his or her behavior), or self-reinforcement (through an internal reward process). In video games external and vicarious reinforcement is mediated through feedback that reinforces the player's behavior by providing confirmatory or corrective information regarding his or her responses (Riegeluth, 1989). Natural feedback provides consequences occurring as a natural result of an action, whereas artificial feedback provides feedback used for purposes of instruction only (Alessi and Trollip, 1991). Prompt and detailed feedback helps the player identify behavioral inadequacies and rectify them thus influencing the player's own outcome expectations. *Outcome expectation* is the belief that enacting a particular behavior will result in the outcome one seeks. By optimizing a player's self-efficacy and outcome expectations the chance that a behavior will be performed is optimized.

Assessing a player's self-efficacy and outcomes expectations regarding a given behavior can be used to tailor the game experience.

Theory of Planned Behavior: Understanding the Behavior in Relation to Perceived Social Influence, Control, and Intentions (Figure 2c)

The Theory of Planned Behavior and the associated Theory of Reasoned Action describe a person's intentions as the critical determinant of their behavior. *Intentions*, in turn, are influenced by an individual's attitudes toward performing the behavior, subjective norms, beliefs about the ease or difficulty of changing, and perceived behavioral control. *Attitudes* toward the behavior are influenced by beliefs about what the behavior entails and outcomes of the behavior. *Subjective norms* are influenced by beliefs about social standards and motivation to comply with those norms, and whether people who are important to the individual approve or disapprove of the behavior. Social influence model approaches like this emphasize behavioral expectations and standards (social norms) present in the environment and prepare the learner to resist pressure to engage in risk-taking behavior. For game developers this theory is particularly useful in consideration of adolescent peer pressure related to sexual risk, substance use, and violence and in developing refusal skills training. *Perceived behavioral control* also influences intentions through the individual's belief that they can control the behavior. Perceived behavioral control is influenced by the presence of things that will make it easier or harder to perform the behavior. Assessing players on attitudes, norms, control, and intentions regarding a particular health behavior provides an opportunity for game developers to tailor the game experience to address these parameters, thus influencing player intentions.

Health Belief Model: Understanding the Health Behavior in Relation to Perceived Susceptibility, Severity, Benefits, and Barriers (Figure 2d)

The Health Belief Model theorizes that people's beliefs about whether or not they are susceptible to a disease, and their perceptions of the benefits of trying to avoid it, influence their readiness to act. The theory is primarily focused on people's decisions about whether to take action to prevent, screen for, and control illness. A person is likely to be ready to take action if they believe they are susceptible (*perceived susceptibility*); believe the condition has serious consequences (*perceived severity*); believe taking action would reduce their susceptibility to the condition or its severity (*perceived benefits*); believe costs of taking action (*perceived barriers*) are outweighed by the benefits; are exposed to factors that prompt action (*cue to action*); and are confident in their ability to successfully perform an action (*self-efficacy*)(Glanz and Rimer, 2005; Skinner and Kreuter, 1997). For game developers the model is very useful in addressing problem behaviors that evoke health concerns (e.g. high fat diets leading to obesity; sexually risky behavior and the threat of HIV infection; screening behaviors for cancer) and to tailor game experiences to player's profiles based on these constructs. Understanding how susceptible a player feels about the health problem, whether they believe it is serious, and whether they believe action can reduce the threat at an acceptable cost can provide guidance on how to frame the gaming experience to impact these determinants.

Transtheoretical Model: Understanding the Processes of Changing Behavior (Figure 2e)

The Transtheoretical Model (TTM) is primarily concerned with an individual's motivation and readiness to change a health behavior. The model has often been used to explain addictive behavior change, in particular, smoking cessation. (DiClemente and Prochaska, 1985; Prochaska et. al., 1988; DiClemente and Prochaska, 1982; Prochaska et. al., 1993) More recently, investigators have applied the model to health behaviors such as weight loss,(O'Connell and Velicer, 1988) adopting healthy diets,(Glanz et. al., 1994) HIV prevention,(Prochaska, 1994) adopting physical activity,(Marcus, 1998; Reed, 1997) mammography screening,(Champion, 1995; Rachowski, 1996a, 1996b) and chronic disease self-management.(Shegog et. al., 2003; 2004) The TTM is based on the premise that health behavior change does not occur automatically with one bold action or effort (Prochaska and DiClemente, 1992). Rather, the *stages of change* represents a quasi-developmental sequence of steps of varied motivational readiness to modify behaviors (Prochaska et. al., 1994). These include: *precontemplation* (not aware or resistant to change), *contemplation* (weighing pros and cons of change), *preparation* (preparing for change and committing to it), *action* (enacting the new behavior), and *maintenance* (adhering to change). *Processes of change* are a set of 10 cognitive, emotional, and behavioral strategies that persons use to move from stage to stage (DiClemente et. al., 1991; Prochaska, 1988). There are two categories of processes: experiential processes (n=5) and behavioral processes (n=5). The *experiential processes* involve cognitive and emotional activities that are most important during the earlier stages of change, i.e., precontemplation and contemplation; that is, using the experiential processes of change during precontemplation and contemplation is associated with a greater chance of successful behavior adoption. Experiential processes include consciousness raising, dramatic relief, environmental reevaluation, self-reevaluation, and self-liberation. The *behavioral processes* involve more concrete, visible activities that are important once the decision to adopt a new behavior has been made, and use of these processes later in the change process (i.e. during action and maintenance) is likely to be associated with greater long-term success. Behavioral processes include stimulus control, counter-conditioning, helping relationships, social liberation, and reinforcement management. During the course of change from one stage to another an individual will engage in *decisional balance*, a weighing of the pros (the perceived positives or benefits) and cons (the perceived negatives or barriers) of a target behavior (Velicer et. al., 1985). Precontemplators have a negative decisional balance, reflecting perceived reasons not to change. Contemplators are closer to the neutral or zero point of equal pros and cons while those in action and maintenance stages have a positive decisional balance (Rakowski et. al., 1996a, 1996b). Serious game developers can consider assessing players in terms of their stage to provide a guide to tailoring gaming experiences. Processes of change represent the "active ingredients" in interventions that lead to successful modification of a problem behavior (DiClemente et. al., 1991). Game developers can use the processes as guides to gaming content to impact cognitive and behavioral change.

Attribution Theory: Understanding the Health Behavior in Relation to Perceptions of Personal Efforts. (Figure 2f)

Attribution theory is concerned with the causal judgments that individuals use to explain events that happen to themselves as well as others in the social and physical domains of life. Attributions can impact thoughts, expectancies, feelings, and behaviors. The causal explanations an individual provides for his or her behavior can be predictive of future morbidity (Peterson & Seligman, 1988), influence success of rehabilitation following disease or distressful events (Gilutz et. al., 1991; Janoff-Bulman, 1979, 1988), impact relapse into unhealthy behavior such as smoking or opiates (Anderson and Anderson, 1990; Grove, 1993; Eiser et. al., 1985; Bradley et. al., 1992), and affect self-management and medication compliance (Morris and Kanouse, 1982, Brubaker, 1988; Hospers et. al., 1990). Typically, when a positive or negative event occurs, individuals assess why this event occurred. People attribute causes to events around them which vary from luck, ability, personal effort, and task difficulty (Weiner, 1985). These explanations can be classified along three primary dimensions including locus, stability, and controllability, or intentionality. *Locus of causality* can be internal (e.g. ability, effort, mood, maturity, and health) or external (e.g. relevant other, the task, or the family) to the individual. Bad health might be perceived as an internal ("I am a sickly person") or as an external ("The flu bug got me") cause of failure. The *stability* dimension can be defined on a continuum ranging from stable (invariant) versus unstable (variant). For example, ability, typical effort, and family would be considered stable causes, while immediate effort, attention, and mood are more unstable. Expectation of success is determined by the perceived stability of the causes for success and failure. A person who attributes success to a stable cause (e.g. ability) will have a higher expectation of success when having to perform the same task again as compared with a person who attributes success on the same task to an unstable cause (e.g. luck). Further, a lower expectation of success leads to less adaptive task behavior. The *intentionality* dimension defines those causes which are controllable (intentional) or uncontrollable (unintentional). Controllable causes include level of effort and physician or observer bias. Uncontrollable causes include innate ability, tiredness, or level of task difficulty. The distinction, here, is not one between control by self or control by other but rather whether the cause is controllable. If health management behaviors and health status is perceived as being in the control of the patient (i.e. subject to his or her intentions) this promotes active involvement in that behavior. A crucial cause of health-related self-management success or failure is the belief that personal effort is an important part of the health care process. Self-management programs instill the understanding that the individual be an active participant in his or her own disease management rather than being at the mercy of external factors. By assessing a player's causal attributions game developers can provide tailored game experiences to modify attributions and instill a sense of the power of personal effort (Lewis & Daltroy, 1990).

UTILITY OF PROCESSING AND MOTIVATIONAL THEORIES FOR PERSUASION AND APPEAL

Information Processing and Messaging Theories: Understanding how to Package the Message

Research on communication, messaging, and information processing provides game developers with guidance on how to most effectively relay information within the game space. The Elaboration Likelihood Model posits that persuasive messages are processed through a central (or systematic) processing route and also through a peripheral (or heuristic) processing route. Central processing involves deliberate and thoughtful consideration of the arguments of the message in a more motivated player. Such weighing of arguments leads to more enduring change (Finnegan and Viswanath, 2008). Peripheral processing represents a response to peripheral cues (e.g. the type of role model used) and tends to occur in less motivated players. The model reinforces the importance of considering both the message and the message context. In this regard Message Effect Theories provide guidance on *message framing* and *exemplification*. Messages can be framed positively or negatively (e.g. advantages of quitting or consequences of not quitting) and through exemplification that are illustrative of a general class of events. Exemplars are most effective when they are simple, emotional, and concrete.

Messages can be embedded within *narratives*, a story with a persuasive element. Narratives have been used in the context of family planning, AIDS education, and healthy lifestyle promotion (Slater, 2002). They generate involvement, absorbing and transporting the audience (Green, 2006). Players who identify and empathize with the character will become more susceptible to the persuasive message.

Motivational Theories: Understanding how to Motivate the Player

While theories of behavior provide explanatory power for behavioral antecedents, theories of motivation provide guidance in how to create interventions that optimize the learner's satisfaction with, and impact of, the learning experience. Of relevance to game developers are Keller's ARC model and Malone's theory of motivation.

ARCS Model

The ARCS model (Keller, 1987) was initially drafted as a method for improving the motivational appeal of educational materials. It was first used in solving motivational problems in instructional materials and methods (Keller and Kopp, 1987), and in computer assisted instruction (Keller and Suzuki, 1987). The model has great facility as a guide for game developers. Features of the model include four conceptual categories of Attention, Relevance, Confidence, and Satisfaction; a set of strategies to use to enhance the motivational appeal of instruction; and a systematic design process (motivational design). The model is aimed at bringing motivational design to the forefront in line with instructional and content design. *Attention.* Attention is typically a first phase in successful information processing and learning. In the instructional context the instructional material must gain the student's attention, sustain it, and focus it on the instruction. This involves attending to the sensation-seeking needs of students (Zuckerman, 1971) and arousing their knowledge-seeking curiosity (Berlyne, 1983). The nature of the player will also influence message effectiveness. Sensation seeking is a personality trait characterized by a search for novelty, thrill seeking, and impulsive decision making that influences the attention to, processing of, and comprehension of messages. Gaming has traditionally been associated with excitement. Sensation seekers prefer intensity and stimulation in their experiences and are more likely to indulge in risky behaviors (Palmgreen and Donohew, 2002). Messages with high sensation value are more likely to be effective among this audience and have been successfully used in discouraging drug use and promoting safe sex (Stephenson and Southwell, 2006).

Keller describes attention strategies including instilling incongruity or conflict, providing concrete yet varied information, applying humor, and encouraging inquiry and participation from learners. *Relevance*. Relevance relates to the meaning of content to the learner, but Keller argues that it is not entirely content driven and that relevance can come from the way something is taught and can relate to the learners need for affiliation (i.e. group process) or the need for achievement (i.e. challenging goals). Fulfilling these goals is pertinent to providing relevance. Relevance strategies include working from the learner's experience, clarifying the present worth and future usefulness of the information, matching the information with needs, modeling its relevance, and providing learner choice in what information is relevant to their needs. *Confidence*. Confidence in the ARCS model relates to causal attributions (see Attribution Theory above) and self-efficacy (see SCT and TTM above). Confident people tend to attribute the causes of success to things such as ability and effort instead of luck or the difficulty of the task (Weiner 1974). They also tend to believe that they effectively accomplish their goals by means of their actions (Bandura, 1977; Bandura and Schunk, 1981). Confidence strategies are designed to help the learner form the impression that some level of success is possible if effort is exerted. Confidence building strategies include clarifying learning requirements, providing appropriate difficulty, setting learning expectations, training positive attributions, and building self-confidence. The implications for providing personal control are similar to Malone's notions for providing challenge and student control (discussed below). *Satisfaction*. Satisfaction relates to the student feeling good about his/her accomplishments. The establishment of external control over an intrinsically satisfying behavior can increase enjoyment of the activity (Lepper & Greene, 1979) To instill satisfaction it is important to define the task and reinforcement schedule but provide appropriate contingencies without over controlling, and to encourage the development of intrinsic satisfaction. Satisfaction strategies include providing natural consequences for the learner's actions as well as unexpected rewards and positive outcomes for appropriate responses, allowing the learner to overcome negative influences, and to provide scheduling. The development model includes 4 principle steps of Define (classify the problem, analyze audience motivation, prepare motivational objectives); Design (generate potential strategies, select strategies); Develop (prepare motivational elements, integrate with instruction); and Evaluate (conduct developmental try-out, assess motivational outcomes) (Keller, 1987).

Malone's Motivation Theory

Malone's early work on motivation suggested three relevant factors: challenge, curiosity, and fantasy (Malone, 1981a. 1981b), to which student control was later added (Lepper and Malone, 1987). The essence of this theory is that instruction will be considered intrinsically motivating if students consider it to be "fun". Intrinsic motivators, things inherent in the instruction, are important in that they motivate the student. Extrinsic motivators, things that are external to the instruction, can diminish student interest because the student's goal can become the reward rather than the learning. *Challenge* relates to task difficulty and is represented by an activity having a goal whose outcome is uncertain. Like goal setting, the element of challenge can increase persistence in learning. The challenge should have a variable difficulty level so that there is uncertainty as to whether it will be achieved or not. The goal should be obvious and personally meaningful. Success over difficult goals is highly reinforcing. However, performance feedback should be constructed so as not to diminish the players self esteem. Challenge must be consonant with student behavioral capability and should increase with mastery (Lepper and Malone, 1987). *Curiosity* is defined as either sensory (e.g. visual or auditory effects) or cognitive (e.g. surprising information that conflicts with the student's existing knowledge, is contradictory, or in some way incomplete) caused by 'surprises' or 'novelty'. This can contribute to arousal or dissonance, encouraging the player to seek new information or mastery to remedy the conflict. The instruction becomes more constructive when it helps the learners remove the misconceptions that caused them to be surprised initially. Progressive complexity can maintain curiosity. *Control* relates to three factors of contingency, choice, and power. Contingency occurs when what the game does is a result of the players actions and responses. Choice is provided when the player can elect sequence and/or parameters and functions from menus, and options. Power relates to the motivation provided when the actions of the player have important or meaningful consequences. Granting control to the learner makes the educational process contingent on the learner's choice of actions and implies an internal locus of instructional control (Milheim, 1991). Providing the learner with moderate control connects the instruction to needs, goals, and motives of the learner and increases his or her expectation of success, thereby contributing to increased motivation (Lepper Malone, 1987; Milheim, 1991). However, providing excessive learner control could run this risk of reducing access to information, increasing time on task for **equivalent achievement** (Pollock, 1990). Reports suggest that control should vary according to individual abilities, with the game initially presenting most material at an appropriate level for novices and students with lower aptitude and giving them extensive opportunity to practice (Alessi and Trollip, 1991). *Fantasy* is provided when players imagine themselves in a situation or which include vivid and realistic images of an imaginary context or event. Malone describes extrinsic fantasies in which the fantasy depends on the use of the skill but not vice versa (e.g. Hangman can use words or sums as the learning subject). Intrinsic fantasies are those that depend on the skill and the skill relies on the fantasy. In this situation problems may be presented in terms of fantasy world elements (e.g. business skills developed in the context of a lemonade stand). These games usually indicate how the skill can be used to accomplish some real-world goal. Malone also emphasized the involvement of other people, either cooperatively or competitively, in increasing motivation of instruction.

INTERVENTION MAPPING: A DEVELOPMENT MODEL TO INCORPORATE THEORY AND EMPIRICAL EVIDENCE INTO GAME DESIGN

The acceptance of theoretical models as providing a practical advantage in designing games for behavioral change is only the first step for the game developer. A fundamental issue in the practice of game design becomes how to successfully apply theory into game development, implementation, and evaluation. Instructional design models have in common a series of production steps that include analysis of the problem, specifying required outcomes, designing the system, creating the system, and testing and revising the system. Commercial

Figure 3. Intervention mapping design sequence illustrating development flow

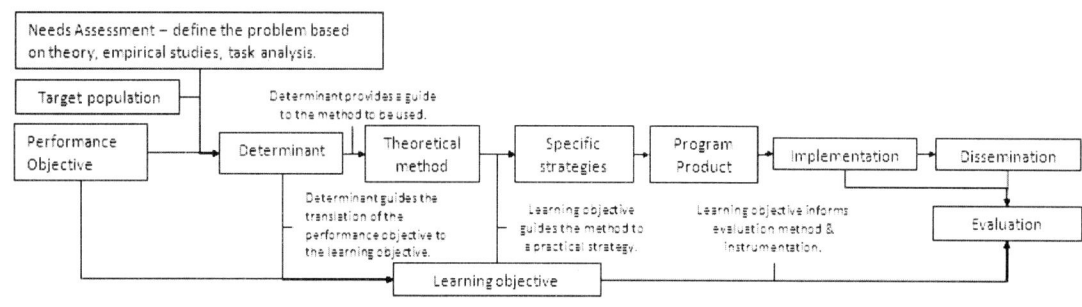

Source: Adapted from Shegog, 1997 and Bartholomew et al. 1998.

game design typically follows a similar sequence including new game proposals, multiple idea generation, concept selection, game development, game testing, and game launch (Thompason et. al., 2007). Varied nuanced and successful frameworks have been drafted for health promotion and instructional design that correspond to this basic approach including PROCEDE (Green and Kreuter, 1980); ARCS (Keller, 1987b); Instructional Design (Gagne et. al., 1988); Tailoring processes (Kreuter et. al, 2000); and simulation development (Reigeluth, 1989). A significant challenge for serious health games is that they focus on health (educational and training) objectives in addition to the gaming (entertainment) objective.

Intervention Mapping is a framework for systematic health promotion program planning, implementation, and evaluation that offers great utility for the development of serious games for health (Bartholomew et. al., 1998, 2000a, 2000b, 2001, 2006; Shegog et. al. 2001; Cullen et. al., 2000). By following the Intervention Mapping framework serious games developers can generate a series of milestone design products including a description of the health problem, critical target behaviors, learner characteristics, setting, theoretical determinants of the target behavior, learning objectives, theoretically derived methods and strategies to impact determinants and optimize player motivation, and plans for implementation, evaluation, and dissemination (Shegog et. al., 2001) [Figure 3] Each intervention element has a clear justification and a logical, conceptual basis. Therefore, it is much more likely that researchers and developers will be able to understand not only who changed, but also and more importantly, *how* the process of change worked. Intervention mapping has been used successfully in the development, implementation, and evaluation of a variety of behavioral health interventions including chronic disease management, disease prevention, and health promotion programs (Bartholomew et. al., 2006).

Case study: 'Watch Discover Think Then Act' (WDTA)

A successful clinic-based tailored asthma management game prototype using an adventure game treatment that simulates asthma management will be used to exemplify the intervention mapping approach. The game was found effective in impacting behavior and behavioral determinants in adolescents (8 – 16 years) and impacting clinic return rates (Bartholomew, 2001; Shegog, 2000). Asthma continues to be a major chronic disease in the United States among children and a primary cause for children's hospital admissions and school absences (Weiss et. al., 1993; Akinbami, Moorman, Garbe, and Sondik, 2009). The medical and social consequences of asthma are more severe in urban, low-income, African American and Hispanic children (Weiss et. al., 1993; Akinbami et. al., 2009; James and Rosenbaum, 2009). Many urban minority children may not get adequate preventive treatment for asthma. Inner-city school children are likely to receive asthma care in emergency rooms rather than from primary care providers (Mak et. al., 1982; Akinbami et. al., 2009). Inner-city minority children are not very likely to be treated with anti-inflammatory medications; the reported rates vary from 11% to 17% of children with moderate to severe asthma (Finkelstein et. al., 1995; Huss et. al., 1994; Homer et. al., 1996). Hispanic and African-American children are less likely to have used either beta-agonist or steroids prior to hospitalization for asthma compared to non-Hispanic whites.

Early studies of computer-based pediatric educational applications through the mid nineties (*Asthma Command©*, *Asthma Control©*, *Air Academy: The Quest for Airtopia* ™, *STARBRIGHT World*™) reported change in knowledge about asthma and its management and, with advances in the theoretical and empirical underpinnings of self-management behavior, determinants of self-management behavior such as self-efficacy (*Bronki the Bronchiasaurus*™)(Shegog and Sockrider, 2009). The goal for WDTA was to provide a computer-based self-management asthma education program for children with the following specifications:

- Stand-alone application that could be implemented in an asthma care medical setting.
- Simulated experience that makes use of behavioral change methods from social and behavioral science theory to elicit change in determinants of self-management behaviors including the child's behavioral capability, self-efficacy, and attributions.
- Individualized educational experience containing real world asthma management scenarios that are responsive to each child's own data.
- Training in asthma-specific skills of medication taking, trigger avoidance, and peak-flow use.
- Training in self-regulatory problem-solving skills.
- Motivational without embedding all asthma management in a fantasy context.
- Output to the physician and family regarding the child's asthma self-management capabilities, to enhance the team approach to asthma care.

WDTA Game Process and Graphic User Interface

Watch, Discover, Think, and Act (WDTA) is a computer-based game that provides tailored self-management education to inner-city upper elementary and middle school-age children with asthma in the context of primary-care clinics and physician offices. After the encounter with the physician a clinic staff member assists the child to input name, age, duration of asthma, and personal asthma symptoms, environmental triggers, medications, and personal best peak flow. This information creates the child's asthma profile. The child chooses a character (a 12 year old child) whose asthma they will manage during the game and a coach who represents a slightly older teenager who has learned to manage asthma. The child can choose whether the character and coach are male or female, and choose between African American and Hispanic ethnicity. The child is introduced to the goal of the game which involves a mission to rescue plans for anti-pollution technology from the castle of Dr. Foulair. The child must move a character through three real-life scenarios with multiple scenes (home, school, and neighborhood) and then to Dr. Foulair's castle. In order to progress from one scenario to the next, the child

Figure 4. 'Watch, Discover, Think, then Act' example screen captures

Example screen scenarios and player functions: Character selection (a); Mission brief (b); Home scenario with health parameters (on right))(c); School scenario with 'Discover' options (on right)) (d); Neighborhood scenario with 'Act' options (on right)(e); Castle scenario with mission tool options (on right)(f).

Source: Watch, Discover, Think, and Act funded by a contract from the National Heart, Lung, and Blood Institute, NIH (Contract No. N01-HO-39220). Lead contractor Macro International, Calverton, MD. For further information contact Dr. Ross Shegog.

must successfully manage the character's asthma by following the four self-regulation steps of watching (monitoring symptoms, environmental triggers, medication taking, and appointment keeping), discovering (deciding if an asthma problem exists and its probable cause), thinking (deciding on a list of possible actions), and acting (choosing an action such as taking medicine, removing or avoiding triggers, getting help). The child must also collect mission handbook pages to complete tutorials on asthma and click and drag tools to be used in the castle. The game scenarios include 18 real-world and 4 castle situations that present the problems that inner-city children with asthma must deal with successfully in order to manage their disease. The game screen shows a scenario scene (left portion of the screen) containing environmental asthma triggers (dust, environmental tobacco smoke, and cleaning chemicals), tools (the roller blades by the coffee table), and mission handbook tutorial pages (Figure 4c). Tutorial components contain information developed de novo as well as drawn from NIH source material and include video segments depicting medicine taking procedures (USDHHS, 1997; NHLBI, 1992). Self-regulatory icons for watch, discover, think and act are vertically aligned at the right of center. Below these, a fifth icon represents a cellular phone with which to contact the coach. The health status boxes on the far right of the screen show the character's symptoms, peak flow, medication taking, appointment scheduling, and personal triggers; they appear during the "Watch" step of the asthma self-management process. WDTA tracks each child's progress so that they can resume the game where they left off in their previous visit. The data output stage provides information for the child, the parents, and the health care provider. Players receive constant feedback from the game as to their progress, and at the end of each session children receive a certificate that congratulates them on the progress made and reiterates the four steps of self-regulation. To guide further asthma education, the parents and physician receive a progress report that indicates the scenario the child reached, the child's use of the self-regulation steps, and tutorial scores. WDTA was originally developed for the Apple platform using MacroMind Director 5.0 ® authoring software.

WDTA Development

Development of WDTA required an identification of asthma self-management behaviors, an understanding of determinants (or antecendents) of these self-management behaviors, identification of appropriate methods to achieve change in these determinants, and identification of appropriate educational strategies to operationalize these methods (Bartholomew, 2000b; Shegog, 2006).

Asthma Self-Management Behavior

For the child with asthma, the aims of management are to control symptoms, prevent exacerbations, attain normal lung function, and maintain normal activity levels. The National Institutes of Health have suggested that the basis of managing persistent asthma in children is control of the inflammatory process that underlies the disease (Halfon, 1993). From the family's perspective, the behaviors required to manage asthma include monitoring symptoms, administering medicine with a metered-dose inhaler or powdered dose inhaler (for both control and exacerbations), and removing and avoiding triggers (USDHHS-NAEP, 1997; 2007). Some physicians suggest the use of a peak flow meter to aid in symptom monitoring. Unfortunately, practice of these asthma-specific behaviors alone are insufficient to provide the child and family with the ability to make independent decisions about what to do under varied circumstances. Asthma's varied clinical pattern makes it impossible for the physician to discuss every contingency when instructing patients, and families must move from adhering to medical regimens to applying complex cognitive-behavioral self-regulatory skills to successfully manage asthma (Clark, 1989; Creer, 1990a; Creer, 1991; Clark, 1993; Clark, 2009)

Asthma education programs have traditionally focused on teaching specific asthma skills to patients and have been able to impact a variety of factors important for asthma management, including improvement in patient's knowledge, feelings of competence, use of self-management behaviors, and adherence to medical regimens. These programs have also been shown to decrease emergency room and unscheduled doctor visits but not hospitalization rates (Klingelhofer,

Figure 5. Framework for asthma self management behaviors and their predisposing, enabling, and reinforcing factors

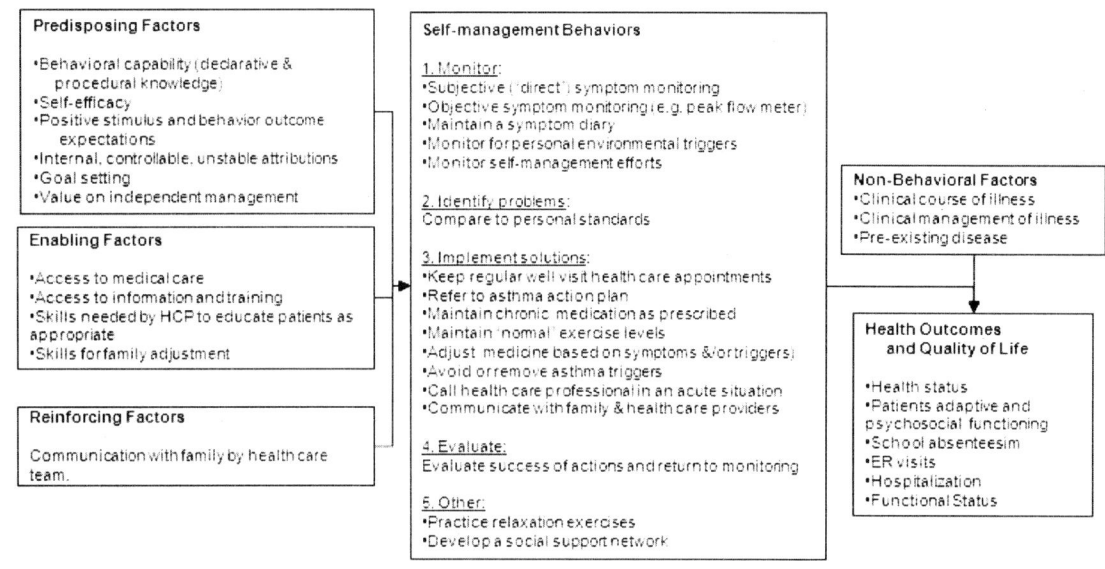

Sources: Adapted from Shegog et. al., 2004 (© 2004, International Review of Asthma. Used with permission.)

1988; Rolnick, 1988; Wigal, 1990). Despite some successes, these programs had shortcomings. For example, instruction has been uniformly presented regardless of individual needs, and self-management training has not reached most children with asthma because it usually has not been offered during regular medical care visits (Creer, 1990b; Kotses, 1991).

Recommendations had been made for a second generation of asthma education programs that are individualized and integrated into the provision of medical care (Creer, 1990). In addition, there had been a call for programs more closely based on the self-management construct of self-regulation (described under Social Cognitive Theory above). The critical processes of self-regulation include goal selection, self-monitoring, self-judgment (information processing and evaluation), decision making, action, and self-reaction (self-evaluation) (Creer, 1991; Clark, 1993; Clark and Zimmerman, 1990; Thorensen, 1983). Self-regulatory skills have become a relevant aspect of patient education because people who manage disease must be able to monitor their own progress and solve disease-related and adjustment problems (Clark and Starr-Schneidfraut, 1993; Clark and Zimmerman, 1990). They must judge medical recommendations in light of their own response to therapy and their efforts to adhere to those recommendations. These self-regulatory behaviors provide a cognitive framework for the performance of asthma-specific behaviors, such as medication taking, using a peak flow meter, or removing and avoiding environmental triggers (Figure 5).

Determinants of Asthma Self-Management

Determinants of a behavior are those factors that must be present for that behavior to occur (Bartholomew et. al., 1998). Determinants identified from the literature related to chronic disease management included behavioral capability, (knowledge of what self-management behavior is required at a given time and the procedural knowledge to perform that behavior), self-efficacy (belief that one has the skill and ability necessary to perform a behavior in a variety of circumstances and in the face of various obstacles) (Bandura, 1986), and attributions (the belief that self-management behavior is controllable and is subject to personal effort)(Weiner, 1985). Determinants of successful game playing were also accounted for with a consideration of the importance of intrinsic motivation (Lepper and Malone, 1987; Lepper and Cordova, 1992; Parker and Lepper, 1992)

Describing Program Objectives

The critical self-management behaviors represent the behavioral outcomes for the program. These are termed "performance objectives" in Intervention Mapping terminology and can be cross-referenced with the determinants of behavior described above. The result is a matrix of "proximal" program objectives (Table 6). The resulting cells of each matrix contain learning objectives. The learning objectives are measurable statements about what the child needs to be able to do in order for the performance objective to be met. These provide a blue-print for the program content. In the abbreviated example matrix provided, to optimize the chance for the child to monitor symptoms of asthma using objective measures (Table 6, row 2) he/she would need behavioral capability, self-efficacy, and positive attributions. The child would need to be able to identify symptoms and state correct procedures for objective monitoring (behavioral capability), state confidence in using a peak flow meter (self-efficacy) and attribute success or failure of peak flow use to personal effort (positive attribution).

Theoretical Methods and Educational strategies

Methods are theoretically derived techniques used to influence behavioral determinants while strategies apply methods in ways that are compatible with the target population and with the context of the intervention (Bartholomew et. al., 1998). For example, theoretical methods to influence the determinant of self-efficacy include symbolic modeling, goal setting, skill training and application (in asthma-specific skills and self-regulation), reinforcement, and verbal persuasion. If symbolic modeling is the method chosen to increase the self-efficacy of a child with respect to monitoring asthma triggers, then an appropriate strategy would be to show older peers successfully performing this skill and receiving positive benefits from it. Methods and strategies in WDTA are herein described (Table 3).

a. WDTA Introduction

Goal Setting: Goal setting is of theoretical importance in providing a measure of self-management success and to subsequently impact self-efficacy. Goals also provide a motivational impetus to self-management (Strecher et. al., 1995). The overriding goal for the child when playing WDTA is to win the game. To do this, however, requires that the child meets the management-related goals of the program: to be able to manage asthma in a variety of circumstances through the application of self-regulatory skills and appropriate decision making. The child's success in the game is linked to achieving these management goals.

Personalized Information: Personalized contexts in education improve learning, facilitate memory, and increase task motivation (Anand and Ross, 1987; Rumelhart and Ortony, 1977). Children given personalized arithmetic materials in computer-based lessons have demonstrated increased retention, problem solving, and information transfer compared with controls using non-

Table 2. Example of the matrix of proximal program objectives with associated items from the measurement instruments created to assess knowledge, self-efficacy, and attribution

	Behavioral determinants		
Performance Objectives	Knowledge	Self-efficacy	Attributions
Monitor symptoms of asthma directly.	Identifies symptoms related to asthma using direct methods. States those symptoms associated with his/her asthma. --- • An upset stomach and dizziness are common symptoms of asthma. (Y/N)	States confidence about assessing asthma symptoms directly. --- How sure am I that I can: • Tell if I am wheezing. • Know that I am having asthma problems before they get bad.	Attributes direct symptom monitoring success or failure to self effort. Attributes direct symptom monitoring failure to unstable causes. --- You haven't had to see the doctor for bad asthma all month. This past month: A. Your parents noticed breathing problems before your asthma got bad. B. You noticed breathing problems before your asthma got bad.
Monitor symptoms of asthma using objective measures (i.e. peak flow).	Identify symptoms related to asthma using objective methods. State the correct procedure to use and interpret a peak flow meter. --- • A "personal best" peak flow is the best peak flow I get when I am sick with asthma. (Y/N) • When I use my peak flow meter, I should blow as hard as I can. (Y/N)	States confidence about assessing asthma symptoms using a peak flow meter. --- How sure am I that I can: • Use my peak flow meter every day. • Tell if my peak flow number is smaller than usual.	Attributes objective symptom monitoring success or failure to self-effort. Attributes objective symptom monitoring failure to unstable causes. You didn't notice your breathing problems until they got bad. This past week: A. Your parents didn't remind you to use your peak flow meter to check your asthma breathing. B. You forgot to use your peak flow meter to check your asthma breathing.
Avoid asthma triggers.	Identify triggers of asthma. States methods to avoid triggers of asthma. --- • Asthma breathing problems can happen if I get infections like colds, flu, or sore throat. (Y/N) • If exercise makes me wheeze I should always stay away from exercise. (Y/N)	States confidence about avoiding asthma triggers. --- How sure am I that I can: • Stay away from things that make me wheeze. • Stay away from cigarette smoke.	Attributes trigger avoidance success or failure to self effort. Attributes trigger avoidance failure to unstable causes. --- Your asthma has been in good control for weeks. Why might this be? A. You try to stay away from things that cause asthma. B. Your parents make sure you stay away from things that cause asthma.
Maintain chronic medication as prescribed.	Identifies asthma medicine (types). States when different asthma medicine is used. --- • A steroid is an asthma medicine that stops swelling in airways. (Y/N) • If I start to have asthma breathing problems I should only take medicine when I start wheezing. (Y/N)	States confidence about taking asthma medicine. --- How sure am I that I can: • Check to make sure I am taking my medicine the way the doctor or nurse has taught me.	Attributes medicine taking success or failure to self effort. Attributes medicine taking failure to unstable causes. --- You have to sit out of the basketball game after you start to wheeze. You might have started wheezing because: A. You don't usually remember to take your asthma medicine before basketball. B. Nobody reminds you to take your medicine before basketball. While walking to school you start to cough. You might be coughing because: A. It's too difficult to take your asthma medicine every day. B. This week it's been too difficult to take your asthma medicine every day.

Source: Adapted from Shegog, 1997.

Table 3. Example learning objectives, behavioral determinants, theoretical methods, and strategies

Learning objectives	Behavioral determinant	Theoretical methods	Operationalization of strategies	Themes & motifs
Player knows how to use the asthma self-regulatory process & asthma specific skills	Knowledge/ behavioral capability	Symbolic modeling Skill building Self-monitoring Self-directed application of acquired skills Reinforcement	Role model character & coach Tutorials Graphic demonstration of asthma specific skills Visual presentation of personal data	Watch-Discover-Think-Act 4 step metacognitive self-regulation process. Mission quest Managing asthma as part of the mission Acquiring information & skills as part of the mission Saving the city by retrieving plans from Dr. Foulair's castle.
Player feels confident in undertaking asthma self-management (self-regulatory behaviors & asthma specific skills).	Self-efficacy	Symbolic modeling Self-evaluation Reinforcement Goal setting Verbal persuasion Self-directed application of acquired skills	Positive feedback (via coach, health status indicators, & success in the game) Goal to maintain character's health Progress pending performance success	
Player perceives asthma self-management as subject to his/her own personal effort.	Attribution	Attribution retraining Symbolic modeling	Coach relates asthma self-management to personal effort (internal, controllable, dimensions). Models are older peers & not authority figures.	

Source: Adapted from Shegog 1997.

personalized materials. As the meaningfulness of the contexts increased from abstract to concrete to personal, performance improved on achievement measures (Anand and Ross, 1987). In addition to improving learning, personalized contexts increase task motivation by describing situations of high interest to learners (Rumelhart and Ortony, 1977). WDTA content was personalized by tailoring the game character to the player's own symptoms, peak flow personal best, and triggers. When the player accesses information about the character during play they see their own health information and are able to alter those health parameters by applying self-management principles.

Fantasy for intrinsic motivation: Fantasy elements of instruction have been shown to promote more efficient and elaborate problem-solving strategies, higher levels of performance, and transfer of learning in 3rd-, 4th- and 5th- grade children (Lepper and Cordova, 1992; Parker and Lepper, 1992). Such effects are attributed to the greater intrinsic motivation such fantasy contexts can provide. The mission to Dr. Foulair's castle provides a fantasy that "book-ends" realistic simulations of home, school, and neighborhood. While players actively manage asthma, they are also picking up items to be used in the final fantasy castle scenario. A motivating fantasy element is thereby woven into the entire application, but even in the fantasy castle, at the game's conclusion, the child still has to manage asthma in order to proceed.

Symbolic Modeling: A central tenet of Social Cognitive Theory is observational learning, by which individuals can acquire cognitive skills and new patterns of behavior by observing the performance of others (Bandura, 1986). The method of symbolic modeling is operationalized by both character and coach. In the introduction stage the child selects from among four characters and chooses a coach. The character and coach models used in WDTA have been chosen with serious consideration to their ability to influence the child's behavior. Using the child's personal asthma data for the character and allowing the child to choose the gender and ethnicity of the character and coach provides role models who can be similar to the learner (Bandura, 1986; Slater, 2002). Models were chosen to be peers rather than authority figures to engender children's beliefs that asthma self-management is within their control (internal attribution for success) and that success is subject to personal effort (controllability of factors leading to success). These attributions can contribute to enhanced self-efficacy for self-management (Bandura, 1986).

Multiple Modalities: Providing information through more than one modality has been shown to increase learner recall (Paivio, 1986). WDTA, like other multimedia games, presents text, audio (narration, character voices, rap song, and sound effects), and visual cues (icons, graphics of environmental triggers, health measures, and tools) to transfer information to learners (Rieber, 1991a, 1991b). These multi-media capabilities also facilitate learning by children whose reading levels may be low.

b. WDTA Game Body

Skills Training Reinforcement: Typically, skill training for the learning of complex behaviors involves demonstration of the skill, opportunity for practice of component parts of the skill and the total skill, feedback and reinforcement, and practice of the skill under varied, challenging circumstances. WDTA includes opportunities to learn both asthma-specific skills and self-regulatory skills through simulations and tutorials. Self-monitoring, self-judgement, and self-evaluation, as previously discussed, are self-regulatory processes that can provide a cognitive framework for using the many specific skills the child with asthma must have to control his or her disease. To provide skills-training in WDTA simulations, buttons for watch, discover, think and act provided visual cues for each step (Figure 4). These self-regulatory buttons are prominently located on the screen, and they separate the graphics that depict the scenario from the health status indicators. This geographic placement helps children consider the appropriate self-regulatory phase prior to choosing specific asthma actions.

The learner must select a self-regulatory button before an action can take place. When learners want to monitor symptoms, peak flow, medication taking, appointment keeping, or triggers, they click on the Watch button. Similarly, to decide if the character has a problem with asthma and to discover possible causes of the problem they click on the Discover button; to choose from a menu of possible solutions, the Think button; to take action, the Act button. With Act, children can have their characters take medicine, remove or avoid asthma triggers, make an appointment with the doctor, change activity, get help from parents or teachers, or go to the emergency room or doctor's office.

Natural feedback occurs with changes in the character's symptoms, peak flow, medicine adherence and appointment scheduling. After the child chooses a self-management action, the health parameters are reset to show the result of the action chosen, and the child can evaluate the effectiveness of the action by clicking on the Watch button. The learner can thus experience immediate feedback for self-management behavior. The character's health parameters are also influenced by the number and type of environmental triggers present in a given scenario, which are triggers for that particular child. These cause-and-effect relationships have important theoretical implications in influencing learner outcome expectations. Stimulus-outcome expectations (the relationship between the environmental stimuli and asthma symptoms) and behavior-outcome expectations (the relationship between specific behaviors and asthma symptoms) are important in the self-management of asthma.(Thorensen and Kirmil-Gray, 1989). The computer program allows the results of self-management behaviors to be experienced immediately. Further, expectations can be tested repeatedly to determine in what situations a certain self-management behavior is effective.

Artificial reinforcement is provided in the form of messages from the coach about the learner's self-regulatory choices and from the score associated with the child's self-management. This score is a composite of self-regulatory logic in arriving at a specific action (i.e., how well the child watches, discovers, thinks, then acts) and the usefulness of the action chosen. Accessing each of the self-regulatory buttons in the correct order provides maximum self-regulatory points. Points are also provided for making correct decisions within each of the four self-regulatory steps. This includes correctly determining severity of symptoms and peak flow, identifying all personal triggers in a scenario and other causes of the character's asthma, choosing all useful solutions, and choosing appropriate actions to control asthma. For example, if preventive medicines have not been taken and the character has symptoms, it is logical to decide that there is a problem, that a possible cause is not having taken preventive medicine, that a possible solution is to take this medicine, and then to choose to act by taking medicine. In such an instance, there is a correspondence across choices that are appropriate and accurate and maximum points are rewarded.

The child is restricted from progressing through the scenarios without self-managing the character's asthma. If the character is not symptom-free, then the child is unable to continue. Further, the child must pick up mission handbook pages (i.e., do tutorials) and pick up tools before being able to move on to the next scenario. The tutorials are didactic instruction segments that provide skill training in measuring peak flow, taking medicine, and other ways to help oneself. In these segments, reinforcement is provided when the player answers multiple-choice questions.

Learner Control, Challenge, and Curiosity: Control, challenge, and curiosity (as well as fantasy) have been suggested as relevant to increasing student interest in the material being learned (Lepper and Malone, 1987). A central tenet of the program is that the child can exert control over the character's health. The program operationalizes challenge and curiosity through the realistic scenarios, which increase in novelty as the child moves from home to school to neighborhood through the tutorials, through the collection of objects in preparation for the castle scene, and through the obstacles present in the castle scene. In WDTA the child has control over pacing, sequence, and content. The child paces movement through the scenarios and spends more time in more difficult content areas until success is achieved. A child must stay within a given scenario until their character is well enough to move on. Children can also choose the sequence in which they perform the self-management steps of watch, discover, think, and act and control the order in which they perform activities in a given scenario: accessing tutorial information on asthma-specific skills, self-managing asthma, and collecting tools.

Persuasion: Verbal persuasion is described in Social Cognitive Theory as one method of influencing self-efficacy (Bandura, 1986). Although it is not the most powerful method, persuasion is simple to provide and can be useful to augment other efficacy-building methods. The coach provides feedback about the child's actions and also encourages his or her management attempts. The written feedback provided to the physician (see below) is designed to enable the physician to encourage the child's performance in specific self-management, translating persuasion beyond the confines of game play into the clinic context.

Attribution Retraining: Attribution, how one attributes success or failure in an endeavor, is an important contributor to self-efficacy (Bandura, 1997). Success will boost self-efficacy expectations if it is attributed to one's own effort rather than to external or uncontrollable factors. Children may see taking care of their health as something that others do for them – parents, doctors, and nurses. However, for children to begin to manage their own asthma, they need to think of caring for their health as something they can do themselves (internal control). Furthermore, when they try to manage asthma, they need to attribute success to internal causes – "I am the kind of person who can manage asthma" – and to effort – "When I work at it, I can manage my asthma". On the other hand, they need to attribute failure or lapses in management to unstable causes (which are subject to change). WDTA provides the child with successes resulting directly from his or her own self-management efforts. The coach also operationalizes this concept because he or she is someone who has managed asthma rather than leaving this task to an external agent.

c. WDTA Conclusion

The development of a team approach to asthma care is an objective outlined in the National Asthma Education Program Guidelines (USDHHS-NAEP, 1997, 2007). The asthma management team typically includes the child with asthma, parents, and a health care provider and can extend to the school to include teachers and school nurses. WDTA is designed to facilitate this team approach by providing information to the child's parents and physician.

Reinforcement: Output to the physician is designed to cue physician behavior in the clinic setting. The program provides a printed report containing information on the learner's self-regulation and knowledge scores. These alert the physician to knowledge or skills areas that need bolstering or reinforcement. WDTA offers common ground on which the child and the physician can discuss asthma. Further, the physician can provide verbal persuasion regarding self-management and positive reinforcement for the asthma self-management that the child has demonstrated in the game. Written output from WDTA to the children and parents congratulates the child on progress and lists the child's symptoms and triggers in the context of the self regulation process. The health care provider can further reinforce the player for their "winning" management and getting through the castle.

Evaluation Studies

WDTA has been evaluated in two research studies, a study of impact on educational variables in a laboratory setting and a study of both impact and outcome variables in a clinical setting (Shegog, 2001; Bartholomew, 2000b). Results of the impact study with 75 children, 9-13 years, suggested that WDTA was motivational and educational (Shegog, 2001). Children who used the simulation were able to reiterate the components of a 4-step problem solving framework, describe behavioral strategies to prevent asthma episodes, describe behavioral strategies to treat asthma symptoms, had confidence to carry out self-management behaviors, and had perceptions of their own self-management autonomy to a significantly greater degree than children who did not use WDTA (all $p<0.05$).

Results of the evaluation study at the clinical sites with 133 children, 8-14 years, indicated that children who were older and those who scored higher at pretest improved their knowledge of how to manage asthma (Bartholomew, 2000b). Among children in the intervention group, both those with conservative estimates of their self-efficacy at pretest and those with higher pretest scores, self-management improved. Children in the intervention group also had a lower rate of hospitalization, and improved functional status. Use of the program was associated with a decrease in symptoms for those children whose symptoms were generally milder (all $p<0.05$).

For children to use a stand-alone intervention like WDTA it is important that it motivate their interest. The motivational appeal of the program has been shown in clinic sites by the way these young asthma patients focus on the program even in bustling hospital clinics, by favorable oral reports about the program from children, and by increases in return rates to the clinic by children using the computer program (p<0.05) (Bartholomew, 2000b).

FUTURE RESEARCH AND DEVELOPMENT FOR SERIOUS GAMES FOR HEALTH

The burgeoning of serious games for health as a viable research and therapeutic / prescription is relatively recent. Early sporadic forays into the development and testing of games has morphed into more unified and collaborative efforts to investigate this medium, backed by research funding opportunities. To date, serious games for health have shown promise to date in impacting behavioral determinants and behaviors in chronic disease management and disease prevention though the number of published studies remains modest (Baranowski et. al., 2008). The use of serious games in the context of health is expected to increase. In fact, research on serious games is likely to lag behind their development and implementation, a common challenge in any rapidly evolving technological field.

In order to truly estimate the effectiveness of serious games it is important that future development and research efforts take into account the importance of (1) theoretical foundations in design and development, (2) rigorous evaluation methods and study designs to determine effectiveness, (3) the importance of contextual exigencies; (4) studies that address cost-benefit and cost effectiveness; (5) studies that elucidate the important aspects of theory application in game design and the importance of the message/receiver/channel relationship.

If one is designing a serious game to impact health behavior then one needs to be serious about the use of behavioral theory to inform development. In the limited literature to date effective programs have been informed by behavioral theory but only a handful of theoretical frameworks are predominant. There remains a great deal of opportunity for developers to employ an array of behavioral, communication, and motivational theories as yet little utilized. Guiding developers on the best way to produce programs that are theoretically and empirically based has become an important research agenda (Alessi and Trollip, 1991; Rhodes et. al., 1997; Revere and Dunbar, 2001). Behavioral theories and models attempt to explain why people behave the way they do by identifying factors that underlie adoption or rejection of a given behavior. Theory can be the basis for specifying program objectives, health behaviors, cognitive determinants of behavior (such as knowledge, attitudes, social perceptions, and self-efficacy), change methods, and evaluation and measurement protocols (Alessi and Trollip, 1985; Lieberman, 1997; Rhodes et. al., 1997; Revere and Dunbar, 2001). Relatively few research papers that describe serious games provide a clear discussion of the theoretical foundations of proposed change methods, in other words matching methods and strategies to the behavioral determinants they attempt to change. While serious games are fun to play and often focus on relevant health issues, publications reporting their evaluation tend to emphasize operationalization of game components or production effects rather than theoretical foundations, methods, and strategies. It is insufficient for serious games for health to impact knowledge alone. This is necessary but not sufficient to determine behavior change. If serious games are going to enhance health behaviors then game design must take into account how behavioral determinants will be impacted and how this will be translated to real world application (Shegog, 2006). Future empirical pursuits include that of user-media-message interactions to understand effective educational gaming strategies rather than comparisons of different media approaches, economic analyses regarding the cost and time benefits of providing serious games in clinic, hospital, and educational settings, studies that examine long term behavioral follow-up in evaluation studies, and diffusion studies that examine the integration of serious games into the educational repertoire of health care providers and curricula (Adler and Johnson, 2000; Street and Rimal, 1997). Baranowksi et. al. (2008) set a research agenda for serious games including what factors provide greater learner attention and information retention, what production specifications can most adequately target different learners, and what engenders greatest motivation to increase "fun" (Baranowski et. al., 2008).

Increased rigor in the design of evaluation studies is also necessary to determine the degree to which serious games can affect behavioral outcomes. Required are designs that can suggest causality such as randomized controlled trials. Studies should also feature reliable measures of determinants of self-management behavior, performance of behavior, and health and functional status indicators. To date, few studies have considered cost-benefit and cost-effectiveness analyses. Serious games may have serious front-end financial expenditure and capital costs and may require greater time on task than traditional programs. Do they, then, provide greater cost benefits in terms of prevention and health maintenance. Also ripe for consideration are implementation barriers to serious games in environments that may not have traditionally accepted this form of educational media such as schools and clinics.

Computer-based instruction has been found motivational and efficacious but studies have indicated that it is a strategy best used as a component of a multi-faceted approach, and not as a replacement to other health educational and clinical management approaches (Revere and Dunbar, 2001; Bartholomew, 2000b; McPherson et. al., 2001; Fall et. al., 1998). This raises the issue of accounting for contextual exigencies. For example, teaching a child health behavior skills through a serious health game may be of little consequence if the enablers of behavior (e.g. parents) do not receive some form of intervention to support and reinforce the behavior or if the triggering environment (e.g. the school) is not modifiable, or if appropriate medical care is not available (Homer, 2000; Rachelefsky, 1987). This represents a potential challenge for serious games in addressing levels of health change: individual, organizational, and community – games generally have been focused on individual level change.

The field of serious games research, development, and evaluation, like that of interactive health communication in general, can call for increased rigor that employs state-of-the-art designs, adequately powered samples, valid and reliable measures of outcomes and mediators, and reporting in a way that is consistent with other randomized clinical trials to allow data for meta-analysis. With such an agenda the potential for serious games to improve health outcomes, decrease healthcare costs, and increase consumer satisfaction might be realized. The application of behavioral theory in this endeavor will contribute to optimal design, development, implementation, and evaluation, guiding the success of games in the serious objectives of a healthier community.

REFERENCES

Adler, M. D., & Johnson, K. B. (2000). Quantifying the literature of computer-aided instruction in medical education. *Academic Medicine, 75*(10), 1025–1028. doi:10.1097/00001888-200010000-00021

Akinbami, L. J., Mooreman, J. E., Garbe, P. L., & Sondik, E. J. (2009). Status of childhood asthma in the United States, 1980-2007. *Pediatrics, 123*(Suppl. 3), S131–S145. doi:10.1542/peds.2008-2233C

Alessi, S. M., & Trollip, S. R. (1991). *Computer-Based Instruction: Methods and Development.* Englewood Cliffs, NJ: Prentice-Hall.

Ammerman, A. S., Lindquist, C. H., Lohr, K. N., & Hersey, J. (2002). The efficacy of behavioral interventions to modify dietary fat and vegetable intake: A review of the evidence. *Preventive Medicine, 35*(1), 25–41. doi:10.1006/pmed.2002.1028

Anand, P. G., & Ross, S. M. (1987). Using computer-assisted instruction to personalize arithmetic materials for elementary school children. *Journal of Educational Psychology, 79*(1), 72–78. doi:10.1037/0022-0663.79.1.72

Anderson, R. C., & Anderson, K. E. (1990). Success and failure attributions in smoking cessation among men and women. *AAOHN, 38*(4), 180–185.

Bandura, A. (1977). Self-efficacy: Toward a unifying theory of behavioral change. *Psychological Review, 84,* 191–215. doi:10.1037/0033-295X.84.2.191

Bandura, A. (1986). *Social Foundations of Thought and Action: A Social Cognitive Theory.* Englewood Cliffs, NJ: Prentice Hall, Inc.

Bandura, A. (1991). Social Cognitive Theory of Self Regulation. Special Issue: Theories of cognitive self-regulation. *Organizational Behavior and Human Decision Processes, 50,* 248–287. doi:10.1016/0749-5978(91)90022-L

Bandura, A. (1997). *Self-efficacy. The Exercise of Control.* New York, NY: W.H. Freeman.

Bandura, A., & Schunk, D. H. (1981). Cultivating competence, self-efficacy, and intrinsic interest through proximal self-motivation. *Journal of Personality and Social Psychology, 41,* 586–598. doi:10.1037/0022-3514.41.3.586

Baranowski, T., Buday, R., Thompson, D. I., & Baranowski, J. (2008). Playing for real. Video game and stories for health-related behavior change. *American Journal of Preventive Medicine, 34,* 74–82. doi:10.1016/j.amepre.2007.09.027

Barry, M. J., Fowler, F. J. Jr, Mulley, A. G. Jr, Henderson, J. V. Jr, & Wennberg, J. E. (1995). Patient reactions to a program designed to facilitate patient participation in treatment decisions for benign prostatic hyperplasia. *Medical Care, 33,* 771–782. doi:10.1097/00005650-199508000-00003

Bartholomew, L. K., Gold, R. S., Parcel, G. S., Czyzewski, D. I., Sockrider, M. M., & Fernandez, M. (2000b). Watch, Discover, Think, and Act: Evaluation of computer assisted instruction to improve asthma self-management in inner-city children. *Patient Education and Counseling, 39*(2-3), 269–280. doi:10.1016/S0738-3991(99)00046-4

Bartholomew, L.K., Parcel, G.S., & Kok, G. (1998). Intervention mapping: A process for developing theory- and evidence-based health education programs. *Health Education & Behavior, Oct 25*(5), 545-63.

Bartholomew, L. K., Parcel, G. S., Kok, G., & Gottlieb, N. H. (2001). *Intervention Mapping: Designing theory- and evidence-based health promotion programs.* Mountain View, CA: Mayfield Publishing Company.

Bartholomew, L. K., Parcel, G. S., Kok, G., & Gottlieb, N. H. (2006). *Planning Health Promotion Programs. An intervention mapping approach.* Hoboken, NJ: John Wiley & Sons, Inc.

Bartholomew, L. K., Shegog, R., Parcel, G. S., Gold, R. S., Fernandez, M. E., & Czyzewski, D. I. (2000a). Watch, Discover, Think, and Act: A model for patient education program development. *Patient Education and Counseling, 39*(2-3), 253–268. doi:10.1016/S0738-3991(99)00045-2

Berlyne, J. (1983). Conceptualizing student motivation. *Educational Psychologist, 18,* 200–215.

Block, G., Miller, M., Harnack, L., Kayman, S., Mandel, S., & Cristofar, S. (2000). An interactive CD-ROM for nutrition screening and counseling. *American Journal of Public Health, 90,* 781–785. doi:10.2105/AJPH.90.5.781

Bogost, I. (2007). *Persuasive Games. The expressive power of videogames.* Cambridge, MA: Massachusetts Institute of Technology.

Bradley, B. P., Gossip, M., Brewin, C. R., Phillips, G., & Green, L. (1992). Attributions and relapse in opiate addicts. *Journal of Consulting and Clinical Psychology, 60*(3), 470–472. doi:10.1037/0022-006X.60.3.470

Brubaker, B. H. (1988). An attributional analysis of weight outcomes. *Nursing Research, 37*(5), 282–287. doi:10.1097/00006199-198809000-00005

Carroll, J. M., Stein, C., Byron, M., & Dutram, K. (1996). Using Interactive Multimedia to deliver nutrition education to Maine's WIC clients. *Journal of Nutrition Education, 28*, 19–25. doi:10.1016/S0022-3182(96)70011-0

Champion, V. L., & Huster, G. (1995). Effect of interventions on stage of mammography adoption. *Journal of Behavioral Medicine, 18*, 169–187. doi:10.1007/BF01857868

Clark, N. M. (1989). Asthma self-management education. Research and implications for clinical practice. *Chest, 95*(5), 1110–1113. doi:10.1378/chest.95.5.1110

Clark, N. M., & Starr-Schneidkraut, N. J. (1993). Management of asthma by patients and families. *American Journal of Respiratory and Critical Care Medicine, 149*, s54–s66.

Clark, N. M., & Zimmerman, B. J. (1990). A social cognitive view of self-regulated learning about health. *Health Education Research, 5*(3), 371–379. doi:10.1093/her/5.3.371

Creer, T. L. (1990a). Strategies for judgment and decision-making in the management of childhood asthma. *Pediatric Asthma Allergy & Immunology, 4*(4), 253–264. doi:10.1089/pai.1990.4.253

Creer, T. L. (1991). The application of behavioral procedures to childhood asthma: Current and future perspectives. *Patient Education and Counseling, 17*, 9–22. doi:10.1016/0738-3991(91)90047-9

Creer, T. L., Wigal, J. K., Kotses, H., & Lewis, P. (1990b). A critique of 19 self-management programs for childhood asthma: Part II. Comments regarding the scientific merit of the programs. *Pediatric Asthma Allergy & Immunology, 4*(1), 41–55. doi:10.1089/pai.1990.4.41

Cullen, K. W., Bartholomew, L. K., Parcel, G. S., & Kok, G. (1998). Intervention Mapping: Use of theory and data in the development of a fruit and vegetable nutrition program for Girl Scouts. *Journal of Nutrition Education, 30*(4), 188–195. doi:10.1016/S0022-3182(98)70318-8

Deardorff, V. W. (1986). Computerized health education: A comparison with traditional formats. *Health Education & Behavior, 13*, 61–72. doi:10.1177/109019818601300107

DiClemente, C. C., & Prochaska, J. O. (1982). Self-change and therapy change of smoking behavior: A comparison of processes of change in cessation and maintenance. *Addictive Behaviors*, 133–142. doi:10.1016/0306-4603(82)90038-7

DiClemente, C. C., & Prochaska, J. O. (1985). Process and stages of self-change: Coping and competence in smoking behavioral change. In S. Shiffman & T.A Wills, (Eds.), *Coping and Substance Abuse*. New York: Academic Press.

DiClemente, C. C., Prochaska, J. O., Fairhurst, S. K., Velicer, W. F., Velasquez, M., & Rossi, J. S. (1991). The process of smoking cessation: An analysis of precontemplation, contemplation, and preparation stages of change. *Journal of Consulting and Clinical Psychology, 59*, 259–304. doi:10.1037/0022-006X.59.2.295

Duncan, T. E., Duncan, S. C., Beauchamp, N., Wells, J., & Ary, D. V. (2000). Development and evaluation of an interactive CD-ROM refusal skills program to prevent youth substance use: "refuse to use. *Journal of Behavioral Medicine, 23*, 59–72. doi:10.1023/A:1005420304147

Eiser, R. J., van der Pligt, J., Raw, M., & Sutton, S. R. (1985). Trying to stop smoking: Effects of perceived addiction, attributions for failure, and expectancy of success. *Journal of Behavioral Medicine, 8*(4), 321–341. doi:10.1007/BF00848367

Fall, A. J., Henry, R. L., & Hazell, T. (1998). The use of an interactive computer program for the education of parents of asthmatic children. *Journal of Paediatrics and Child Health, 34*(2), 127–130. doi:10.1046/j.1440-1754.1998.00178.x

Finkelstein, J. A., Brown, R. W., Schneider, L. C., Weiss, S. T., Quintana, J. M., & Goldmann, D. A. (1995). Quality of care for preschool children with asthma: the role of social factors and practice setting. *Pediatrics, 95*, 389–394.

Finnegan, J. R., & Viswanath, K. (2008). Communication theory and health behavior change. In K. Glanz, B.K. Rimer, & K. Viswanath, (Eds.), *Health Behavior and Health Education. Theory, Research, and Practice,* (4th ed). San Francisco, CA: Jossey-Bass, John Wiley & Sons, Inc.

Gagne, R. M., Briggs, L. J., & Wager, W. W. (1988). *Principles of Instructional Design.* New York: Holt, Rinehart, and Winston, Inc. Games for health, available at http://www.gamesforhealth.org/

Gee, J. P. (2007). Good video games + Good Learning. In *Collected essays on video games, learning, and literacy.* New York: Peter Lang Publishing Inc.

Gerbert, B., Berg-Smith, S., Mancusco, M., Caspers, N., McPhee, S., & Null, D. (2003). Using innovative video doctor technology in primary care to deliver brief smoking and alcohol intervention. *Health Promotion Practice, 4*(3), 249–261. doi:10.1177/1524839903004003009

Gilutz, H., Bar-On, D., Billing, E., Rehnquist, N., & Cristal, N. (1991). The relationship between causal attribution and rehabilitation in patients after their first myocardial infarction. A cross cultural study. *European Heart Journal, 12,* 883–888.

Glanz, K., Patterson, R. E., Kristal, A. R., DiClemente, C. C., Heimendinger, J., & Linnan, L. A. (1994). Stages of change in adopting healthy diets: Fat, fiber, and correlates of nutrient intake. *Health Education Quarterly, 21,* 499–519.

Glanz, K., Rimer, B. K., & Viswanath, K. (2008). The scope of health behavior and health education. In Glanz, K., Rimer, B.K., & Viswanath, K., (Eds.), *Health Behavior and Health Education: Theory, Research, and Practice,* (4th ed.). San Francisco, CA: Jossey-Bass, John Wiley & Sons, Inc.

Glanz, K., & Rimer, K. (2005). *Theory at a glance. A guide for health promotion practice,* (2nd ed.). Washington, DC: NIH Pub. No. 05-3896.

Gochman, D. S. (1982). Labels, systems, and motives: Some perspectives on future research. *Health Education Quarterly, 9,* 167–174.

Gochman, D. S. (1997). Health Behavior Research: Definitions and Diversity. In D.S. Gochman (Ed.), *Handbook of Health Behavior Research,* (Vol I: Personal and Social Determinants). New York: Plenum Press.

Green, L. W., & Kreuter, M. W. (2005). *Health Promotion Planning: An Educational and Environmental Approach,* (4th Ed.). New York: McGraw-Hill.

Green, M. C. (2006). Narratives and cancer communication. *The Journal of Communication, 56*(s1), S163–S183. doi:10.1111/j.1460-2466.2006.00288.x

Grove, R. J. (1993). Attributional correlates of cessation self-efficacy among smokers. *Addictive Behaviors, 18*, 311–320. doi:10.1016/0306-4603(93)90032-5

Halfon, N., & Newacheck, P. W. (1993). Childhood asthma and poverty: differential impacts and utilization of health services. *Pediatrics, 91*(1), 56–61.

Herz, J. C. (1997). *Joystick Nation: How computer games ate our quarters, won our hearts, and rewired our minds*. Boca Raton, FL: Little, Brown, and Company.

Homer, C. J., Szilagyi, P., Rodewald, L., Bloom, S. R., Greenspan, P., & Yazdgerdi, S. (1996). Does quality of care effect rates of hospitalization for childhood asthma? *Pediatrics, 98*(1), 18–23.

Hornung, R. L., Lennon, P. A., Garrett, J. M., DeVellis, R. F., Weinberg, P. D., & Stretcher, V. J. (2000). Interactive computer technology for skin cancer prevention targeting children. *American Journal of Preventive Medicine, 90*, 781–785.

Hospers, H. J., Kok, G., & Strecher, V. J. (1990). Attributions for previous failures and subsequent outcomes in a weight reduction program. *Health Education Quarterly, 17*(4), 409–415.

Hsu, S., Lee, F., & Wu, M. (2005). Designing action games for appealing to buyers. *Cyberpsychology & Behavior, 8*, 585–591. doi:10.1089/cpb.2005.8.585

Huss, K., Rand, C. S., Butz, A. M., Eggleston, P. A., Murigande, C., & Thompson, L. (1994). Home environmental risk factors in urban minority asthmatic children. *Annals of Allergy, 72*, 173–177.

James, C. V., & Rosenbaum, S. (2009). Paying for quality care: Implications for racial and ethnic health disparities in pediatric asthma. *Pediatrics, 123*(Suppl. 3), S205–S210. doi:10.1542/peds.2008-2233L

Janoff-Bulman, R. (1979). Characterological versus behavioral self-blame: Inquiries into depression and rape. *Journal of Personality and Social Psychology, 37*(10), 1798–1809. doi:10.1037/0022-3514.37.10.1798

Janoff-Bulman, R., & Lang-Gunn, L. (1988). Coping with disease, crime, and accidents: The role of self-blame attributions. In L.Y. Abramson (Ed.), *Social Cognition and Clinical Psychology: A synthesis* (pp. 116-147). New York: Guilford Press.

Kahn, G. (1993). Computer-based patient education: A progress report. *Patient Informatics, 10*, 93–99.

Kasl, S. V., & Cobb, S. (1966a). Health behavior, illness behavior, and sick-role behavior: I. Health and illness behavior. *Archives of Environmental Health, 12*, 246–266.

Kasl, S. V., & Cobb, S. (1966b). Health behavior, illness behavior, and sick-role behavior: II. Sick-role behavior. *Archives of Environmental Health, 12*, 531–541.

Keller, J. M. (1987b). Development and use of the ARCS model of instructional design. *Journal of Instructional Development, 10*(3), 2–10. doi:10.1007/BF02905780

Keller, J. M., & Kopp, T. (1987). Application of the ARCS model of motivational design. In C.M. Reigeluth (Ed.), *Instructional Theories in Action: Lessons Illustrating Selected Theories and Models*. Hillsdale, NJ: Lawrence Erlbaum Associates.

Keller, J. M., & Suzuki, K. (1987a). Use of ARCS motivation model courseware design. In D.H. Jonassen (Ed.), *Instructional Designs for Microcomputer Courseware*. Hillsdale, NJ: Lawrence Erlbaum Associates.

Kirreimuir, J., & McFarlane, A. (2006). *Literature review in games and learning*. [FutureLabs Report 8]. Accessed February 2009, available from http://www.futurelab.org.uk/resources/documents/lit_reviews/Games_Review.pdf

Klingelhofer, E. L., & Gershwin, M. E. (1988). Asthma self-management programs: Premises, not promises. *The Journal of Asthma, 25*(2), 89–101. doi:10.3109/02770908809071359

Kotses, H., Stout, M. A., Wigal, J. K., Carlson, B., Creer, T., & Lewis, P. (1991). Individualized asthma self-management: A beginning. *The Journal of Asthma, 28*(4), 287–289. doi:10.3109/02770909109073386

Kreuter, M., Farrell, D., Olevitch, L., & Brennan, L. (2000). *Tailoring Health Messages. Customizing Communication with Computer Technology*. Mahwah, NJ: Lawrence Erlbaum Associates.

Krishna, S., Balas, E. A., Spencer, D. C., Griffin, J. Z., & Boren, S. A. (1997). Clinical trials of interactive computerized patient education: implications for family practice. *The Journal of Family Practice, 45*, 25–33.

Legler, J. (2002). The effectiveness of interventions to promote mammography among women with historically lower rates of screening. *Cancer Epidemiology, Biomarkers & Prevention, 11*(1), 59–71.

Lepper, M. R., & Cordova, D. I. (1992). A desire to be taught: instructional consequences of intrinsic motivation. *Motivation and Emotion, 16*(3), 187–208. doi:10.1007/BF00991651

Lepper, M. R., & Greene, D. (1979). *The Hidden Costs of Reward*. Morristown, NJ: Lawrence Erlbaum Associates.

Lepper, M. R., & Malone, T. W. (1987). Intrinsic motivation and instructional effectiveness in computer-based education. In R.E. Snow & M.J. Farr (Eds.), *Aptitude, Learning, and Instruciton, III. Conative and Affective Process Analysis*. Hillsdale, NJ: Lawrence Erlbaum Associates.

Lepper, M. R., & Malone, T. W. (1987). Intrinsic motivation and instructional effectiveness in computer-based education. In R.E. Snow and M.J. Farr (Eds), *Aptitude, Learning, and Instruction, III. Conative and Affective Process Analysis*. Hillsdale, NJ: Lawrence Erlbaum Associates.

Lewis, F. M., & Daltroy, L. H. (1990). How causal explanations influence health behavior: Attribution theory. In K. Glanz, F.M. Lewis, & B.K. Rimer, (eds.), *Health behavior and health education*. San Francisco, CA: Jossey-Bass Publishers.

Lieberman, D. A. (1997). Interactive video games for health promotion: Effects on knowledge, self-efficacy, social support, and health. In R.L. Street, W.R. Gold, & T. Manning (Eds), *Health Promotion and Interactive Technology*. Mahwah, NJ: Lawrence Erlbaum Associates.

Mak, H., Johnson, P., Abbey, H., & Talamo, R. C. (1982). Prevalence of asthma and health utilization of asthmatic children in an inner city. *The Journal of Allergy and Clinical Immunology, 70*, 367–372. doi:10.1016/0091-6749(82)90026-4

Malone, T. W. (1981a). Towards a theory of intrinsically motivating instruction. *Cognitive Science, 5*, 333–369.

Malone, T. W. (1981b). What makes computer games fun? *BYTE Publications*.

Marcus, B. H., Emmons, K. M., Simkin-Silverman, L. R., Linnan, L. A., Taylor, E. R., & Bock, B. C. (1998). Evaluations of motivationally-tailored vs. standard self-help physical activity interventions at the workplace. *American Journal of Health Promotion, 12*, 246–253.

McGuire, W. (1972). Social psychology. In P.C. Dodwell (Ed.), *New Horizons in Psychology*, (pp. 219-242). Middlesex, UK: Penguin Books.

McPherson, A., Glazerbrook, C., & Smyth, A. (2001). Double click for health: the role of multimedia in asthma education. *Archives of Disease in Childhood, 85*, 447–449. doi:10.1136/adc.85.6.447

Milheim, W. D., & Martin, B. L. (1991). Theoretical bases for the use of learner control: three different perspectives. *Journal of Computer-Based Instruction, 18*(3), 99–105.

Mokdad, A. H., Marks, J. S., Stroup, D. F., & Gerberding, J. L. (2000). Actual causes of death in the United States. *Journal of the American Medical Association, 10*(291), 1238–1245.

Morris, L. A., & Kranouse, D. E. (1982). Informing patients about drug side effects. *Journal of Behavioral Medicine, 5*, 363–373. doi:10.1007/BF00846163

National Institute on Media and the Family. *Sixth Annual Video and Computer Report Card.* (2001). Retrieved September 12, 2002, from http://www.mediaandthefamily.org/research/vgrc/2001-2.shtml

NHLBI. (1995). *Global initiative for asthma: global strategy for asthma management and prevention.* NHLBI/WHO Workshop Report. 1995; Pub. No. 95-3659.

O'Connell, D. O., & Velicer, W. F. (1988). A decisional balance measure and the stages of change model for weight loss. *The International Journal of the Addictions, 23*, 729–740.

Owen, N., Fotheringham, M. J., & Marcus, B. H. (2002). Communication technology and health behavior change. In Glanz, K., Rimer, B.K., & Lewis, F.M. (Eds.), *Health Behavior and health Education. Theory, research, and Practice,* (3rd ed.). San Francisco, CA: Jossey-Bass, John Wiley & Sons, Inc.

Paivio, A. (1986). *Mental Representations: A Dual Coding Approach.* Oxford, UK: Oxford University Press.

Palmgreen, P., & Donohew, L. (2002). Effective mass media for drug abuse prevention campaigns. In Z. Sloboda & W. Bukoski (eds), *Effective Strategies for Drug Abuse Prevention.* New York: Plenum Press.

Parker, L. E., & Lepper, M. R. (1992). Effects of fantasy contexts on children's learning and motivation: Making learning more fun. *Journal of Personality and Social Psychology, 62*(4), 625–633. doi:10.1037/0022-3514.62.4.625

Peterson, C., Seligman, M. E. P., & Vaillant, G. E. (1988). Pessimistic explanatory style is a risk factor for physical illness: A thirty-five year longitudinal study. *Journal of Personality and Social Psychology, 55*, 23–27. doi:10.1037/0022-3514.55.1.23

Pollock, J. C., & Sullivan, H. J. (1990). Practice mode and learner control in computer-based instruction. *Contemporary Educational Psychology, 15*, 251–260. doi:10.1016/0361-476X(90)90022-S

Prochaska, J. J., Zabinski, M. F., Calfas, K. J., Sallis, J. F., & Patrick, K. (2000). PACE+: Interactive communication technology for behavior change in clinical settings. *American Journal of Preventive Medicine, 19*, 127–131. doi:10.1016/S0749-3797(00)00187-2

Prochaska, J. O., & DiClemente, C. C. (1992). Stages of change in the modification of problem behaviors. In M. Hersen, R.M. Eisler, P.M. Miller (Eds.), *Progress in Behavior Modification* (pp 184-218). Sycamore, IL: Sycamore Publishing Company.

Prochaska, J. O., DiClemente, C. C., Velicer, W. F., & Rossi, J. S. (1993). Standardized, individualized, interactive and personalized self-help programs for smoking cessation. *Health Psychology, 12*, 399–405. doi:10.1037/0278-6133.12.5.399

Prochaska, J. O., Redding, C. A., Harlow, L. L., Rossi, J. S., & Velicer, W. F. (1994). The Transtheoretical Model of Change and HIV prevention: A review. *Health Education Quarterly, 21*, 471–486.

Prochaska, J. O., Velicer, W. F., DiClemente, C. C., & Fava, J. L. (1988). Measuring processes of change: Applications to the cessation of smoking. *Journal of Consulting and Clinical Psychology, 56*, 520–528. doi:10.1037/0022-006X.56.4.520

Rachelefsky, G. S. (1987). Review of asthma self-management programs. *The Journal of Allergy and Clinical Immunology, 80*, 506–511. doi:10.1016/0091-6749(87)90087-X

Rakowski, W. A., Dube, C. A., & Goldstein, M. G. (1996b). Considerations for extending the transtheoretical model of behavior change to screening mammography. *Health Education Research, 11*, 77–96. doi:10.1093/her/11.1.77

Rakowski, W. A., Ehrich, B., Dube, C. A., Pearlman, D. N., Goldstein, M. G., & Peterson, K. K. (1996a). Screening mammography and constructs from the Transtheoretical Model: Associations using two definitions of the stages-of- adoption. *Annals of Behavioral Medicine, 18*(2), 91–100. doi:10.1007/BF02909581

Reed, G. R., Velicer, W. F., & Prochaska, J. O. (1997). What makes a good staging algorithm: Examples from regular exercise. *American Journal of Health Promotion, 12*, 57–67.

Reigeluth, C., & Schwartz, E. (1989). An instructional design theory for the design of computer-based simulations. *Journal of Computer-Based Instruction, 16*(1), 1–10.

Reis, J., Riley, W., Lokman, L., & Baer, J. (2000). Interactive multimedia preventive alcohol education: a technology application in higher education. *Journal of Drug Education, 30*, 399–421. doi:10.2190/LWMQ-9CQA-B78H-9MA7

Revere, D., & Dunbar, P. J. (2001). Review of computer-generated outpatient health behavior interventions: clinical encounters "in absentia." . *Journal of the American Medical Informatics Association, 8*, 62–79.

Rhodes, F., Fishbein, M., & Reis, J. (1997). Using behavioral theory in computer-based health promotion and appraisal. *Health Education & Behavior, 24*(1), 20–34. doi:10.1177/109019819702400105

Rieber, L. P. (1991a). Animation, incidental learning, and continuing motivation. *Journal of Educational Psychology, 83*(3), 318–328. doi:10.1037/0022-0663.83.3.318

Rieber, L. P., & Kini, A. S. (1991b). Theoretical foundations of instructional applications of computer-generated animated visuals. *Journal of Computer-Based Instruction, 18*(3), 83–88.

Rolnick, S. J. (1988). Self-management of pediatric asthma: Four programs being studied. *Journal of Pediatric Health Care, 2*(5), 264–266. doi:10.1016/0891-5245(88)90159-9

Rumelhart, D. E., & Ortony, A. (1977). The representation of knowledge in memory. In R.C. Anderson, R.J. Spiro, & W.E. Montague (Eds), *Schooling and the Acquisition of Knowledge.* Hillsdale, NJ: Lawrence Erlbaum Associates.

Sawyer, B., & Smith, P. (2008). *Serious games taxonomy*. Serious Games Initiative. Available from www.seriousgames.org/index2.html.

Schild, E. O. (1968). The shaping of strategies. In S.S. Broocock & E.O. Schild (Eds.), *Simulation Games in Learning.* Beverley Hills, CA: Sage Publications.

Sechrest, R. C., & Henry, D. J. (1996). Computer-based patient education: observations on effective communication in the clinical setting. *The Journal of Biocommunication, 23*, 8–12.

Shegog, R. (1997). *Computer-assisted instruction for self-management education in pediatric asthma*. Unpublished doctoral dissertation, University of Texas, Houston.

Shegog, R., & Bartholomew, L. K. (2004). Computer-assisted asthma education for children: Impact on self-management behavior. *International Review of Asthma, 6*(2), 70–86.

Shegog, R., Bartholomew, L. K., Craver, J., Sockrider, M. M., Mullen, P. D., & Pilney, S. (2004). Development of an expert system knowledge base: A novel approach to promote guideline congruent asthma care. *The Journal of Asthma, 41*(4), 385–402. doi:10.1081/JAS-120026098

Shegog, R., Bartholomew, L. K., Gold, R. S., Pierrel, E., Parcel, G. S., & Sockrider, M. M. (2006). Asthma management simulation for children: Translating theory, methods, and strategies to effect behavior change. *Simulation in Healthcare, 1*(3), 151–159.

Shegog, R., Bartholomew, L. K., Parcel, G. S., Sockrider, M., Czyzewski, D., & Masse, L. (2001). Impact of a computer-assisted education program on variables related to children's asthma self-management behavior. *Journal of the American Informatics Association, 8*(1), 49–61.

Shegog, R., Conroy, J., Murray, N. G., Agurcia, C., Kelder, S., & Prokorov, A. (2003). Process evaluation of ASPIRE: A CD_ROM based smoking curriculum for high school students. *American Public Health Association*, San Francisco, CA.

Shegog, R., McAlister, A., Hu, S., Ford, K., Meshak, A., & Peters, R. (2005). Using the web to impact smoking intentions in middle school students: A pilot test of the "Headbutt" risk assessment program. *American Journal of Health Promotion, 19*(5), 334–338.

Shegog, R., & Sockrider, M. M. (2009). Computer-based applications for asthma education and management. In Harver, A. (Ed.), *Asthma, Health, and Society*. New York: Springer.

Skinner, C. S., & Kreuter, M. W. (1997). Using theories in planning interactive computer programs. In: R. Street, W. Gold, and T. Manning (Eds.), *Health Promotion and interactive technology: Theoretical applications and future directions*. Mahwah, NJ: Lawrence Erlbaum Associates.

Slater, M. D. (2002). Entertainment education and the persuasive impact of narratives. In M.C. Green, J.F. Strange, & T.C. Brock (Eds), *Narrative Impact: Social and Cognitive Foundations*. Mahwah, NJ: Lawrence Erlbaum Associates.

SocialGames. (n.d.). Available at http://www.socialimpactgames.org

Stephenson, M. Y., & Southwell, B. (2006). Sensation-seeking, the activation model, and mass media health campaigns. *The Journal of Communication, 56*, S38–S56. doi:10.1111/j.1460-2466.2006.00282.x

Strecher, V. (1999a). The role of interactive strategies in cancer risk communication. *Journal of National Cancer Institute*.

Strecher, V. J. (1999b). Computer-tailored smoking cessation materials: A review and discussion. *Patient Education and Counseling, 36*, 107–117. doi:10.1016/S0738-3991(98)00128-1

Strecher, V. J., Greenwood, T., Wang, C., & Dumont, D. (1999b). Interactive multimedia and risk communication. *Journal of the National Cancer Institute. Monographs, 25*, 134–139.

Strecher, V. J., Seijts, G. H., Kok, G. J., Latham, G. P., Glasgow, R., & DeVellis, B. (1995). Goal setting as a strategy for health behavior change. *Health Education Quarterly, 22*(2), 190–200.

Street, R. L., & Rimal, R. N. (1997). Health promotion and interactive technology: A conceptual foundation. In R. Street, W. Gold, & T. Manning (Eds.), *Health Promotion and Interactive Technology: Theoretical Applications and Future Directions*. Mahwah, NJ: Lawrence Erlbaum Associates.

Thompson, J., Berbank-Green, B., & Cusworth, N. (2007). *The Computer Game Design Course. Principles, Practices, and Techniques for the Aspiring Game Designer*. London: Thames and Hudson Ltd.

Thoresen, C. E., & Kirmil-Gray, K. (1983). Self-management psychology and the treatment of childhood asthma. *The Journal of Allergy and Clinical Immunology, 72*(5), 596–610. doi:10.1016/0091-6749(83)90487-6

Tripp, M. K., Herrmann, N. B., Parcel, G. S., Chamberlain, R. M., & Gritz, E. R. (2000). Sun Protection is Fun! A skin cancer prevention program for preschools. *The Journal of School Health, 70*(10), 395–401. doi:10.1111/j.1746-1561.2000.tb07226.x

USDHHS. (1997). National Asthma Education Program Expert Panel Report. Expert panel report 2. *Guidelines for the diagnosis and management of asthma*. Bethesda, MD: National Asthma Education Program, Office of Prevention, Education and Control, National Heart, Lung and Blood Institute, U.S. Department of Health and Human Services, Public Health Service; Report No.: 97-4051.

USDHHS. (1999). Science Panel on Interactive Communication and Health. In T.R. Eng & D.H. Gustafson (Eds.), *Wired for Health and Well-Being: The Emergence of Interactive Health Communication*. Washington, DC: US Department of Health and Human Services, US Government Printing Office.

USDHHS. (2007). National Asthma Education and Prevention Program, Expert panel report 3: *Guidelines for the Diagnosis and Management of Asthma*. Bethesda, MD: National Asthma Education Program, Office of Prevention, Education and Control, National Heart, Lung and Blood Institute, U.S. Department of Health and Human Services, Public Health Service.

USDHHS. (n.d.). *Healthy People2010*. Office of Disease Prevention and Health Promotion. U.S. Department of Health and Human Service. Retrieved February 2008, from http://www.healthypeople.gov/

Velicer, W. F., DiClemente, C. C., Prochaska, J. O., & Brandenburg, N. (1985). A decisional balance measure for assessing and predicting smoking status. *Journal of Personality and Social Psychology, 48*(5), 1279–1289. doi:10.1037/0022-3514.48.5.1279

Walters, S., Wright, J. A., & Shegog, R. (2006). A Review of Computer- and Internet-Based Interventions for Smoking Behavior. *Addictive Behaviors, 31*, 264–277. doi:10.1016/j.addbeh.2005.05.002

Watkins, S. A., Hoffman, A., Burrows, R., & Tasker, F. (1994). Colorectal cancer and cardiac risk reduction using computer-assisted dietary counseling in a low-income minority population. *Journal of the National Medical Association, 86*, 909–914.

Weiner, B. (Ed.). (1974). *Achievement motivation attribution theory*. Morristown, NJ: General Learning Press.

Weiner, B. (1982). An attribution theory of motivation and emotion. In H.W. Krohne and L. Laux, *Achievement, stress, and anxiety*. New York: McGraw-Hill.

Weiner, B. (1985). An attributional theory of achievement motivation and emotion. *Psychological Review, 92*(4), 548–573. doi:10.1037/0033-295X.92.4.548

Weiss, K. B., Gergen, P. J., & Wagener, D. K. (1993). Breathing better or worse? The changing epidemiology of asthma morbidity and mortality. *Annual Review of Public Health, 14*, 491–531.

Wigal, J. K., Creer, T. L., Kotses, H., & Lewis, P. (1990). A critique of 19 self-management programs for childhood asthma: Part I. Development and evaluation of the programs. *Pediatric Asthma Allergy & Immunology, 4*(1), 17–39. doi:10.1089/pai.1990.4.17

Wood, P. R., Hidalgo, H. A., Prihoda, T. J., & Kromer, M. E. (1993). Hispanic children with asthma: morbidity. *Pediatrics, 91*, 62–69.

Zuckerman, M. (1971). Dimensions of sensation seeking. *Journal of Consulting and Clinical Psychology, 36*, 45–52. doi:10.1037/h0030478

Chapter 12
Avatars and Diagnosis:
Delivering Medical Curricula in Virtual Space

Claudia L. McDonald
Texas A&M University-Corpus Christi, USA

ABSTRACT

Medical education faces a host of obstacles in coming decades requiring that it rethink the way it delivers medical curricula, especially with regard to critical thinking and differential diagnostics. Traditional didactic curricula must be coupled with emerging technologies that provide experiential learning without risk to patients while not eroding the clinical effectiveness of advanced medical learners. Virtual-world technologies have advanced to a level where they must be considered as a method for delivering medical curricula effectively and safely; moreover, research must establish that such systems are reliable and valid means for delivering medical curricula; otherwise, they are of no use to the medical community, regardless of their technical sophistication. Pulse!! The Virtual Clinical Learning Lab is a project designed to explore these issues by developing a reliable and valid learning platform for delivering medical curricula in virtual space, space usually reserved for entertainment videogames.

INTRODUCTION

Medical education must reform to stay abreast of rapid change. Baby-boom retirements from academic faculties and other demographic factors are creating looming shortages of medical personnel, especially physicians and nurses (Rasch, 2006). Shorter hospital stays and medical residents' workweeks are reducing clinical training opportunities and expertise development (e.g., Verrier, 2004). Continuously changing warfare and terrorist technology and methods drive a need for rapid deployment of training for continuously evolving medical treatment (Zimet, 2003). Battlefield fatalities from potentially survivable injuries might have been prevented with consistent application through more effective training of U.S. armed forces' Tactical Combat Casualty Care (TCCC) guidelines (Holcombe, et al., 2006).

DOI: 10.4018/978-1-61520-739-8.ch012

Simulation is an established, effective means for providing medical training and education. Issenberg et al. (2005) conclude "that high-fidelity medical simulations facilitate medical learning under the right conditions" (p. 24). This first comprehensive review of 34 years of research on the efficacy of simulation in medical training found that "high-fidelity medical simulations are educationally effective and simulation-based education complements education in patient care settings" (Issenberg et al., 2005, p. 10). Issenberg notes that the medical community's increasing interest in simulation and virtual reality (VR) is driven in part by "the pressures of managed care" that lead to fewer clinical opportunities for medical learners, including practicing physicians constrained by shrinking financial resources to spend less time keeping abreast of developing medical topics. Issenberg cites a 1997 study by Mangione & Nieman, which found a decline in bedside acumen and urged "the use of simulation systems for training" (Issenberg et al., 2005, p. 12).

Reznick and MacRae (2006) observe that well-established motor learning theory (Fitts & Posner, 1967), as it applies to surgical training, ratifies the use of simulators in the acquisition of skills through three stages – cognitive, integrative, and autonomous. Ericsson (1996) applies a thicker description directly to surgical training with the concept of "deliberate practice" defined as "repeated practice along with coaching and immediate feedback on performance" (Reznick & MacRae, 2006, p. 2665).

Watters et al (2006) propose that substantial learning occurs through what has become standard entertainment-game architecture, including instantaneous feedback, a rising scaffold of challenges, visible goal indicators, personalization and customization, fluidity and contextual grounding. Watters and Duffy (2004) propose "a framework of motivational constructs" (p. 16) found in games that are applicable in developing interactive health-related software: self-regulation, or autonomy; relatedness, which includes role-playing, narrative and personalization; and competency, or self-efficacy built upon completion of meaningful tasks.

Virtual-world technologies are emerging as the next big step for medical simulation. "We are in a complete revolution in surgical education," according to Richard Satava, a professor of surgery at the University of Washington. "If history serves us well, such a revolution occurs only once every hundred years, as evidenced by the fact that the last revolution began in 1908 with the Flexner Report. Whatever is established during these next 10 years is likely to endure for the next century" (Satava, 2008, p. 2).

Entertainment has been the main use of virtual reality, but virtual training in various formats also has been explored. In the academy, there has been incidental development of human-simulation trainers anchored to computer-assisted case presentations for surgeons, U.S. Army combat medics engage electronic role-playing games aimed at clinical soft-skills education. The military originally developed the state-of-the-art *America's Army* online war game for recruitment, using the 'first-person shooter' game genre. A modified version of *America's Army* became a standard aspect of some training at the U.S. Military Academy after research showed that those who used virtual-reality training achieved better scores in a in ground-maneuver warfare course than those who had used traditional print materials. Farrell et al. (2003) said in conclusion: "The result is excited and motivated cadets who take ownership of their military development, in terms of what they study, when they study it, and how both cadets and instructors receive feedback on performance" (p. 1).

This chapter will address specific issues based on the development of Pulse!! The Virtual Clinical Learning Lab, including funding and organization; the interface among various fields of expertise brought into play by Pulse!! development, and the process of development that evolved from the project. This chapter also will recommend a model for developing medical education in virtual space with regard to funding, management, case-based development, curricular standards and embedded evaluation.

BACKGROUND

Military and civilian health-care delivery systems are at risk from medical errors, their related costs, and rapidly diminishing educational and training resources. The Institute of Medicine (IOM) reports that errors are responsible for 44,000 to 98,000 deaths annually; and the Agency for Healthcare Research and Quality estimates that related costs are between $17 billion and $29 billion annually. Patients' length of stay has shortened dramatically, giving students significantly fewer and briefer opportunities for clinical experience, including exploring further diagnostic options and examining treatment results (Kohn, Corrigan, & Donaldson, 1999).

A growing shortage of nurses and looming shortage of physicians contribute to the safety issue, which is compounded by case-mix intensity at health-care facilities – more and sicker people requiring more attention from a dwindling base of caregivers. A number of demographic factors, including baby-boom retirements – 55 percent of nurses and a third of physicians intend to retire between 2011 and 2020 (Hader et al., 2006; AAMC, 2006) – and increased numbers of women physicians since the 1970s, which due to child-rearing reduce overall full-time equivalents (FTEs), are driving an estimated shortage of 200,000 physicians, beginning in 2010. Even conservative estimates, such as that reported in 2005 by the Council on Graduate Medical Education (COGME), indicate that the domestic shortage of physicians will be no less than 85,000 (COGME, 2005). The U.S. Census Bureau predicts that the number of elderly will increase 104 percent from 2000 to 2030; meanwhile, first-year medical-school enrollment per 100,000 has steadily declined since 1980 (Campbell, 1996; AAMC, 2006). The U.S. Department of Health and Human Services warns that, despite increased numbers of nursing programs, projected demand for nurses substantially exceeds projected supply. Capacity in nursing education programs is at a maximum, faculty are retiring in unprecedented numbers, care patterns are changing, and according to the National League for Nursing (2009), 99,000 qualified nursing students, at all levels, were denied admission in 2006.

The development of three-dimensional virtual simulation comes at a moment in the history of American health care when the treatment paradigm is shifting to include greater emphasis on care as the population ages. The current system is built on the concept of cure, but that goes against the grain of U.S. demographics and society, which are dominated by aging baby boomers living longer than any previous generation due to pharmaceutical intervention, surgical advances, and the successful treatment of chronic disease. U.S. medicine is evolving a new concept of care that has implications for medical education and underscores the need for virtual learning space. Medical practice in the future will become more complex as the concept of care for older patients matures and technology continues to advance. Virtual simulation of such complexity makes possible the kind of education and training that will rise to these challenges.

Modeling and simulation are generally recognized in the medical community as performance enhancers. A U.S. Food and Drug Administration panel has recommended virtual-reality training for carotid artery stenting: "Given the advances in technology and the accruing evidence of their effectiveness, now is the time to take stock of the changes we can and must make to improve the assessment and training of surgeons in the future" (Reznick & MacRae, 2006, p. 2668). An Accreditation Council on Graduate Medical Education (ACGME) "toolbox" of clinical assessment methods recommends simulation and models as the most desirable means for evaluating residents' competence in medical procedures and the next best method for the development and execution of management plans, investigatory and analytic thinking, knowledge. and application of basic science and ethical practices. Simulators and models are deemed potentially applicable to physicians' self-analysis of performance and where improvements might be made (AGCME Outcomes Project, 2000).

Medical-education is shifting, perforce, from traditional process evaluation to performance outcome. The Association of American Medical Colleges (AAMC) notes that "measurement of performance outcome is gradually replacing process evaluation alone" due to "increasing regulatory and payor influences [that] constrain the conduct and governance of clinical teaching activities" (2005, p. 6). The ACGME performance now requires outcomes evaluation in six domains of competency: medical knowledge, patient care, practice-based learning and improvement, systems-based practice, professionalism, and interpersonal and communication skills (2006).

Responding to the IOM report, a partnership between the National Academy of Engineering and the Institute of Medicine formed to conduct a study of health care mistakes and to identify future remedies. The academy's 2005 report, *Building a Better Delivery System: A New Engineering/Health Care Partnership*, is a culmination of that team's research. The report concludes that the "U.S. health care industry neglected engineering strategies and technologies that have revolutionized quality, productivity, and performance in many other industries," and calls for an array of powerful new tools in medicine (p. 1). Virtual environment simulations are being explored for use in various formats and are one solution for a necessary paradigm shift in clinical health-care education.

Virtual-reality learning takes into account the remarkable development of computer technologies as tools for teaching the "Net Generation," born since 1982, of which 89.5 percent are computer literate, 63 percent are Internet users, and 14.3 percent have been using the Internet since age four (U.S. Department of Commerce, 2002). Research by the Kaiser Family Foundation found that American children four to six years of age "are living richly media-centric lives"—90 percent using "screen media" two hours a day; 43 percent using a computer and 24 percent playing video games several times a week (Rideout, 2006, p. 33).

DESIGNING MEDICAL EDUCATION IN VIRTUAL SPACE

The Pulse!! platform design is consistent with the Whitelock-Brna-Holland (1996) model for conceptual learning in virtual reality, which posits three essential properties of such domains for "implicit" or experiential learning: representational fidelity, "the extent to which the environment that is simulated is familiar to the user;" immediacy of control through a continuum of media from imitative haptics to command-line coding; and presence, viewed "as subjectively reported phenomenon . . . and as a set of repeatable objective measures" (p. 6). The authors conclude that "a high presence value and a high degree of immediacy of control (i.e., autonomy and interaction) leads to a high degree of implicit learning . . . the ability to perform tasks consistent with an improved understanding" as opposed to mere "conceptual understanding," which is the projected outcome of low-value immediate control (p. 7).

Thus conceived, the Pulse!! project models a strategy for providing cross-disciplinary expertise and resources to educational, governmental and business entities engaged in meeting looming health-care crises with three-dimensional virtual learning platforms that provide unlimited, repeatable, immersive clinical experience without risk to patients anywhere there is a computer. Pulse!! has been in development since 2005 by an international team of academicians, subject matter experts in medicine and human-factors research and top-flight simulation industry professionals.

The project has been supported so far by more than $14 million in federal grants through the Office of Naval Research. Based on industry standards, development from scratch of the base layer of a platform such as Pulse!! requires $25 million to $30 million. Government investment is the most likely source of funding for virtual-world research and development, which, at least for the time being, competes for private investment with the lucrative entertainment industry.

Pulse!! products will be grounded in research findings and equipped with tools and generators that enable clients to author cases and create scenarios within a variety of virtual environments. The Pulse!! platform is being rigorously researched, developed and tested extensively for reliability and validity, yielding a product for delivering medical and health-care curricula with confidence.

ESSAY: ON AVATARS AND DIAGNOSIS

In the beginning, Pulse!! was about establishing meets and bounds. In general, the project was to (1) *develop* a (2) *training platform* using (3) *game-based technologies* for (4) *health-care providers*. Decision-making formed around those four general topics.

Pulse!! was envisioned as a high-fidelity virtual-world platform from the beginning, because, in the author's view, only intensively realistic virtual reality provides the immersive experience required for problem-based medical education in real time. Pulse!! thus distinguishes itself from "computer-assisted" learning, which delivers didactic material electronically, and interactive platforms that render cases in abstract imagery or so-called "stick figures." Pulse!! is a wholly new approach in curriculum-delivery and virtual-world development: Every case is generated from standard medical curricula, rendered in virtual space and evaluated for learning effectiveness and the capability to measure/assess outcomes against objectives; moreover, it's portable and thus cost-effective. The Pulse!! virtual learning lab can be run wherever there is a properly configured and sufficiently powered computer or network.

Pulse!! was a research project from the beginning because the virtual-reality platform had to be shown to be a reliable and valid delivery system for medical education – a first in the field of VR simulation. Evaluation was the "third leg of the milking stool" that gave the project its stability amid the demands of academic medicine. It wasn't enough to depict cases in an immersive, interactive virtual environment. Medical educators want to know whether simulators actually teach critical thinking in the appropriation of medical curricula. It was critical that innovation in the delivery system not alter the substantive content of the accredited curricula. Pulse!! is a teaching strategy not a curriculum generator. The curriculum decision, moreover, was a major decision in developing the platform: Everyone got around the table – medical subject-matter experts, evaluators and game producers – to determine what cases actually would be presented in virtual space from the infinite spectrum of all possible standard medical cases.

Collaboration

There is no one way to structure management and execution of a virtual-world project such as Pulse!!, but the following discussion touches upon some major issues that may confound those seeking to develop and produce virtual-reality (VR) educational platforms that can be used with confidence to deliver complex curricula to high-level, motivated learners. These issues fall under the general topic of collaboration and include funding, technology and evaluation. This discussion will conclude with what lies ahead for the Pulse!! research and development project.

Collaboration is a polite term for the inevitable culture clash that makes for spirited conversation around the conference-room table. Academics, subject-matter experts (SMEs) and business executives don't always operate from the same point of view; in fact, they seldom see the world – or their collaborative project – in quite the same way. The product of their collaboration is a synthesis of often antithetical standpoints.

Pulse!! experience indicates that there must be a "decider" at the heart of the process. Achieving consensus is improbable in a project that brings three such diverse fields into collaboration. When the project is research and development, the principal investigator (PI) – and, in this case, the chief fundraiser – is the most likely decision-maker if for no other reason than to ensure the integrity of the research. In corporate language, the PI is the chief executive officer (CEO), the captain of the ship, whose decisions make or break the project. Still, the project's ultimate success depends on collaborative performance. The team is interdependent such that when one piece falters, all falter. Each piece of the project is continually modified as other pieces provide input, feedback, and expertise. The PI may be the captain of the ship, but midcourse corrections are numerous and inevitable, especially when the course is uncharted and the seas unknown.

As a public/private venture – funded by government but driven in part by private-sector entrepreneurs – Pulse!! actually leans toward being a small business. The Pulse!! project lays a foundation through learning research and development of the core engines and features of the learning platform. As a business venture, however, the Pulse!! collaboration also must be aware of market developments, technological shifts, and medical trends. The three-way conversation that informs Pulse!! development, production, and research never ends, and neither does the role of the PI/CEO. There always must be a captain/decider at the helm.

Business executives and their employees are constrained by costs and are wont to put parameters on the PI/CEO's vision. The PI/CEO, meanwhile, pushes back with expectations that the product will be technologically innovative, even trailblazing, as well as pedagogically sound. SMEs bring deep knowledge to the depiction of medical cases in virtual reality, and their desire for realism – without which the integrity of the Pulse!! vision would be compromised – often reaches far beyond what the business executive thinks is either necessary or affordable. The evaluation team, meanwhile, sets a high standard for the platform's interface, tutorials, and instructional materials to ensure that learners are not inhibited by the platform itself in their appropriation of the all-important curricula. The military, meanwhile, wants bang for its buck and so does the U.S. Congress. The clock is always ticking, and everyone feels the pressure.

The Pulse!! project's private-sector partner describes stumbling blocks for VR educational products based on years of experience in the interface of academe and the "serious games" industry (McDonald, 2009). Some things academics think are difficult are actually easy; such as creating demonstrations with off-the-shelf VR engines. On the other hand, some things academics think are easy are actually difficult; such as publicly showing a product in development, developing user interfaces and depicting the world "exactly as it looks in real life" (McDonald, 2009, p. 29). And the hardest things of all? "Getting a subject matter expert to tell you what you should put into the game" and "getting all the SMEs to agree" (McDonald, 2009, p. 31).

The collaborative relationship is symbiotic, but there must be a single authority responsible for the outcome of the whole emulsion of virtual-world technologies, medical curricula, and learning evaluation. For Pulse!! the single authority is the principal investigator/producer, whose position is defined at least in part by her pivotal role in fund-raising and in recruiting and coordinating subject-matter experts. It is a collaborative partnership with genuine respect on both sides – but someone must be in charge, at whose desk the buck stops.

Terminology Matters

Sorting out nomenclature issues among collaborators is a significant undertaking. It matters in how the project's story is told to the outside world wherein resides customers, critics, and potential investors. One nomenclature issue, for example, has been whether the Pulse!! project is about "training" or "education." The medical field regards training as something one can do with simians, while "education" is what must occur for medicine to be practiced critically as well as skillfully. "Training," however, is something taken seriously for human beings by those in the academy whose *field* is *education*. It's not just for simians; ergo, there is a bit of creative tension in how to speak internally and externally about Pulse!! It's a full-spectrum "learning platform," from training to continuing education for healthcare professionals, terminology that also finesses the education-training debate.

And then, there's the term "game," which is the bread and butter of the project's private-sector partner but which is anathema in selling Pulse!! to congressional committees and military review boards. Games are not serious business and education, except to gamers and companies that develop games for entertainment or serious simulation. Games are "played," while simulations are "learned." Pulse!! collaborators have crafted the phrase "learning platform," yet there has been a steep learning curve, especially in private conversation, in not referring to Pulse!! as a game.

How to overcome the nomenclature issues? Clarify, clarify, clarify, and be consistent as possible, especially in public conversation but also in private consultation among collaborators.

Case Development

There's a scene in the film "A River Runs Through It," in which a young man is being taught to write by his exacting father, an educated Presbyterian pastor. It's a study in refinement, as each effort is returned to the young writer with the instructions to use "half-again" as many words in the next draft. Because few VR learning projects have unlimited resources, they must accomplish as much as possible with no superfluous features, which requires a similar kind of "half-again" refinement.

It is traditional in medical education to teach by body system – endocrine, respiratory, muscular – or by clinical area – pediatrics, geriatrics, emergency medicine and the like. Pulse!!, however, had to take a different approach – "outside in" – because the level of visual fidelity required in a whole-body model, especially internal fluid dynamics, is currently beyond reach of virtual-world technologies, given available funding and the state of development of programming and animation.

Pulse!!, moreover, had something to prove – to the military and to Congress but also among its collaborators. The project was conceived as research, after all, and even though everyone at the table anticipated the outcome, the data had to speak for themselves. The Pulse!! learning platform had to show it was more than computer-assisted learning, that it was a trailblazing application of virtual-world technologies used by millions in the entertainment game industry. The original case design for Pulse!! was rooted in the need to show the platform's diversity, the width, breadth and depth of it as a delivery system for medical curricula and that it could deliver at every level of medical training and education, from entry-level first-responder skills to residents and practicing physicians doing continuing education.

First-generation case selection was tempered by the availability of financial and technical resources and putting together the talent team, as well as the need to deliver results for government and military stakeholders. Pulse!! collaborators decided not to develop virtual patient physiology by body-systems or medical discipline; instead, collaborators went with a whole-body physiology approach, from outside in, and cases that initially would not require technically difficult invasive procedures. The first four cases developed were for various types of shock common to battlefield trauma.

Consequently, Pulse!! case-development delivers problem-based curricula stemming from trauma and transmitted through the experience of the project's medical subject-matter experts. Every aspect of the virtual-world medical facility and operation depicted on-screen is shaped by experience and established medical practice. All interactive components of the platform reflect real-world possibilities and probabilities; moreover, the patient's physiology is programmed to be responsive to treatment in real time. If the problem-based case is not properly understood, diagnosed, and treated, the patient suffers; therefore, learners are required not only to know what procedures must be performed but must think critically and differentiate amid several possible diagnoses. Meanwhile, an on-screen clock ticks off the minutes, cardio-vascular monitors operate realistically, and sometimes the virtual patient's condition deteriorates rapidly.

Pulse!! problem-based cases in virtual reality offer repeated practice for gaining expertise in the simulated conduct that prevents real-world errors. In theory, iterative training in VR leads to a level of expertise in which there's less hesitation and less chance for error on procedures that have been done many times in virtual space. Pulse!! is predicated in part on the assumption that opportunities for practicing a wide range of clinical skills will continue to decline in coming years, such that the number and range of opportunities offered won't offer sufficient opportunity to achieve optimum learning outcomes. VR technology is poised to fill that void with immersive, experiential, iterative learning experience. VR also offers the possibility of training practicing health-care professionals in new techniques that require repeated practice to move from novice to expert. Moreover, it's not just a matter of physical dexterity but also mental dexterity, which makes the immersive VR environment all the more applicable not only to medical education but also continuing education.

Engines

Whether to build your own engines or modify commercially available products off-the-shelf was a major decision for the Pulse!! project. The collaborators decided it was better to start from scratch, because it made little difference whether project resources were invested in purchasing and modifying someone else's product or developing something new; moreover, existing engines could not produce soft-tissue and fluid effects without major, time-consuming modifications. Time and money considerations for both strategies came down to a wash, which doesn't mean there wasn't a grand debate within the development team.

Those in favor of purchasing commercial engines argued that case-development would begin sooner. Another group also postulated that Pulse!! could be developed as an open-source modification of an existing core engine with which others could create their own cases. Quality-control, however, would have required an enormous expenditure of time and resources. There would have been no guarantee that others would create Pulse!! cases in compliance with accreditation standards; moreover, it would be difficult to establish the platform's reliability and validity had it been an open-source project built by modifying existing virtual-world development engines. The fact that no engine could emulate human physiology was the capstone argument in favor of building from scratch.

Funding

Funding is a major issue in developing virtual-world educational platforms. It requires upward of $30 million to develop a complex commercial entertainment videogame, and that level of funding is simply not available from academic granting sources. Private investment is an unlikely source for educational learning platforms, because entertainment is much more profitable and the educational genre is not yet a proven product. Public investment has driven Pulse!! development through the Office of Naval Research, but it has come incrementally – almost $15 million since 2005 – so development has been relatively slow compared with the entertainment industry. Eventually, Pulse!! is expected to pay for itself through licensing and client arrangements for case development, but not before the learning research is complete – and not before the platform includes robust case-authoring tools and assets.

Congressional funding drove some significant strategic decisions for the Pulse!! project; for example, the project's base of operations; money needed to flow to the congressional district in Texas where Pulse!! was conceived. That meant creation of a new production studio far removed from the private-sector partner's corporate headquarters in Maryland. Logistical issues have sometimes been inconvenient but generally not insurmountable; and in any case, the project's "geography" has not inhibited Pulse!! from achieving its strategic goals.

Funding for ventures such as Pulse!! does not match the level of general interest in virtual-reality as a simulation strategy for enhancing medical education. There are few domestic programs similar to that recently announced by the South Korean government, which will provide the equivalent of $63.5 million in seed money for the so-called serious-games market. The government hopes that market will grow to almost $400 million by 2012 (Korea IT Times, 2009). Nowhere near enough private capital has been attracted to educational serious games such as Pulse!! Hefner's market study (2007) concludes that while VR technology will be in higher demand as a medical-training strategy over the next 10 to 15 years, development costs will continue to be prohibitive for the private sector. (The study suggests that establishing technological standards for VR health-education systems would lead to more rapid commercial development of learning platforms such as Pulse!! [p. 107].)

It is likely, then, for the foreseeable future, that government funding similar to that of the South Korean plan will be key to seeding the "serious-games" market with sufficient resources to establish, among other things, that VR medical education is reliable and valid and that technologies and conventions common to entertainment videogames are sufficiently playable to be of value to emergent generations of health-care learners, from first responders to medical residents.

Interest in these developments, meanwhile, remains high, as evidenced by the Robert Wood Johnson Foundation's sponsorship of the Games for Health program. The American College of Surgeons (ACS) has demonstrated its general interest in VR technologies as well as its particular interest in Pulse!!, which was demonstrated by invitation in 2007 at the ACS Clinical Congress in New Orleans, La. U.S. House and Senate caucuses have formed around modeling and simulation technologies with annual demonstrations on Capitol Hill. Both major U.S. political parties have supported congressional resolutions calling for public-private partnerships in developing these technologies, including virtual-world simulation, toward their having a more significant role in military and civilian training and education. The U.S. military has a longstanding interest in simulation technologies, which are used extensively not only for training but also for strategic planning. Pulse!! is a logical extension of the military's continued exploration of computer-based simulation, which since 2002 has supported the remarkably popular multi-player online game *America's Army*, an adaptation of which has proved effective in training cadets at the U.S. Military Academy (Farrell, 2003).

To date, no definitive cost-benefit analyses have been completed on the use of virtual-reality systems in health-care training and education. Rizzo & Kim (2005) observe: "Without such cost-benefit proofs, healthcare administrators, and mainstream practitioners who are concerned with the economic 'bottom-line' may have little motivation to spend money on high tech solutions if there is no expected financial gain" (pp. 15-16).

Future Directions

Pulse!! research indicates that the platform is a viable delivery system for medical curricula and that users not only enjoy using it but believe it is an effective educational strategy they would recommend to others. Three generations of problem-based cases have been developed, including first-responder training in tactical combat casualty care (TCCC). The project has reached a plateau from which to launch its next major phases, including the following:

- A case-authoring system is in preliminary stages of development. These tools will enable medical trainers and educators to develop problem-based cases from the accumulated assets of the Pulse!! learning-platform database. Case-authoring development also includes creation of new assets for the database, including virtual personnel, equipment, environments and conditions that are not associated with early generations Pulse!! cases.
- Pulse!! will be integrated with standard medical reference curricula that learners will be able to access from the platform. This component will be one with an integrated "smart tutoring" system capable of assisting learners before, during or after engaging a problem-based case.
- Long-range Pulse!! development calls for creating a server-based, multiuser system within which teams of learners will assume roles and work cases. It will not be unlike "America's Army" in this regard. This development also will include inter-user voice communication, and will not require that users be in the same location to train together.
- Pulse!! collaborators are exploring whether voice-recognition technology is sufficiently advanced that it can be integrated with the platform such that users can exchange oral conversation with virtual medical personnel.

CONCLUSION

The Pulse!! project is predicated on the belief that virtual-world technologies, which have in recent years achieved a remarkably high level of realism, present medical education with an opportunity to appropriate these media not only to convey curricula but to make them come alive, as it were, presented as problem-based cases that induce critical thinking and differential diagnostics. Continued technological development, moreover, will only enhance the educational community's ability to present curricula in dynamic, realistic ways that improve performance and reduce medical errors due to insufficient clinical training. These technologies never will replace other means of providing medical curricula, but will so dramatically enhance medical-education experience that the whole spectrum of medical and paramedical practitioners, from first-responders to practicing physicians will become more proficient in practice. In effect, over time, Pulse!! collaborators believe that this cutting-edge project is but one of many that will change the face of medical education.

Pulse!! collaborators are absolutely persuaded that VR medical-education learning platforms must rise to the high standard of reliability and validity through rigorous field testing and evaluation that is embedded in the development process from the inception of a learning platform. The medical-education community *will* hold VR technologies to this high standard, so it must be an assumption for any who would contribute to this significant paradigm shift in training and educating medical and paramedical practitioners. Adoption of the technology alone will require a quantum leap for the medical-education community. It simply must be able to assume that VR technologies have *demonstrated* value as an educational strategy.

The Pulse!! project provides a model based on experience since 2005 of the collaborative process required to achieve the high standards and goals of medical education in virtual reality. It is a collaboration fraught with natural tensions and competing interests, none of which are insurmountable but all of which must come into play as each takes ownership of its piece of a complex development process aimed at producing VR learning platforms that will enhance and improve U.S. military and civilian health care for decades to come.

NOTE

Texas A&M University-Corpus Christi technical writer Ronald E. George contributed to this chapter.

REFERENCES

AAMC. (2006). *Help wanted: More U.S. doctors.* Retrieved June 1, 2009, from the AAMC Web site, http://www.aamc.org/workforce/help-wanted.pdf

AAMC Task Force On the Clinical Skills Education of Medical Students. (2005). *Recommendations for clinical skills curricula for undergraduate medical education.* Washington, DC: Author.

Accreditation Council for Graduate Medical Education (ACGME) Outcomes Project. (2006). *Educating physicians for the 21st century.* Chicago, IL: Author.

AGCME Outcomes Project. (2000). *Toolbox of assessment methods.* Chicago, IL: Author.

Association of American Medical Colleges (AAMC). (2006). *Questions and answers about the AAMC's new physician workforce position.* Retrieved June 1, 2009, from the AAMC Web site, http://www.aamc.org/workforce/workforceqa.pdf

Council on Graduate Medical Education (COGME). (2005). *Physician Workforce Policy Guidelines for the U.S. for 2000 – 2020.* Rockville, MD: U.S. Department of Health and Human Services.

Ericsson, K. A. (1996). The acquisition of expert performance: an introduction to some of the issues. In K.A. Ericsson, (Ed.), *The road to excellence: the acquisition of expert performance in the arts and sciences, sports, and games.* Mahwah, NJ: Lawrence Erlbaum Associates.

Farrell, C. M., Klimack, W. K., & Jaquet, C. R. (2003). Employing interactive multimedia instruction in military science education at the U.S. Military Academy. In *Proceedings of the Interservice/Industry Training, Simulation, and Education Conference* (I/ITSEC); Orlando, FL, Dec. 1-4, 2003.

Fitts, P. M., & Posner, M. I. (1967). *Learning and skilled performance in human performance.* Belmont, CA: Brock-Cole.

Hader, R., Saver, C., & Steltzer, T. (2006). No time to lose. *Nursing Management, 37*(7), 23–29, 48.

Heffner, S. (Ed.). (2007). *Virtual Reality Applications in the U.S. Healthcare Market.* New York: Kalorama Information.

Hobbs, F. B., & Damon, B. L. (1996). *65+ in the United States: U.S. Bureau of the Census, current population reports, special studies, P23-190.* Retrieved June 1, 2009, from the U.S. Census Bureau Web site, http://www.census.gov/prod/1/pop/p23-190/p23-190.pdf

Holcomb, J. B., Caruso, J., McMullin, N. R., Wade, C. E., Pearse, L., Oetjen-Gerdes, L., et al. (2006). *Causes of death in U.S. special operations forces in the global war on terrorism: 2001-2004.* MacDill AFB, Tampa, FL: U.S. Special Operations Command.

Issenberg, S. B., McGaghie, W. C., Petrusa, E. R., Lee, G. D., & Scalese, R. J. (2005). Features and uses of high-fidelity medical simulations that lead to effective learning: A BEME systematic review. *Medical Teacher, 27*(1), 10–28. doi:10.1080/01421590500046924

Kohn, L., Corrigan, J., & Donaldson, M. (Eds.). (1999). *To Err is Human: Building a Safer Health System.* Committee on Quality of Health Care in America, Institute of Medicine. Washington, DC: National Academy Press.

Korea, I. T. *Times.* (n.d.). Government likes serious games, May 18, 2009. Retrieved June 1, 2009 from the Korea IT Times Web site, http://www.koreaittimes.com/story/1403/government-likes-serious-games

McDonald, C., Cannon-Bowers, J., Whatley, D., & Dunne, J. (2009). *The Pulse!! collaboration: Academe & industry, building trust.* Panel tutorial presented at the Medicine Meets Virtual Reality annual conference. Long Beach, CA: January 19-22, 2009.

National League for Nursing. (2009). *NLN annual nursing data review.* Retrieved June 1, 2009, from http://www.nln.org/newsreleases/annual_survey_031609.htm

Rasch, R. F. R. (2006). *The nurse educator shortage: Opportunities for education and career advancement.* Unpublished paper. Nashville, TN: Vanderbilt University School of Nursing.

Reid, P. P., Compton, W. D., Grossman, J. H., & Fanjiang, G. (Eds.). (2005). *Building a Better Delivery System: A New Engineering/Health Care Partnership.* Committee on Engineering and the Health Care System. Geneva: IOM Press.

Reznick, R. K., & MacRae, H. (2006). Teaching surgical skills – changes in the wind. *The New England Journal of Medicine, 355*(25), 2665–2669. doi:10.1056/NEJMra054785

Rideout, E., & Hamel, E. (Eds.). (2006). *The media family: Electronic media in the lives of infants, toddlers, preschoolers and their parents.* Menlo Park, CA: The Henry J. Kaiser Family Foundation.

Rizzo, A., & Kim, G. J. (2005). A SWOT analysis of the field of virtual reality rehabilitation and therapy. [Cambridge, MA: Massachusetts Institute of Technology Press.]. *Presence (Cambridge, Mass.), 14*(2), 119–146. doi:10.1162/1054746053967094

Satava, R. (2008). Competency, proficiency, and the next generation of skills training and assessment curricula using simulators. *Laparoscopy Today,* Aug. 13, 2008. Retrieved June 3, 2009, at http://www.laparoscopytoday.com/2008/08/competency-prof.html

U.S. Department of Commerce, Economics and Statistics Administration, National Telecommunications and Information Administration. (2002). *A nation online: How Americans are expanding their use of the internet,* Washington, DC. Retrieved May 11, 2007, from http://www.ntia.doc.gov/ntiahome/dn/anationonline2.pdf

Verrier, E. D. (2004). Who moved my heart? Adaptive responses to disruptive challenges. *Journal of Thoracic and Cardiovascular Surgery, 127*(5), 1235-1244. Retrieved May 4, 2007, from http://dx.doi.org/10.1016/j.jtcvs.2003.10.016

Watters, C., & Duffy, J. (2004). Metalevel analysis of motivational factors for interface design. In K. Fisher, S. Erdelez, & E.F. McKechnie, (Eds.), *Theories of Information Behavior: A Researcher's Guide*. Medford, NJ: ASIST (Information Today, Inc.).

Watters, C., Oore, S., Shepherd, M., Azza, A., Cox, A., Kellar, M., et al. (2006). Extending the use of games in health care. In *Proceedings of the 39th Hawaii International Conference on System Sciences*, (pp. 1-8).

Whitelock, D., Brna, P., & Holland, S. (1996). What is the value of virtual reality for conceptual learning? Towards a theoretical framework. In *Proceedings of EuroAIED*, Lisbon.

Zimet, E., Armstrong, R. E., Daniel, D. C., & Mait, J. N. (2003). Technology, transformation, and new operational concepts. *Defense Horizons, 31*. Retrieved May 4, 2007, from http://www.ndu.edu/inss/DefHor/DH31/DH_31.htm

Chapter 13
Using Serious Games for Mental Health Education

Anya Andrews
Novonics Corporation, Training Technology Lab (TTL), USA

Rachel Joyce
University of Central Florida, USA

Clint Bowers
University of Central Florida, USA

ABSTRACT

A significant number of research and experimentation efforts are currently underway to identify the effective ways of leveraging advanced gaming technologies towards the development of innovative training and education solutions for the mental health domain. This chapter identifies mental health training and education needs of modern "at risk" populations and discusses the potential of serious games as instructional interventions for addressing those needs. Special attention is paid to the importance of prevention training and ways to facilitate prevention by using serious games. Within the chapter, the authors cite a number of specific mental health-related serious game efforts and discuss design considerations for effective serious games.

INTRODUCTION

Using games within a medical domain has been a subject of a continuous debate during the last decade due to a rather fragmented nature of systematic research in this area as well a diverse spectrum of community perspectives on the effectiveness and potential implications of serious games applied for heath education and training. Considering that medical community primarily relies on tradition within its training and education methods, the medical domain has been one of the toughest areas for serious games to break into. However, new learning paradigms and modern gaming technology options continue to foster innovation within the education and training industry, and now we can observe games and immersive learning simulations being applied to a wide variety of medical training and education needs, ranging from medical equipment operations to mental health training. The latter one remains of a particular interest to researchers due to the significantly increasing demand for mental

DOI: 10.4018/978-1-61520-739-8.ch013

healthcare services in the U.S., specifically being attributable to high military deployment rates to Iraq and Afghanistan during the last few years.

This chapter is focused on discussing the use of serious games for psychological health education, with a particular emphasis on the prevention of mental health issues by way of providing effective game-based learning interventions targeted at the development of appropriate coping skills associated with certain sets of mental health risks. Starting with the review of the modern mental health training and education needs, the chapter will explore the value of prevention and how serious games can facilitate it. Special attention is paid to the design and development considerations of using gaming technology to develop effective immersive learning solutions for the mental health training and education purposes.

MODERN MENTAL HEALTH TRAINING AND EDUCATION NEEDS

The modern mental health training and education needs are very diverse, ranging from addiction-based disorders (e.g. alcoholism, gambling, and substance abuse) to psychological stress disorders, acute and post-traumatic, resulting from military and non-military stress-inducing events, such as combat stress, domestic violence, traumatic loss, sexual assault, economic hardship, etc. Although virtually everyone can be considered at risk of any of the above-mentioned stress inducers, the current composition of "at risk" populations for mental health disorders is largely represented by the military personnel and their families. Research has shown that deployment stressors and exposure to combat result in considerable risks of mental health problems, including Post-Traumatic Stress Disorder (PTSD), major depression, substance abuse, impairment in social functioning and in the ability to work, and the increased use of health care services.

According to a report issued by the Department of Defense Task Force on Mental Health in June 2007, the mental healthcare system of the U.S. armed forces is currently unable to meet the mental health needs of its service members. The task force concluded that military health system lacks the resources and fully trained staff to meet the mental healthcare needs for troops and their families. Social stigma represents the primary barrier to care with the current psychological screening procedures still unable to mitigate the bias against seeking mental health services. Family members have poor access to psychological health training and education due to insufficient availability of relevant training programs and general lack of coordination between military organizations dealing with mental health care of military personnel. Too much emphasis is currently placed on short-term treatment models, while not enough support is being provided for prevention purposes and long-term management of chronic disorders. The key recommendation made by the task force calls the Department of Defense to build a "culture of support for psychological health" by updating research and providing access to effective training and education about mental health to military personnel and their families throughout the entire military life continuum.

The Army Surgeon General's Mental Health Advisory Team IV (MHAT IV) determined that combat stress, family separation, and multiple deployments have a significant impact on the mental health of the U.S. troops. During a series of structured evaluations, the MHAT IV (2006) concluded that mental health of troops was significantly undermined, with combat exposure and the length of deployment being the greatest influencing factors on mental health status. Troops facing high levels of combat were two to three times more likely to screen positive for anxiety, depression, acute stress, and other mental health problems. The report rules out a common belief that a previous deployment experience 'inoculates' soldiers against further increases in mental health issues. The more times troops are deployed and the longer the duration of the deployment, the higher would be their rates of mental health and marital problems. The report recommended extensive mental health awareness training for troops, non-commissioned officers, and junior officers before, during, and after deployment as well as improved education and training for military mental health care providers in the areas of combat and operational stress control.

PREVENTION IS KEY

The research links the absence of effective mental health training and education programs to the lack of critical coping skills that could safeguard "at risk" populations against potential psychological health issues. Focused on providing relevant skills, such as recognizing signs and symptoms of stress, identifying sources of stress, seeking professional help and/or social support, these programs could help prevent psychological disturbances resulting from traumatic stress events and minimize the severity of existing mental health disorders. Research focused on the specific mental health needs of various "at risk" populations, barriers to accessing care, and the efficacy of existing prevention and intervention programs is critical to making mental health care more relevant, available, and effective.

In response to the frightening suicide rates among military personnel resulting from long deployments, continuous separations from family, and the perceived stigma associated with seeking help, several military organizations initiated a series of training and education efforts for military suicide prevention, specifically focused on helping eliminate the perceived stigma and shame associated with asking for help. The focus is on prevention and early intervention versus treatment of psychological disorders.

What is prevention? Feldner, Monson, and Friedman (2007) emphasized that prevention exclusively focuses on reducing the *incidence* (rather than prevalence) of a disorder. Thus, three categories of prevention programs have been distinguished:

- A *universal intervention* is applied to all members of the population, regardless of their risk for developing a disorder.
- A *selective intervention* targets only persons at risk for developing, but showing no signs of, a disorder.
- An *indicated intervention* is aimed at individuals demonstrating aspects of a disorder but who are sub-syndromal.

According to Cannon-Bowers and Bowers (2007), prevention training for populations with a potentially high-risk of Post-Traumatic Stress Disorder (PTSD) will "minimize risk, minimize cost, minimize impact on healthcare system, and overall improve the quality of life." The need for working prevention interventions not only has life in literature and theories, but in reality. Funded by the U.S. Department of Defense, a study by Stetz, Long, Wiederhold, and Turner (2008) took a close look at investigating ways to mitigate the negative psychological effects warfighters develop as a result of combat experience. Critical Incident Stress Debriefing (CISD) is a common tool used to treat traumatic stress conditions, such as Acute Stress Disorder (ASD) and PTSD (Feldner et al., 2007; Stetz et al., 2008). The CISD method of discussing the details of the traumatic event individually or in groups with a facilitator is typically prescribed and applied immediately after a person has already been exposed to a traumatic event. Nevertheless, there is a consensus within the findings of Feldner's and Stetz' research teams regarding the fact that CISD does not necessarily reduce PTSD, and, in some cases, may actually increase post-event arousal along with amplifying traumatic event-related memories. Based on these findings, it is not a leap of faith to surmise that CISD methods should not be the benchmark of success for future, improved prevention interventions of PTSD and related conditions.

Cognitive Behavioral Therapy (CBT) is another method used in PTSD treatment and prevention that has been widely described in literature and empirically evaluated. Feldner et al. (2007) summarize a series of studies using CBT in comparison to repeated assessment, delayed assessment, and supportive counseling. Results of these studies appear to be promising in the short term, with significantly lower rates of PTSD, at two months in one study and six months in another. Additionally, psychobiological treatments using pharmaceuticals to amend the psychobiological responses to traumatic events and stress have been extensively used alone or in combination with CISD and CBT. Results of these studies also have minimal efficacy in the prevention of disorders such as PTSD and ASD.

In their quest to find novel approaches to facilitate prevention of mental health disorders, Stetz et al. (2008) conducted a series of experiments using Virtual Reality Stress Inoculation Training (VR-SIT) and Coping Training (CT) via relaxation behavior techniques, such as Progressive Muscle Relaxation (PMR) and Controlled Breathing (CB) administered to 63 volunteers attending a combat medical class designed to psychologically 'harden' this population to combat stress to face pre- and post-traumatic events during deployment. The VR-SIT consisted of two scenarios where (1) medics had to decide to shoot or treat in a combat situation, and (2) transporting a victim via helicopter to their next care destination while under fire and with effects of turbulence. The VR-SIT and the CT tools were tested separately and together with the overall results of the study indicating that "VR-SIT may be particularly effective way to raise stress levels in individuals who have already experienced combat, " and " the combination of these two techniques might be instrumental to producing the 'hardening' effect against combat stress." (Stetz et al., 2008, p. 244). While this study blazes a trail for the use of virtual environments for stress prevention interventions, the absence of long-term data for the participants' ability to cope with their traumatic exposure requires further investigation and validation of the effectiveness of these two treatment techniques for prevention purposes.

USING SERIOUS GAMES TO FACILITATE PREVENTION

During the last decade, the quest for innovative approaches to military training has brought considerable attention to Serious Games. While the decades of traditional video games with their aggressive contexts and competition schemes have added a rather tainted slant to the term 'serious games', there is a growing body of research looking into positive impacts of play and leveraging gaming technology for a wide range of instructional treatments. The notion of Serious Games (SG) today refers to the uses of state-of-the-art gaming technologies for developing adaptive instructional solutions. According to the Serious Games Forum (2009), "a serious game may be a simulation which has the look and feel of a game, but corresponds to non-game events or processes, including business operations and military operations." There is a growing trend to look at serious games as Immersive Learning Simulations (ILS) that represent "an optimized blend of simulation, game element, and pedagogy that leads to the learner being motivated by, and immersed into, the purpose and goals of a learning interaction. Serious games use meaningful contextualization, and optimized experience, to successfully integrate the engagement of well-designed games with serious learning goals" (Wexler et al., 2008, p.3).

Gaming environments offer exceptional potential for teaching behavioral skills by providing opportunities for simulating the real-life situations

Table 1. Psychological health issues, stress management skills, and instructional elements

Psychological Health Issues	Relevant Stress Management Skills	Instructional Techniques	Game Elements
• Fear and anxiety • Intrusive thoughts • Sleep disturbance • Hypervigilance • Grief & depression • Guilt feelings • Social isolation • Anger & aggression • Substance Abuse	• Resilience • Self-assessment • Reflection • Interpretation • Understanding sources of stress • Strategy formation • Seeking support • Routine adjustment	• Digital Storytelling • Learning by Doing • Case Study • Guided Discovery • Coaching • Situated Learning • Reflective Dialogue • After-Action Review	• Challenge • Rules/Controls • Fantasy/Mystery • Sensory Stimuli • Cues/Feedback • Character Interaction • Humor

and conditions, under which the development of these skills occurs. Considering that knowledge acquisition and the development of cognitive and behavioral skills rely heavily on learner engagement and motivation, the major benefit of using simulations and games for learning relates to their potential to engage the learner with the instructional content, making the learning experience active rather than passive. Multiple studies have concluded that problem-solving and decision-making skills are natural by-products of educational games resulting from inductive discovery, exploration, trial and error, drawing conclusions and other activities (Rieber, 1996; Prensky, 2001; Gorriz & Medina, 2000). Games motivate learners to take responsibility for their own learning, which leads to intrinsic motivation contained by the game-based learning method itself (Rieber, 1996). A properly constructed game can train the learners to adapt to novel situations, thereby, increasing cognitive agility. While engaged in a serious game, the learners tend to retain a significant amount of information by building cognitive maps. This process fosters adaptive decision making, a critical competency for managing mental health.

Simulations and games have been successfully applied to the medical training domain for a number of years. In most cases, however, the primary focus of those games has been procedural training. It is important to explore the use of advanced gaming technologies for creating effective instructional treatments that would provide an optimal blend of education and training and focus on the mental health knowledge acquisition as well as stress management skills and behaviors. Behavioral health screening is another avenue where serious games can be applied to facilitate early detection and prevention of mental health concerns. Table 1 below summarizes the common types of psychological health issues and associated stress management skills that can be fostered via an appropriate combination of instructional techniques and elements of game-based learning.

Due to their ability to deliver engaging empathy-infused content, serious games have also been recognized for their persuasiveness, a quality, which would be a significant benefit to any kind of mental health training. In fact, there is a new genre of serious games which claim to boost users' mental health and self-esteem. Inspired by the success of Nintendo's "Brain Age" focused on math and word puzzles, some developers attempt to harness the persuasive abilities of video games for psychological health purposes.

A Tokyo-based Dimple Entertainment has released a game that promises to deliver a daily measurement on your mental and emotional health based on a series of questions ranging from love to money under the unconventional title "DS Therapy." Dr. Mark Baldwin, a psychology professor at McGill University in Montreal, Canada, and his research team designed a game to address insecurity and stress based on social psychology research. Titled "MindHabits Booster," this game is supposed to make people feel good about themselves by having players repeatedly pick an agreeable face from a group of frowning faces with the objective to train the players to look for acceptance and ignore rejection. These stress busting, confidence boosting games use simple, fun-to-play exercises that help players develop and maintain a more positive state of mind.

To address the mental health education and training needs of the military community, Novonics Training Technology Lab (TTL) and Institute for Simulation and Training (IST) at the University of Central Florida have teamed to develop an interactive game-based system for psychological health education under the sponsorship of the Telemedicine Advanced Technology Research Center (TATRC) of the U.S. Army Medical Research and Materiel Command. The goal of the effort is to explore the effective and efficient ways of leveraging advanced gaming technologies to develop an interactive learning solution for the Psychological Health (PH) domain. Within the context of this effort, Novonics Corporation and the IST are developing a theoretical model to serve as a map for linking specific stress management skills to the respective psychological health issues commonly encountered within military environments. This theoretical model will illustrate a Soldier's decision-making process for approaching stress-related issues and inform the design and development of the interactive instructional solution for military personnel that could be expanded to the veteran and civilian populations. The scenario-based instructional system will allow the learners to practice identifying different types of psychological issues, determining the appropriate options within the framework of the military PH process, and applying proven coping strategies within a virtual environment. Focused on the military PH process, available PH options, PTSD prevention and coping strategies, the instructional solution is envisioned to include closely intertwined educational and training components and provide opportunities not only to raise the learner awareness regarding the PH-related issues and available options, but also help develop stress prevention, stress recognition, and stress management skills.

The Virtual Reality (VR) Assessment and Treatment of Combat-Related Post-Traumatic Stress Disorder (PTSD) project focuses on using immersive virtual environments for the treatment of Iraq War veterans diagnosed with PTSD. Sponsored by the U.S. Office of Naval Research (ONR), this project is being conducted as a multi-year effort by the Center for Virtual Reality and Computer Simulation Research at the Institute for Creative Technologies (ICT). Within the VR environment, a veteran with PTSD can experience a combat-relevant scenario in a low-threat context designed to facilitate therapeutic processing of emotions and ultimately de-condition the effects of the disorder. The main goal of the project is to discover innovative means to help restore quality of life for veterans and their families by overcoming the effects of PTSD. The development team implemented a cost-effective technology solution by recycling virtual graphic assets from the combat tactical simulation training game, Full Spectrum Warrior and other ICT project assets (ICT, 2009).

In the case of addiction recovery and prevention, there are existing methods of Relapse Prevention Therapy (RP) currently being used by many addiction therapy professionals. The RP model is a cognitive-behavioral model asserting that a relapse to substance abuse can be prevented through a set of skills that allow the patient to recognize and cope with situations that are likely to initiate it (Larimer, Palmer, & Marlatt, 1999). Currently, RP treatment is regularly delivered using psychoeducation methods either to individuals in a therapy session with a trained specialist or as part of a group session. While recovering addicts have RP treatment available, it is not readily accessible outside of a therapy session when patients are most likely to face a

Figure 1. Guardian angel game interface

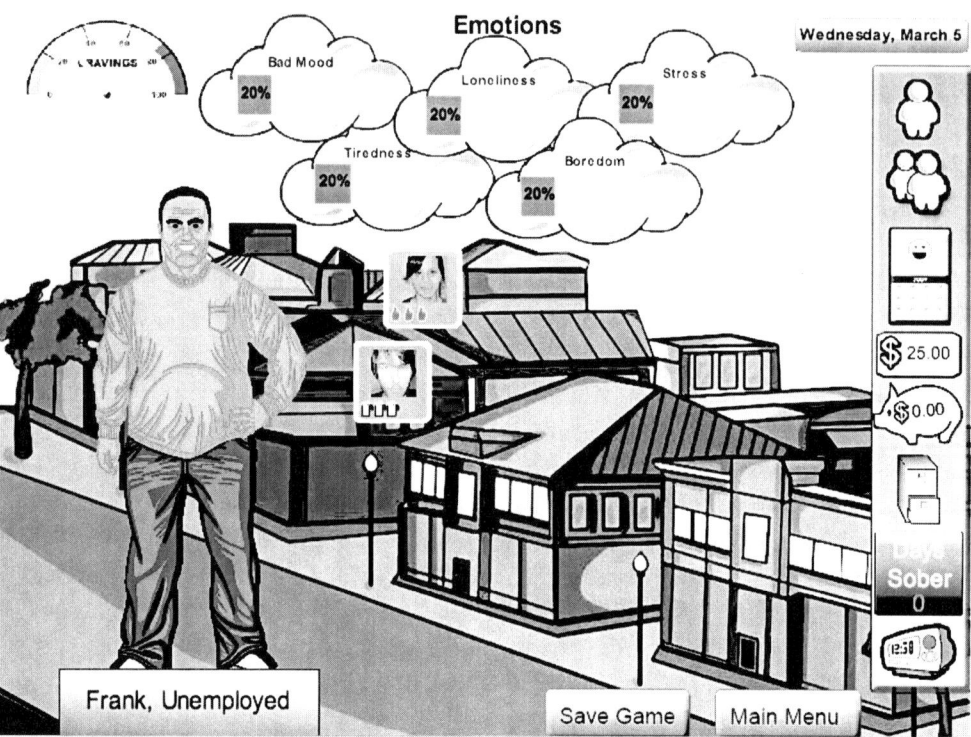

relapse situation at home or on a social setting. The unfortunate result of this is a high relapse rate among patients after receiving treatment. In an effort to minimize relapse rates, researchers at the University of Central Florida and the Ralph H. Johnson VAMC have been contracted by The Robert Wood Johnson Foundation to create a serious game based on the principles and lessons of RP treatment for patients currently receiving treatment for alcohol addiction at the Ralph H. Johnson VAMC in Charleston, SC.

Currently in its last phases of development, this game is a decision-making simulator where a player is an angel who is on the fast-track to becoming a guardian angel, provided he or she 'watches over and guides' a soul to sobriety for one full year. The player has a choice of four characters to guide on their way to recovery. The Guardian Angel game is composed of two interrelated parts of a core game where the player-angel must assist in choosing their character's activities for the day using RP craving-management techniques as well as choosing relationships to invest in by selecting either a positive, recovery-supporting person or an addiction-supporting person, both of whom are supported by the narrative and interface of the game. Four of the approximately thirty activity choices will generate a minigame, or part two of game play. These four minigame activities are directly derived from the following RP intervention techniques: (1) drink refusal, (2) identification of high-risk situations, (3) lifestyle-balancing, (4) assertiveness training, (5) cognitive restructuring, and (6) stimulus control (Larimer, Palmer, & Marlatt, 1999). Minigames are subgames designed like a casual or puzzle game. While minigame game play is intentionally simpler, it provides a more robust game experience in comparison to a

Figure 2. Route planner minigame interface

strictly psychoeducational method of reading RP treatment material or having it delivered in group or one-on-one sessions.

Each activity, including the minigames, constitutes one day in the game. The objective of the game is to play for 365 days without a relapse by investing in healthy personal relationships and choosing positive outlets as your daily activities to ultimately create a balanced, simulated life for a person recovering from an addiction. If a player does encounter a relapse, the in-game consequences, such as a loss of income or a job, will occur. However, opportunities to try again via remedial activities, such as job training, driving lessons, and Alcoholics Anonymous (AA) classes, do emulate the real-life redemption options available to recovering addicts to assist in continuing to build-up their skill sets for staying sober.

The four minigames inside Guardian Angel are Route Planner, Clean House, Drink Refusal, and Balance Out. In Route Planner, players have to avoid establishments where alcohol is available as they navigate home by plotting out their 'route' in an interactive map. Players have a designated amount of money and amount of time to get home and win. Route Planner is designed to practice RP training in stimulus-control and identification of high-risk situations.

In Clean House, a stimulus-control exercise, players must throw out all alcohol and related items in a time-pressure situation while being careful to preserve items that are still necessary and safe.

In Drink Refusal, players must use their RP drink refusal skills, assertiveness training, and identification of high-risk situation skills when participating in social activities, such as bowling, birthday parties, camping etc. This conversation game allows players to choose who joins them in a given activity. The scenario that transpires between the player and an in-game friend while socializing will include one or two drink offers providing the opportunity for players to engage in drink refusal, assertiveness practice, or high-risk detection by disengaging from the activity.

The final minigame Balance Out elicits practice in many areas of RP, but specifically aims to accomplish overall cognitive restructuring and lifestyle balance. When a player chooses to attend therapy as their activity for the day, Balance Out

Figure 3. Clean house minigame interface

deploys as a large scale surrounded by a wheel of Individual Resources (IRs). The player spins the wheel as if it was the "wheel of fortune", and a random IR gets selected, for example, body language. In RP treatment, recovering addicts learn how to communicate via body language as an individual resource for their continued sobriety. Once the IR is selected, an information description of the IR is narrated to the player followed by a narrated scenario where the player must choose their body language or other response. Each response is weighted as gold, silver, and/or bronze and after a player chooses their response, a gold, silver, or bronze coin is deposited onto the scale via animation. The objective of this minigame is to balance the scale with the coins awarded based on the player's IR decisions.

The Guardian Angel game has been piloted through informal, observation-based focus groups and is scheduled to be used in empirical studies in 2009–2010.

Serious Games for Mental Health Domain: Key Design Considerations

Serious games have proved to be effective both from the pedagogical and business perspectives affording the possibility of developing sophisticated training at a fraction of the costs traditionally associated with interactive and immersive courseware. To date, however, most serious games remain outcome driven and fail to provide the learner with effective instructional supports. In order to maximize the instructional value of serious games, we recommend referring to the good old principles of Instructional Systems Design (ISD) and game-based learning and viewing the technology as the enabler and not the goal in itself.

Figure 4. Drink refusal minigame screenshot

Figure 5. Balance out minigame interface

Advanced gaming technologies do lend themselves to develop visually appealing instructional solutions, but appropriate learner supports, engaging storyline, dynamic content sequencing, and assessment tools require a skilled instructional designer's touch. As instructional design practitioners, we strongly believe that serious games must go beyond simply raising the learner's awareness about coping strategies for a given set of mental health issues. In order to make sure a serious game results in skill development, it is necessary to create opportunities within a game to walk the learner through all of three phases of skill acquisition: cognitive, associative, and autonomous.

Figure 6. Balance out minigame interface of an individual resource (IR)

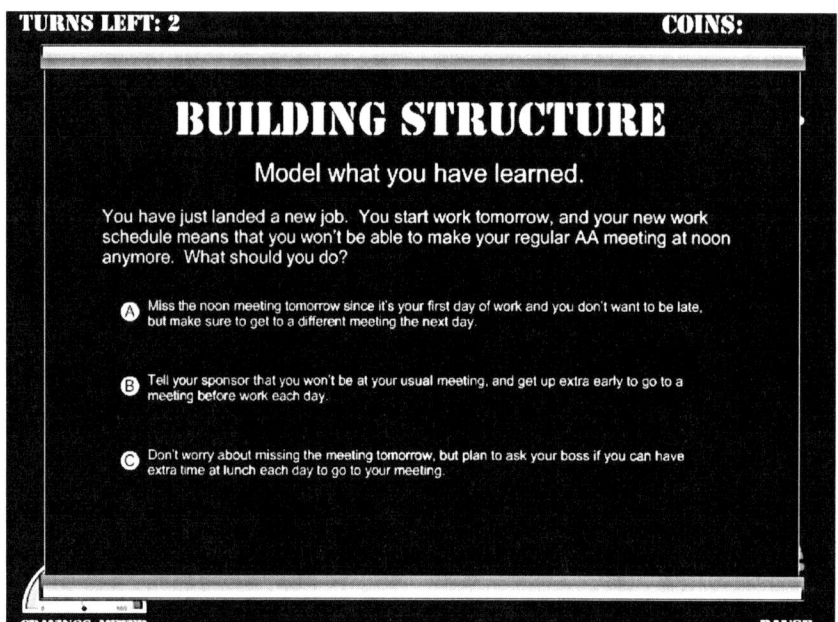

Based on our experience with using serious games for behavioral training purposes, we recommend the following design ingredients as a recipe for successful serious games:

1. ***Instructionally sound serious game concept.*** Similarly to any training development effort, a serious game concept must be based upon a series of analyses to determine, the training goals, learner characteristics, desired learning outcomes, an assessment strategies. Designing a serious game is a creative process that must be based upon a number of calculated decisions with the ultimate focus on maximizing the learning value.

2. ***A solid construct of explicit and implicit instructional events and activities that support clearly defined learning objectives.*** It is important to remember that the learning objectives must be driving any serious game design effort. Identify your goals before starting the design and development. Unfortunately, there are too many examples of serious games where the learning goals and objectives appear to be more of an afterthought.

3. ***Use of state-of-the-art gaming technologies to ensure realism and immersion and provide sensory stimuli.*** With the recent advances in the gaming technology field, serious game developers have endless possibilities for selecting the appropriate technology solution for any training context and budget. It is up to the designers to determine the degree of justified immersion and sensory stimuli. Game modding and machinima remain amongst the most popular techniques to develop quality serious games while keeping development costs low.

4. ***A robust storyline that is both relevant and engaging.*** Contextual relevance is the key enabler for learner motivation and engagement. The ability to "hook" the learner from the very beginning is an unwritten guarantee that he or she will follow through the entire instructional treatment. A powerful storyline can also compensate for low-fidelity sensory

experience due to its ability to engage the learner on the emotional level to the extent that the visual fidelity may become irrelevant.

5. *The "Fun Factor."* This ingredient has always been subject to interpretation and divided opinion due to diverse perspectives on what constitutes "fun." Does it relate to learner engagement, competition, use of humor within the game, or other entertainment elements? Our answer would be "all of the above." Effective serious game play must be constructed to engage the learner via creative use of entertainment means, which, if used appropriately, will encourage the learner to make not only a time investment in this game, but also take away the critical nuggets of experience from it.

FUTURE RESEARCH DIRECTIONS AND CONCLUSIONS

The APA has recently called for a "figure-ground reversal" in professional psychology, emphasizing the need to view it as a health profession, with mental health as a subset of its expertise. Such a change in perspective would require a rather dramatic change in psychology training and education programs. It is interesting to note that 50-75% of all visits to primary care medical personnel are for problems with a psychological origin or for problems with a psychological component.

There is a growing body of empirical evidence supporting the effectiveness of mental health interventions in ameliorating a wide range of physical health problems, including both acute and chronic disease affecting literally every organ system and encompassing pediatric, adult and geriatric populations (Levant, 2008). In addition to being clinically effective, these interventions are dramatically less expensive than alternative somatic interventions across a wide variety of illnesses and disorders, including cardiovascular disease, hypertension, diabetes, neoplasms, and traumatic brain injury. One way to reduce the costs associated with psychological health care is to provide the public with effective PH education and training programs.

Considering the cost-effectiveness of PC-based training solutions and their ability to reach wide learner audiences, research on using serious games to address specific mental healthcare needs of general populations is needed. Serious games could potentially be used to reduce the stigma of seeking mental health care, educate patients about the behavioral nature of their problems, and help during initial patient evaluations. Although there are numerous examples of serious games with a promise of being effective within the medical domain, the first and most important thing they lack are the actual users who could help validate these games as training or medical interventions and build a research base to facilitate cultural change and promote wider use of serious games within the medical domain.

Games for Health is a project conducted under the overarching umbrella of the Serious Games Initiative, a Woodrow Wilson International Center for Scholars effort that applies game technologies to a wide range of public and private policy, leadership, and management issues. Games for Health serves as a platform for the games specifically focused on health care applications. The project brings together researchers, medical professionals, and game developers to share information about best practices, research results, and ideas that might improve health care administration and policy.

REFERENCES

Cannon-Bowers, J., & Bowers, C. (2007). *Learning and Technology-Based Solutions for PTSD Prevention: An Example of Future Medical Simulation.* Paper presented at the Interservice/Industry Training, Simulation & Education Conference (IITSEC), Orlando, FL.

Department of Defense Task Force on Mental Health. (2007). *An achievable vision: Report of the Department of Defense Task Force on Mental Health.* Falls Church, VA: Defense Health Board.

Feldner, T. M., Monson, C. M., & Friedman, M. J. (2007). A Critical Analysis of Approaches to Targeted PTSD Prevention Current Status and Theoretically Derived Future Directions. *Behavior Modification, 31*(1), 80–116. doi:10.1177/0145445506295057

Gorriz, C. M., & Medina, C. (2000). Engaging girls with computers through software games. *Communications of the ACM, 43*, 42–49. doi:10.1145/323830.323843

Hoge, C. W., Castro, C. A., Messer, S. C., McGurk, D., Cotting, D. I., & Koffman, R. L. (2004). Combat Duty in Iraq and Afghanistan, Mental Health Problems and Barriers to Care. *The New England Journal of Medicine, 351*(1), 13–22. doi:10.1056/NEJMoa040603

Institute for Creative Technologies. (2009). *Post-Traumatic Stress Disorder Assessment and Treatment: Project Description.* Retrieved on June 2, 2009, from http://ict.usc.edu/projects/post_traumatic_stress_disorder_assessment_and_treatment_ptsd/C45

Krupnick, J. L., & Green, B. L. (2008). Psychoeducation to Prevent PTSD: A Paucity of Evidence. *Psychiatry, 71*(4), 329–331.

Larimer, M. E., Palmer, R. S., & Marlatt, G. A. (1999). Relapse prevention: An overview of Marlatt's cognitive-behavioral model. *Alcohol Research & Health, 23*(2), 151–160.

Levant, R. F. (2008). *Evidence-Based Practice in Psychology.* Paper presented at the Culturally Informed Evidence-Based Practices (CIEBP) National Conference, Bethesda, MD.

Mental Health Advisory Team (MHAT IV) (2006). *Operation Iraqi Freedom 05-07 Final Report.* Office of the Surgeon Multinational Force-Iraq and Office of the Surgeon General Unites States Army Medical Command.

Prensky, M. (2001). *Digital Game-Based Learning,* New York: McGraw-Hill.

Rieber, L. P. (1996). Seriously considering play: Designing interactive learning environments based on the blending of microworlds, simulations, and games. *Educational Technology Research and Development, 44*(2), 43–58. doi:10.1007/BF02300540

Serious Games Forum. (2009). *What are Serious Games?* Retrieved June 02, 2009, from http://seriousgames.ning.com/

Stetz, M., Long, C., Wiederhold, B. K., & Turner, D. (2008). Combat Scenarios and Relaxation Training to Harden Medics Against Stress. *Journal of CyberTherapy & Rehabilitation, 1*(3), 239–246.

Wexler, S., Corti, K., Derryberry, A., Quinn, C., & Van Barnveld, A. (2008). *Immersive Learning Simulations: A 360-Degree Report,* eLearning Guild.

ADDITIONAL READING

Ahlers, R., Driskell, J. E., & Garris, R. (2002). Games, motivation, and learning: A research and practice model. *Simulation & Gaming, 33*(3), 441–467.

Anderson, J. R. (1982). Acquisition of cognitive skill. *Psychological Review, 89*, 369–406. doi:10.1037/0033-295X.89.4.369

Belanich, J., Sibley, D., & Orvis, K. (2004). *Instructional Characteristics and Motivational Features of a PC-Based Game.* Research Report # 1822; Army Research Institute.

Bergeron, B. (2006). *Developing Serious Games*, NY: Charles River Media.

Bonk, C., & Dennen, V. (2005). *Massive Multiplayer Online Gaming: A Research Framework for Military Training and Education*. ADL Technical Report # 2005-1.

Chapman, B. (2004). *E-learning Simulation Products and Services*. Brandon Hall.

Cianciolo, A., et al. (2007). *Using Digital Storytelling to Stimulate Discussion on Army Professional Forums*. Paper presented at the Interservice/Industry Training, Simulation, and Education Conference (IITSEC) Conference, Orlando, FL.

Hayes, R. T. (2005). *The Effectiveness of Instructional Games: A Literature Review and Discussion*. Technical Report, Orlando, FL: Naval Air Warfare Training Systems Division.

Hoyt, C. L., & Cuilla, J. B. (2004). *Using Advanced Gaming Technology to Teach Leadership Skills*. Woodrow Wilson International Center for Scholars, Foresight and Governance Project Report.

Iuppa, N., & Borst, T. (2007). *Story and Simulations for Serious Games: Tales from the Trenches*. Burlington: Elsevier, Focal Press.

Laff, M. (2007). Serious Gaming: The Trainer's New Best Friend. *Training & Development, 01*, 52–57.

Mayo, M. J. (2007). Games for Science and Engineering Education. *Communications of the ACM, 50*(7), 31–35. doi:10.1145/1272516.1272536

McDowell, P. (2007). Serious Games: Why Today? Where Tomorrow? *MS&T Magazine, 2*, 26-30.

Rieber, L. P. (1996). Seriously considering play: Designing interactive learning environments based on the blending of microworlds, simulations, and games. *Educational Technology Research and Development, 44*(2), 43–58. doi:10.1007/BF02300540

Royle, K., & Clarke, R. (2003). *Making the Case for Computer Games as a Learning Environment*. University of Wolverhampton, School of Education.

Salen, K., & Zimmerman, E. (2003). *Rules of Play*, Boston, MA: MIT Press.

Salen, K., & Zimmerman, E. (2005). *The Game Design Reader: A Rules of Play Anthology*. Boston, MA: MIT Press.

Chapter 14
Pervasive Health Games

Martin Knöll
University of Stuttgart, Germany

ABSTRACT

Prevention and therapy of emerging lifestyle diseases are strongly linked to daily behavior, physical activity, and knowledge of healthy life. The potentials of serious game applications in a health context for user's motivation, education, and therapy compliance is investigated and so far widely accepted. Pervasive Health Games (PHGs) combine pervasive computing technologies with serious game design strategies, in order to unfold user's playground to the city and therefore to their everyday life. The following article presents the typology of PHG within Games for Health as an interdisciplinary working field consisting of health care, psychology, game design, sports science, and urban research. A brief introduction to the theme is illustrated with a conceptual "showcase," a pervasive game concept for young diabetics.

URBANISM AND HEALTH

The relationship between healthy living and the city is a "critical one." Since the 19th Century, "urban city life" has been considered as a primary cause for various epidemical outbreaks of diseases. Cholera, typhus, tuberculosis, and "infectious fevers" had been closely linked to the massive growth of population and the "insanitary" conditions in urban living areas of the industrialized city. Various political parties used the "picture of the unhealthy city" as an instrument in the health discussion in order advocate their own differing interests (Rodenstein, 1988, p. 81).

In September 2006, the New York Times featured an article series called "Bad Blood," headlining that one of eight adult persons in New York City has diabetes (Santora, 2006). Authors brought up the word of a "social epidemic" referring to the fact that the diabetes type 2 epidemic is not only caused by a genetically disposition but by a certain lifestyle. Modern "lifestyle diseases" affect all social milieus, but especially the socially deprived. The western way of life, such as physical inactivity, stress, and an unhealthy diet, takes place in the city and as shown in the case of New York City, is adopted in cities by millions of immigrants every year.

The *19th century picture* of the "unhealthy city" influenced modern town planning mainly in two ways. On the one hand the emerging hygiene and sanitary movement transferred the health discourse from a social to an engineer terrain. Within the political philosophy of the time, any problematic symptom of the industrial city would be addressed by technical progress. Hygiene and public health policies therefore midwifed the "scientific" carrier of the discipline of modern town planning (Benevolo, 1967, p. 32). On the other hand most of the modern Avant-garde architects of the early 20th century responded to this picture with an explicit "anti-urban" idea of healthy living, which still effects contemporary urban designs (c.f. Fehl & Rodriguez-Lores, 1997, p. 51).[1] To put it in a nutshell, current health issues are fundamentally interwoven with both - spatial *and* social aspects of city life. Their epicenters are situated in the post-urban megapolises. Facing the big challenges of modern lifestyle diseases such as obesity, type 2 diabetes or chronic heart diseases, health orientated urban research must focus on urban lifestyle, rather than merely investigating, analyzing and influencing the built environment. Seen as a research *and* design discipline, urban planning cannot be seen anymore as restricted to the production of buildings, streets, and parks. With the references *cybernetics, pervasiveness,* and *game* urban planning must face a postmodern, computer-assisted of health promoting systems. In subsequent chapters, the author would therefore discuss possibilities to *re-use* rather than to re-build cities within the health context.

SERIOUS GAMES ARE ABOUT TO LEAVE THEIR ELECTRONIC SHELLS...

In order to develop mechanism and strategies to re-use our cities, the cultural technique of serious games comes into play. The "Serious Games Initiative" founded by the Woodrow Wilson International Centre for Scholars has defined the term "serious game" as "digital games with non-entertainment purposes such as health care, security, management or learning" in 2002. Since 2004 the sub-group "Games for Health" focuses on "the impact games and game technologies can have on health care and policy" ("Games for Health - About," n.d., Welcome section). [2] Current trends include video games for rehabilitation and therapy issues and the emerging field of "Exergaming," motivating players for more physical activity.

Considering the increase of chronic diseases and prevention projects, we seek for game design strategies, which integrate the game play into the everyday life of its players. A new generation of computer games, called "Serious Pervasive Games" therefore overlay the physical space with a virtual game zone. According to Borries, Walz, & Böttger, (2006):

"They [pervasive games] not only serve as a new type of gaming, but also as a new form of using and experiencing the city. In pervasive games, the city transforms into a playground that can be played every time and everywhere. And this functional assignment does not depend any longer on the building structures but on the available technology." (Borries, Walz, & Böttger, 2006, p. 41)

Several prototypes for Serious Pervasive Games (SPGs) have been developed in the fields of health care, security, tourism, management, or learning in the last ten years.[3]

One of the first pervasive projects that focus on "the relationship between art, technology and health" is the Mixed Reality Game "Ere Be Dragons" by Active Ingredient and the Middlesex University in the UK. "The player wears a heart rate monitor, and inputs his or her age into a pocket PC." An optimal heart rate is calculated and the player starts to walk where ever or however he or she wishes. During the walk an on-screen landscape is built, which uses GPS and corresponds to the real environment surrounding the player and his or her measured heart rate. If players do well for example, adequately exercising their hearts, the landscape flourishes, "while overexertion leads to the growth of a dark, forbidding forest (Davis, Jacobs, Moar, & Watkins, 2007, p. 296)."

Apart from that scientific and artistic approach, doctors and managers in the public health sector begin to see that healthcare is no longer restricted to the hospital or the clinician's practice. Researchers begin to use new technical possibilities and pick up for example the growing interest of users in collecting private physical data. The Aarhuis-based "Centre for Pervasive Healthcare" (CfPH) works on several case studies for a so-called "HealthyHome" (Centre for Pervasive Healthcare, n.d.). Pervasive Healthcare provides a technical and mental support for patients in their living space in special situations (e.g. for the period of a pregnancy, for elderly or isolated people).

TOWARDS PERVASIVE HEALTH GAMES

Facing the diverse challenges of preventing and caring for chronic "lifestyle" diseases, the user's potential field of action should be discussed: from the scale of a "healthy home" to the scale of the city. By combining technological developments in pervasive computing and modern communication methods with serious game design strategies, a new interdisciplinary working field within "Games for Health" is unfolded. In order to highlight this topic, the term *Pervasive* Health Games (PHGs) is proposed, referring to its principal possibility of being played everywhere and every time. Before discussing the particular conceptual challenges of design, development, and implementing of this sort of Health Games, let us have a short look on potential impacts:

- *Increase of motivation and user compliance:* Games package complex serious content into playable, entertaining, and "not-so-serious" units. Games for Health therefore certainly will have the short-term effect of increasing user's motivation and compliance to their therapy. As a long-term effect of the proposed overlay of gaming and healthy everyday life, we see a potential development of the self-perception by the user - emphasizing the figure of a self-confident, active "Health Player."
- *Educational games:* The individual learning of healthy lifestyle and habits in our everyday life becomes more important within preventive and therapeutically treatment. Using pervasive technologies, PHGs can specify this content with integrated therapeutically specifications (e.g. personal risks, chronic diseases) directly to end-users' needs.
- *An extension of therapeutically education:* In general, patients of chronic diseases learn under clinical circumstances how the disease affects their body. For example when and how they need to apply their medicine and what an appropriate diet should look like. They need to be instructed how to use their personal technical devices and how to do documentation of all the important parameters. After this period the patients return to their everyday lives facing its whole complexity. The patients then often realize that their therapy was adjusted under the so-called "white-coat effect." In this point a PHG can support patients as an extension

of the clinical education into their everyday routine.
- *Change of lifestyle:* As an extension of educative or learning games and one possible further direction of research, the potentials of PHGs in supporting an active change of lifestyle must be investigated. Pervasive game technologies working on biological sensors, deliver direct feedback for mobile environments. This "virtual augmentation" of the body bears potentials for new strategies of serious games, objecting education for lifestyle changes.

Features

Conceptually, PHGs would be composed composed of the following features. For a start a PHG consists of mobile, "medical" devices. They are the central sensorial devices of the body's biological parameters. Sensors can alter according to the medical content provided, from heart rate monitors to blood glucose measurement systems or "low-tech" pedometers. Additionally, "communication" between the sensorial device and the interface as a mobile phone must be provided. Possible technical solutions vary from mobile Internet access to "body near networks." The third conceptual feature deals with the "medical content." This would imply the specification of the serious content, such as therapy goals and parameters as well as educative information. This feature has to be developed by the game designer in alliance with an interdisciplinary team of physicians, nutritionists, psychologists, or sport scientists. The game design process would aim to reduce the complex scientific perspective on the theme. Finally, we face the aspect of "game play," as Salen and Zimmerman put it: "The focus of a game designer is designing *game play,* conceiving and designing rules and structures that results in experience for players" (Salen & Zimmerman, 2004, p. 1).

SHOWCASE: PLAYING WITH DIABETES?

The subsequence division presents versions of an educative game for diabetics. All three descriptions focus on the possible experience, e.g. the game play for the users. They therefore instinctively integrate and anticipate features introduced above. In order to visualize possible materializations of a Pervasive Health Game, the author would like to present the project "DiabetesCity" as a showcase of a "research by design" project.

Motivation

Diabetes is a chronic disease likely becoming a major epidemic in the developed countries. Today 200 million live with the disease and the World Health Organization (WHO) predicts an increase to 350 million diabetics until 2025. Type 2 diabetes, which comprises 90% of the diabetes cases, is largely the result of excess body weight and physical inactivity. Moderate changes in lifestyle, an adequate diet, more physical activity and loss of weight positively influence the therapy of type 1 and type 2 diabetes.

Documenting daily medical data is one essential part of an education for diabetics pointing out an adaptation of these lifestyle factors. Diabetics collect their data in so called "diabetes-diaries." These documents, often hand-written note books, are the basis for a discussion between the doctors, diabetes-assistants and the patients in order to improve the therapy set up.

Digital documentation tools, for example applications running on a mobile phone or a personal computer are already available, but hardly accepted by diabetics. Further more the documentation is merely focused on medical aspects. Even if the positive effects on diabetes by a sound lifestyle, e.g., healthy diet and physical activity, is proven by many studies, there is no documentation tool available so far that detects the connection between the medical data and the daily personal behavior.

Objectives

"DiabetesCity" therefore aims to develop a game design concept, which playfully motivates children and young adults with diabetes to document their medical data. In their age it is of particular importance to combine an active lifestyle with a well-adjusted diabetes treatment to avoid the long-term complications of the disease. Combining medical treatment with daily activities, we seek to encourage a modern form of daily recording therapeutic data. So-called "spatial diaries" should help to visualize deeply inscribed personal behavior, such as physical activity, stress, or inadequate nutrition - factors, which have a grave impact on the patient's sugar levels. As a result, the serious claim of the game DiabetesCity would be to document as much (medical) data as possible for the game period of one week. The collected data could be visualised within in a therapy tool, on which patients and doctors would discuss necessary therapy improvements.

Developing Ideas of Gameplay

The subsequent three proposals adress possible versions of game play for DiabetesCity according to the objectives framed above, which we discussed with patients in Stuttgart. All concepts are developed in order to run on a mobile platform for a test period of about one week.

Version I: "The Zuckerschloss"

In *The Zuckerschloss* ("Candy Castle") the players take on the role of creative builders. A team consists of one diabetic and his or her parents. Both jointly document a part of their daily routines and create a "spatial diary" - the virtual Castle. The diabetic for example, measures his or her blood sugar and takes a photo of his lunch. As soon as he or she puts his data into his or her mobile phone, the game starts asking to design a new room. These rooms can be filled up with information, photos, texts, stories, and sugar levels or insulin doses. During the game play users begin to classify situations and can link them with certain parts of the castle. The rooms can be individually designed, atmospheres and surroundings can be added and chosen. The player's teammate, who is not diabetic, would build his own Castle, documenting for example the meals he or she is eating during the game period. Players can go for a virtual walk through their "diaries." In further levels, players can exchange their spatial diaries, can explore their teammate's castle and invite other players to their version of "The Zuckerschloss."

Version II: "Capture the Flag"

Players (one diabetic and up to three of his friends) form a gang and are filming, photographing and describing their everyday life in school. By documenting data, as measured blood sugar levels, insulin doses, meals, or physical activities, the team may set claims throughout the physical school grounds. After having uploaded a "post" consisting of clips, picture and text-messages, the team may mark the spot with a tag symbolizing that this area had been captured. When other teams pass the spot, players can see the uploaded documentation by holding their mobiles close to the Near Field Communication (NFC)-tag and download the content. Teams can conquer spots by creating and uploading a new documentation to the "Capture the Flag" server. The team capturing and defending most of the school claims during the game period, wins the match.

Version III: "Sugar Spore"

The *Sugar Spore* is a single player version of DiabetesCity. In contrast to the first two concepts the user plays with a virtual companion by "feeding" it with information. Entering his or her glucose levels the game is started and the player can create and watch its companion. The virtual companion develops from a simple, friendly Spore to a complex character as it does in "Spore," developed by Will Wright (Wright, 2008), for which it could work as an extension. Every time the player enters data (blood sugar levels, carbohydrates, or insulin doses) he earns credit points for creating and developing his virtual companion. Moreover, the virtual friend begins to mirror all therapeutic actions. For example, every time the player measures his blood sugar, the digital pet is doing a strange gesture, like scratching his neck. The more parameters the player enters, the better the player gets to know his pet. The interactive storytelling uses therefore a psychological trick: the pet is diagnosed as diabetic as the player is. Since it obviously needs a hand in managing disease, the player begins to take over responsibility for his friend and subconsciously for himself.

PLAYTESTING

In this section, several design techniques are presented based on the general design proposals by Ballagas and Walz for "Player-Centered Iterative Design" (Ballagas & Walz, 2007, p. 255). In the early stages we produced one-page conceptual design treatments as the three presented above. "Early Concept Prototyping" supported us in two ways. For a start, it started a transparent design process, which is not focused on one single solution track. Developing different versions of game play, storytelling, and even technical set up has the big advantage of releasing ideas rather than absorbing them.

Moreover, we are able to analyze and identify the several versions, which is an important part of theory development, as follows later. Finally, the short descriptions of formal and dramaturgical elements of the prospective game are primarily used for the communication with design partners. For example, we could perfectly outline several Do's and Don'ts by sending them to the doctors of the Center for Diabetes Care and Education at the Olgahospital in Stuttgart in very early design stages. Secondly, we used a technique of "visual brainstorming," initiating a sort of "correspondence of conceptual sketches" with befriended game designers. First step is to send them your concept treatment. Your correspondence partner spend one hour of his or her time to read and simultaneously put together a rough conceptual drawing, including his or her first impressions about emotions, characters, and the mood of the game.[4] Moreover, we built board game prototypes based on the three different versions presented. "In addition to being a demonstration tool, a board game prototype provides a world-in-miniature that allows game play to be easily tested" (Ballagas & Walz, 2007, p. 268). Designers know this kind of prototyping very well by constructing countless models and design studies. As a Pervasive Game Design tool, the three-dimensional board game gives an impression of spatiality and travel time in the game. Apart from that board game prototypes worked as interactive demonstration tools for the potential end users. After presenting board game versions and sketches of the three versions to patients of the Olgahospital in Stuttgart we asked for their first impression and asked questions about their general diabetes management.

RESEARCH BY DESIGN

The comparison of the three game versions presented above, should give a brief insight to the challenges of Pervasive Health Game projects. The author lines up randomly keywords of Serious Game Design & Development and stresses their materialization in the DiabetesCity project.

Social Interaction

Social interaction is one of the key points of long-term motivation in pervasive computing games. Regarding the two multi-player concepts ("The Zuckerschloss" and "Capture the Flag"), we face the contest between several teams. Additionally, we have an interaction between the playmates within the team, which is of particular importance for a Game for Health. It is important to note that patients of chronic diseases often deal with their handicap on their own. Therefore they are in danger of feeling isolated in society and even from their friends. Two game concepts in particular, encourage social interaction between the diabetic person and its social entourage. As a result of this proposition, we focused on two questions when we presented our concepts to the patients: First, with whom and where patients would like to play a PHG? And second, how can a non-diabetic player interact within a game including therapeutic care?

In "The Zuckerschloss", we try to integrate the parents into the game play. We know by interviews, that „monitoring" of the diabetic child by the parents is a problematical theme for both of them in everyday live. We decided incorporate exactly this habit as a possible interactive and playful feature of the game. The kids might like the idea that their parents as well as themselves lead a diary and the two of them together build up "The Zuckerschloss". Having presented the three concepts to the children, we asked 11 patients between 8 and 13 years: "With whom you would like to play DiabetesCity?" Eight out of Eleven answered "with my friends." Three of them answered "with other diabetics" and "with my parents" and two children would prefer to play "alone." It is important to note that all children (four out of the whole group, aged between 8 and 10) who specified that their parents still do their diabetes documentation did not favor playing DiabetesCity "with their parents" at all. Even though two of them liked the general concept of the game. On the one hand it is a common place that children prefer to play with partners of their own age. On the other hand we suppose that escaping from parental monitoring is an essential attraction of any game situation, especially within a game for health.

The design challenge that results by this assumption is how to involve non-diabetics into the game play. To include measuring blood sugar levels for non-diabetics is imaginable, but will loose its fascination after the first tries and bears certain issues for uneducated users. What we did know is that patient's friends are extremely curious about the disease, about its therapy and the featured "medical gadgets." A further idea was to implement a quiz-module, in which diabetic and non-diabetic players would answer questions. This feature would provoke friends to talk with each other about the disease and its consequences, which would be a positive side effect to the increase of motivation. Since it remains questionable whether the quiz-feature would work over a game period of a whole week, we decided to focus on the documenting facet of interaction. Diabetics and non-diabetics are interested in documenting and sharing their experiences and activities, as can be seen by the enormous success of social networks. The particular challenge is, how to implement the very personal, individual need of e.g., documenting blood sugar levels into the group's game play. This issue is addressed within the version of Capture the Flag, in which the diabetic child teams up with its friends.

Audience

Every game has its audience, which is an important aspect to social interaction within pervasive games, as their playgrounds include public spaces. Since the game play of DiabetesCity incorporates very personal actions for the patient to do like measuring blood sugar or injecting an insulin doses, the game design has to consider who could and who *may* be watching the game. We asked the patients therefore "[...] Who should be the audience to DiabetesCity?" Most of the answers (7 out of 11) stated "Friends," followed by "Parents" (5), "No One" (4) and "Other Diabetics" (2). Only one 10-year-old boy seemed not to seek for any restrictions, saying "anyone, who cares!"

On and Off Playgrounds

Talking about playgrounds of Pervasive Games, it is important to note that spatiality is layered with the sociology of the people using the spaces. The strong interconnection becomes obvious in patients' answers regarding places, in which they "*would not* like to play DiabetesCity" or would not like to measure their blood sugar. Eight out of eleven patients denied that there are places, where they would not like to play the game. Only three said: "Yes, there are places, in which we do not want to play DiabetesCity." Two of the three with rather technical explanations, as it "was forbidden to play during school lessons" (Capture the Flag) or it is "too loud to play in my own or my friend's room." The third child answered, that he would not like to play DiabetesCity in school, because there, "he prefers to talk with his friends." In contrast to that, the latter question, whether there would be places, where they would not measure their blood sugar was denied by only six of eleven. The places, where patients, would not like to measure their blood sugar varies from "on a party" over "on the stage or during presentation" to "swimming lessons in school" and "in the bus." Most of the time explanations were linked to the audience of these places. They span from "because there are too many people (party)" to "embarrassment (e.g. being on stages)."

Storytelling/Educative Features/Immersion

For all three of our game play versions, storytelling is an essential feature. Especially, if you are working in a health context, it is very important to provide a different, "playful" perspective onto the serious content. Transferring the "hard", scientific, and medical notion of a therapy of chronic diseases into a softer and *edutaining* concept is one key point of developing a story within Games for Health. Secondly, a good story would "absorb" the player and therefore can be considered as one technology of immersion in serious games. Regarding *Sugar Spore*, storytelling might be even used as an education and therapy tool. By switching the role from a patient to a "mentor" of his digital friend, the player could be motivated for certain habits. Doctors and diabetes educators appreciated this feature. Incidentally as well as the patients: Six of them proposed that they "favor the version of *Sugar Spore*" in particular.

PROJECT SET UP DIABETESCITY/ FURTHER STEPS

For the development of complex games for health, it is a prerequisite to use diverse competences of an interdisciplinary design team. Therefore the author has been working together with doctors, psychologists, sport scientists, game designers, and technical developers. The concept of DiabetesCity has been developed in cooperation with Martin Holder and the team of the Centre for Diabetes Care and Education at the Olgahospital in Stuttgart, Germany. Conceptual and technical guidance has been provided by the Stuttgart-based game designer Steffen P. Walz. Stephen Boyd Davis and Magnus Moar provided additional support from the Lansdown Centre for Electronic Arts at Middlesex University in London, UK. The project has been supported by the MFG Stiftung Baden-Württemberg, and was supervised by Gerd de Bruyn, head of the Institute of Modern Architecture and Design (IGMA) at University of Stuttgart, Germany.

Further investigations will address the relationship between health and urban planning with the focus on the development of participative, user-centered design processes. Digital health services and the proposed Pervasive Health Games form an innovative, interdisciplinary working field. An investigation on theories of health orientated planning *for* and *in* the city could be of particular interest for current health projects and are addressed in the current work on the author's PhD thesis. Next step for the DiabetesCity project the evaluation of the board game prototypes with patients of the Centre for Diabetes Care and Education in Stuttgart, which could be provided for this publication. The evaluation work together with prospective end-users aims the development of an application prototype reflecting on the presented challenges of game play, added value for the patient and technical feasibility. During this "research by design" project aspects as possible financial models, cooperation with health insurances and health care suppliers as well as judicial questions were being revealed and require further investigations. Further research within the "Games for Health" topic might discuss challenges and potentials of *Pervasive* Health Games in detail.

REFERENCES

Ballagas, R., & Walz, S. P. (2007). REXplorer: Using Player-Centered Iterative Design Techniques for Pervasive Game Design. In C. Magerkurth & C. Röcker (Eds.), *Pervasive Gaming Applications - A Reader for Pervasive Gaming Research Vol. 2*. Aachen, Germany: Shaker Verlag.

Benevolo, L. (1967). *The origins of modern town planning*. London, United Kingdom: Routledge and Kegan Paul.

Centre for Pervasive Healthcare (CfPH). (2007). *HealthyHome: Supportive technology for pregnant women with diabetes*. Retrieved from http://www.healthyhome.dk/

Davis, S., Jacobs, R., Moar, M., & Watkins, M. (2007). *Exploring the Subjective City*. In F. von Borries, S.P. Walz, and M. Böttger (Eds.), *Space Time Play: Computer Games, Architecture and Urbanism: The Next Level*. Basel, Germany: Birkhäuser Publishing.

Fehl, G., & Rodriguez-Lores, J. (Eds.). (1997). *Die Stadt wird in der Landschaft sein und die Landschaft in der Stadt - Bandstadt und Bandstruktur als Leitbilder im modernen Städtebau* (transl. by MK). Basel, Germany: Birkhäuser Publishing. Games for Health/The Serious Games Initiative (n.d.). *Games for Health – About*. Retrieved from http://www.gamesforhealth.org/about2.html

Lake, A., & Townshend, T. (2006). Obesogenic environments: exploring the built and food environments . *The Journal of the Royal Society for the Promotion of Health*, (June): 2006.

Rodenstein, M. (1988). *Mehr Licht, Mehr Luft - Gesundheitskonzepte im Städtebau seit 1750*. New York: Campus Verlag.

Salen, K., & Zimmerman, E. (2004). *Rules of Play – Game Design Fundamentals*. Cambridge, MA: MIT Press.

Santora, M. (2006, January 12). Bad blood: East meets west, adding pounds and peril. *The New York Times*. Retrieved from http://www.nytimes.com/2006/01/12/nyregion/nyregionspecial5/12diabetes.html von Borries, F., Walz, S.P., & Böttger, M. (2006). Ausweitung der Schiesszone. *Archithese, April 2006.* (transl. by MK).

Wright, W. (2008). *Spore* [computer/video game]. Emeryville, CA: Maxis/EA Games.

ENDNOTES

[1] Current research on the relationship between the built environment and the prevalence of obesity shows that low populated, "car-friendly" and "less walkable" living areas suggest a negative influence on physical activities and calorie intake of their inhabitants. See Lake, A., & Townshend, T. (2006), Obesogenic environments: exploring the

2   built and food environments, The Journal of The Royal Society for the Promotion of Health, June 2006, p. 262 ff.
3   Please see as well the article "Games for Health" by Debra Lieberman in this book.
4   See for example the tourist information game "Rexplorer" for the city of Regensburg, Germany on http://archiv.ethlife.ethz.ch/e/articles/sciencelife/rexplorer.html.
5   The website www.intothepixel.com provides an overview of the art of conceptual sketches within computer games.

Chapter 15
Influencing Physical Activity and Healthy Behaviors in College Students:
Lessons from an Alternate Reality Game

Jeanne D. Johnston
Indiana University, USA

Lee Sheldon
Indiana University, USA

Anne P. Massey
Indiana University, USA

ABSTRACT

Physical inactivity is largely preventable through education, individual, and/or community-based interventions. Yet, in the college-age population, traditional interventions (e.g., lecture-based academic courses) may not fully meet their social and learning needs. Here, the authors report on a study regarding the effectiveness of an Alternate Reality Game (ARG) – called The Skeleton Chase – in influencing physical activity and wellness of college-age students. A growing game genre, an ARG is an interactive narrative that uses the real world as a platform and involves multiple media (e.g., game-related web sites, game-related blogs, public web sites, search engines, text/voice messages, video, etc.) to reveal a story. The authors' initial results are extremely promising relative to the impact on physical activity, as well as tangential learning such as teamwork and problem-solving. They also report students' reactions to the game itself, highlighting game design strengths and weaknesses that may inform game designers.

INTRODUCTION

There is growing interest in the use of games to encourage physical activity (Singh & Mathew, 2007; Goran & Reynolds, 2005; Anderson et al., 2007). Examples include physically interactive computer games such as Dance Dance Revolution (DDR) and web-based games such as Fish'n'Steps (Lin, et al., 2006). Early results provide evidence that games such as these have the potential to motivate

DOI: 10.4018/978-1-61520-739-8.ch015

physical activity and influence healthy behaviors (Hoysniemi, 2006; Lieberman, 2001).

In this chapter, we describe a game designed to influence physical activity and wellness in the *college-age population*. In the transition to college, an alarming decrease in the percentage of individuals participating in physical activity has been found. Simultaneously, a significant weight gain during early college years has been shown to increase the risk of obesity and associated diseases later in life such as diabetes and coronary heart disease (National Center for Health Statistics, 2005; CDC, 2007). Despite these realities, college-age students are significantly under-represented in national research studies (Kahn et. al., 2002).

Physical inactivity is largely preventable through education, individual, and/or community-based interventions. Yet, traditional interventions (e.g., lecture-based academic courses) may *not* fully meet the social and learning needs of college-age students. Today's "millennium students" exhibit a learning preference that tends toward teamwork, experiential activities, structure, and the use of technology (Oblinger, 2003; Oblinger, 2004). Their strengths include multi-tasking, goal orientation, and a collaborative style (Raines, 2002; Howe & Strauss, 2000). All these factors must be considered when designing an intervention aimed at college students (Keating, Guan, Pinero, & Bridges, 2005). Characteristics of effective learning paradigms – including experiential learning, social learning, and goal-setting – may be found in games (Bransford, Brown, & Cocking, 2000); thus suggesting a potential role as an intervention with college students.

In this study, we sought to explore the effectiveness of a prototype Alternate Reality Game (ARG) – called *The Skeleton Chase* – in influencing physical activity and wellness of college-age students. A growing game genre, an ARG is an interactive narrative that uses the real world as a platform, often involving multiple media (e.g., game-related web sites, game-related blogs, public web sites, search engines, text/voice messages, video, etc.) to reveal a story (Kim, Allen, & Lee, 2008; Szulborski, 2005; http://www.argn.com).

Over time, players engage in a complicated series of puzzles and challenges that not only involve the players with emerging story, but also with fictional characters, each other, and with the real world. Puzzles and challenges can expose players to new knowledge and ideas, facilitate the development of critical thinking/problem-solving skills, and promote collaboration and cooperation. Many also require players to "get up" from their computers, "move" from one location to another to find clues or other planted assets in the real world, and/or participate in a live event. Influencing physical activity requires that players *"get up and move"*! In designing *The Skeleton Chase*, our goal was *not* to teach students about wellness or force physical activity on them; rather our goal was to enable tangential learning with physical activity as a backdrop to gameplay.

In the following section, we provide background on the context of our efforts, the college population, and associated intervention considerations. We then describe *The Skeleton Chase* and how its design was informed by the learning preferences and strengths of today's college students. Following this, we report on the results of an eight-week pilot study involving 17 competing teams comprised of 58 college freshmen. Here, we examine the effectiveness of the ARG in influencing physical activity. We also report students' reactions to the gameplay experience, highlighting game design strengths and weaknesses. We conclude this chapter with a discussion of our preliminary findings, lessons learned, and directions for future work.

MOTIVATION AND CONTEXT

Despite national recommendations, the incidences of obesity and chronic diseases associated with physical inactivity continue to increase (Must & Anderson, 2003). These changes can be seen across the lifespan; however, they are particularly important in the younger generation as risk factors and lifestyle behaviors established during the formative years are likely to carry forward throughout an individual's life. An alarming decrease in the percentage of individuals meeting physical activity recommendations during the transition period from high school to college has been found, dropping from 68% to 44% (American College Health Association, 2007; Leslie et al., 1999). In addition, significant weight gain (ranging from 1–3 kg) have been consistently noted during the first year of college (Hivert et al., 2007), and obesity prevalence has been consistently increasing (35%) in the young adult population (Lowry et al., 2000; Wang & Beydoun, 2007).

The target population in our study is freshman-level college students. At Indiana University (IU), we have a unique opportunity to assess a game intervention relative to existing interventions already in place. At IU, Living Learning Centers (LLCs) are sections of university residence halls where students with common interests live together. IU has been a leader in the development of LLCs, including thematic centers such as the Fitness & Wellness LLC (FWLLC). The FWLLC offers an environment where annually 350 freshmen explore and apply the six dimensions of wellness: intellectual, emotional, social, physical, life planning, and spiritual. Residents have a dedicated fitness facility, a quiet room, and a collection of health-related literature. New residents must enroll in Foundations of Fitness and Wellness, a sixteen-week development course covering topics related to physical activity, nutrition, stress, and substance abuse. The course consists of a weekly 50-minute lecture and a 50-minute lab. During the lab, students are exposed to the on-site fitness center and provided opportunities to participate in physical activity. Collectively, the residential community, physical environment, dedicated staff, and focused programming provide a holistic experience intended to encourage residents to make healthy choices during and after college.

Thus, the FWLLC reflects several interventions, including: (a) providing safe, convenient place for physical activity (e.g., Fitness Center), (b) educational programming (e.g., the course); and, (c) using community to influence behaviors (e.g., residence LLC).

In the fall of 2007, preliminary research was conducted to track and compare physiological, anthropometric, and health outcomes of FWLLC students enrolled in the Foundations of Fitness & Wellness course (n = 60) to a non-resident control group (n = 40) over the course of their first year of college. Our findings indicated a significant ($p < 0.05$) increase in body weight and associated decrease in physical activity during the first semester of college regardless of group affiliation. However, the magnitude of the shift of students from active to insufficiently active (9.1% and 18.8% for FWLLC and control, respectively) was smaller than has been reported in previous literature (Bray & Born, 2004). These preliminary results suggested that the combination of environmental, social, and educational interventions the FWLLC students were exposed to did not have a significant impact on physical activity and health-related outcomes. It may be that the FWLLC, and specifically the educational course, are not sufficient interventions to alter (at least in the short term) physical activity and health-related outcomes within this population. Thus, these findings provided the impetus to explore alternative interventions to reach the college student population during the critical transition period from high school to college.

Designing effective interventions requires an understanding of college students' physical activity patterns and the determinants of behavior within this specific population. Moreover, interventions must address the learning preferences and strengths of today's students, i.e., teamwork/collaborative style, experiential activities, structure, goal orientation, multi-tasking, and the use of technology (Oblinger, 2003; Oblinger, 2004; Raines, 2002; Howe & Strauss, 2000). Traditional interventions

Figure 1. Fictional crop circle with IU logo

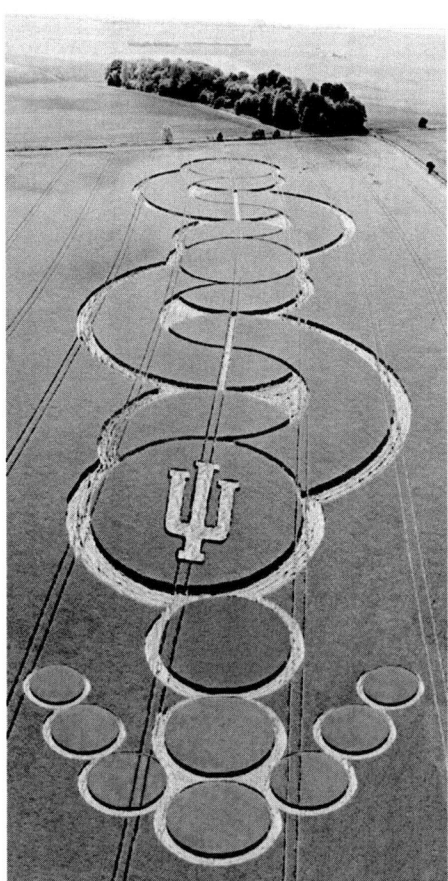

(e.g., "tell and test" lecture-based courses) may not meet the needs of these students. All these factors must be considered when designing interventions aimed at college students (Keating, Guan, Pinero, & Bridges, 2005), and they also suggest a strong role for games as interventions. Characteristics of serious games such as goal-attainment and social/experiential learning offer the potential to create interventions where students can be engaged in experiences not possible in the traditional classroom (Mayo, 2007; Brandsford et al., 2000). In the next section, we describe the design of *The Skeleton Chase*, a prototype alternate reality game, intended to serve as a gameplay complement to the existing FWLLC interventions.

THE SKELETON CHASE ALTERNATE REALITY GAME

The Skeleton Chase is an Alternate Reality Game or ARG (pronounced either ā·ar·jē or ărg). As noted earlier, an ARG is a narrative with multiple gameplay elements that takes place in *real time* in the *real world*. In its simplest form, an ARG can be a single website as in *World without Oil* (http://worldwithoutoil.org) where students from around the world projected in a variety of media how a future global oil crisis would affect their lives; or an event such as *Last Call Poker* (http://42entertainment.com/poker.html) where players moved from an online poker site to actual cemeteries where they used the dates of births and deaths on gravestones they could simultaneously touch to create the best poker hands.

ARGs are convergent, utilizing a wide range of media to tell a story. The gameplay in *The Skeleton Chase* featured websites both real and fictional, text messaging, phone calls, emails, images, audio, video, real-world landmarks and locations (some staged specifically for the game), and live performances by actors. Images, such as Figure 1, were created using Photosphere and included in the fictional websites.

While many ARGs to date have used such real-world interactions – and a few featured characters and objects experienced directly in the real-world – *The Skeleton Chase* took place in the real world to an unprecedented degree to accomplish its goal of influencing physical activity and, in turn, overall wellness. To complete many of the puzzles and challenges presented by the game, the narrative was constructed so that players would need to traverse the large university campus from one corner to another on foot, as well as scale stairs in the football stadium and to a tower in the student union.

The Skeleton Chase was designed to run for eight weeks during the fall 2008 semester. Its fictional narrative centered on Sarah Chase, a young

Figure 2. Sarah chase

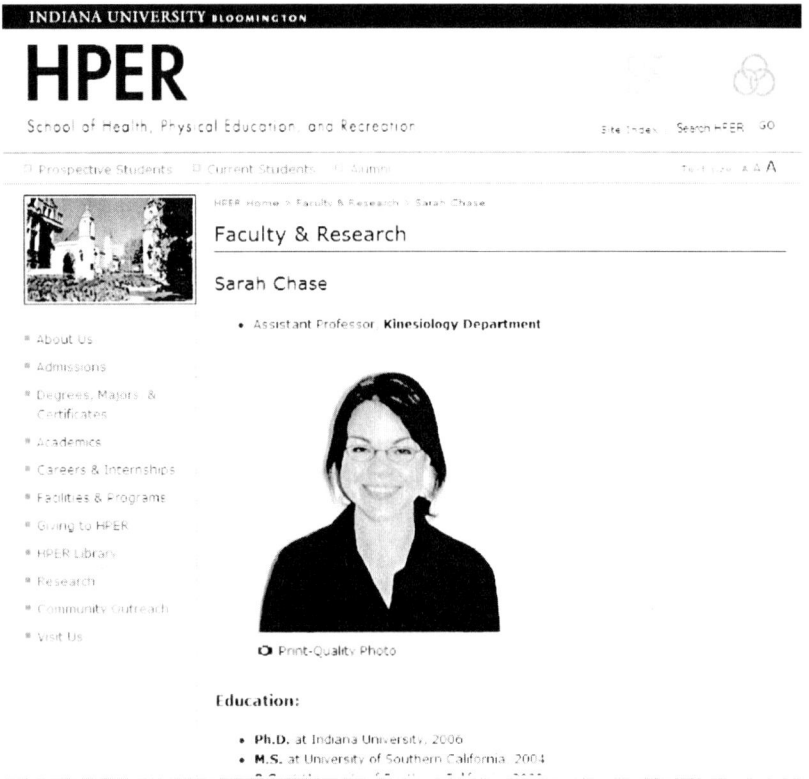

assistant professor in kinesiology and folklore who had disappeared not long after announcing a breakthrough discovery. The (fictitious) faculty web site of Sarah, played by a local actress, is shown in Figure 2.

In the first week, gameplay began with the appearance of an actor portraying Steven Cartwright, a public relations representative from the Source Corporation. He explained that Source was a company supporting university research into health and nutrition. Cartwright announced a series of fitness challenges, distributed a worksheet for students to keep track of their progress in the first challenge, and handed out free unmarked bottles of vitamin water.

The actual rabbit hole that launched gameplay was the URL of the Source Corporation web site printed at the bottom of the worksheet. From Lewis Carroll's *Alice's Adventures in Wonderland*, a rabbit hole is the point of entry into the world of the ARG. The website appeared to be a legitimate corporate website with a welcome video by Cartwright, announcement of new testing facilities being built, and stories on nutrition and fitness. However, one link on the web site labeled "Internal Site" was password protected. Players, using clues gleaned from Cartwright's talk to the class and the welcome video, were able to discover the password.

Upon entry into Source Corporation's internal website, players found various links including one to the (fictitious) IU Security public web site where a request was posted for information concerning the whereabouts of Sarah Chase. Links from that article led them to Sarah's blog (see Figure 3), and the blog of her former associate instructor, a graduate student named Sam Clemens. Videos on the blogs introduced these major characters

Figure 3. Sarah's blog

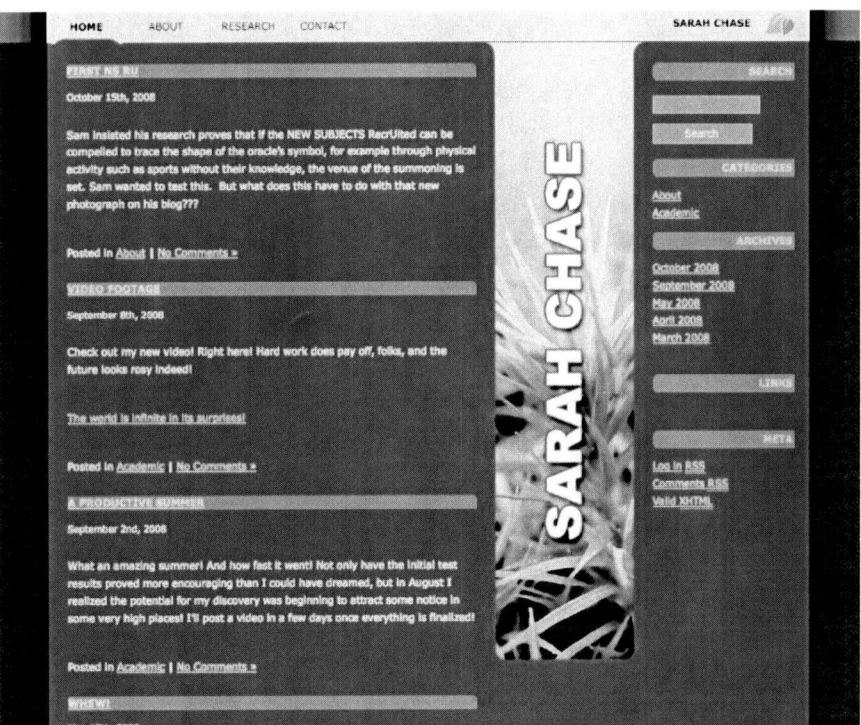

to players, and began to fill in some plot details. The game was now fully underway.

Each week, gameplay consisted of physical challenges either presented overtly by the Source Corporation to encourage students or elaborate puzzles students had to solve in their quest to find out what had become of Sarah. In the first few weeks, players had direct communication with the major game characters only via their cell phones and email. Sarah (apparently kidnapped) and Sam (also missing, but hiding somewhere on campus) were able to contact them with text messages and phone calls.

Through the course of these first few weeks, students searched Sarah's office (staged with planted assets including Sarah's diplomas and research notebook); hacked into the (fictional) IU Security internal website (see Figure 4); found Sam's hiding place (but not Sam); and were able to uncover a wide-ranging conspiracy tied to a formula that may or may not retard aging.

In the process, they learned of a third person's disappearance, were alerted to flying saucers sited near IU's Cyclotron facility, and investigated appearances of a creature dubbed the "Bloomington Bigfoot" in some campus woods. At the heart of the mystery was a wild flower – a purple dandelion, also known as a *skeleton* plant. This flower not only appeared to have remarkable properties as an herbal remedy, but it was also believed it could be used somehow as an oracle to predict the future.

On Halloween Day, players were finally able to track Sam Clemens to a jungle-like campus greenhouse. A week later they discovered an unmarked office in the Business School used by the Source Corporation and learned the full extent of a diabolical experiment that included the students themselves. Remember that vitamin water they drank that first evening? In the final week of gameplay, they were brought face to face with Sarah at last. And the solution to the final mystery was revealed.

Figure 4. Fictitious campus security web site

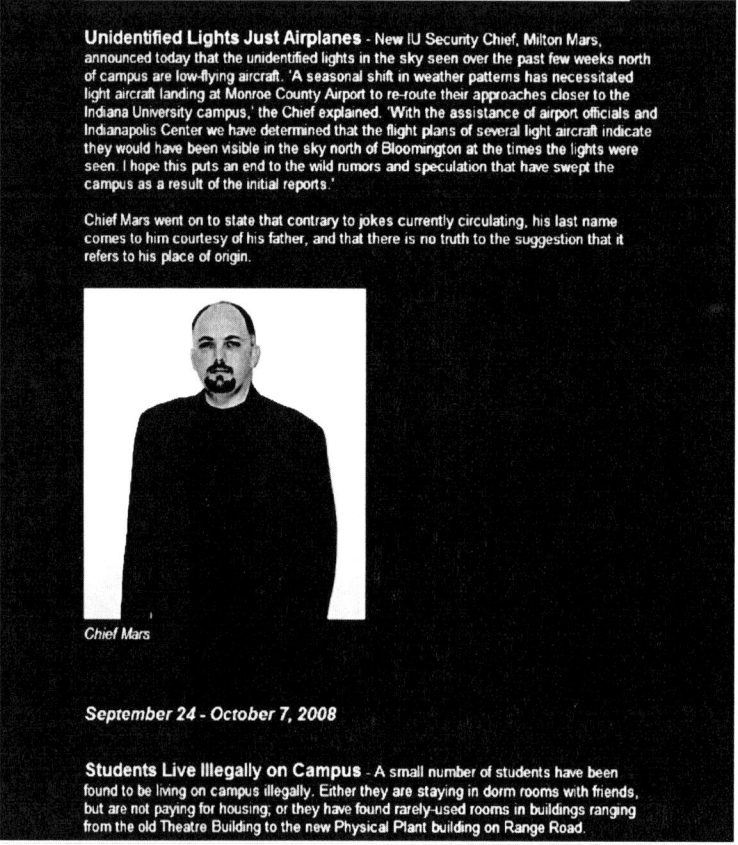

The primary goal of the game was to provide an engaging alternative to structured weekly workouts. For that reason, the following were critical elements of the game's design:

1. The narrative must be intriguing enough for the students to want to follow it and experience it as it unfolded. To that end, it deliberately avoided instructional elements, leaving those to the Foundations of Fitness & Wellness class, and focused instead on being as entertaining and immersive as possible.
2. The physical challenges were not simply Source Corporation fitness challenges, even though these fit into the fiction. Other challenges were initiated by characters as part of the story with no reference to fitness whatsoever. For example, one week students followed an elaborate "treasure map" left by Sam for Sarah that directed her (and ultimately the students) to his lair. In another week, Sam played "phone tag" with students, similar to what kidnappers or blackmailers have done to make sure those trying to contact them will not be followed. The students logged many steps by simply pursuing the game's narrative.
3. The physical challenges should be such that accelerometers could track student movement. This unfortunately eliminated such activities as climbing, swinging on ropes, swimming, etc.

There were a number of outside factors the game design needed to take into account:

1. The game was restricted to the campus, so that students could play in their free time between classes.
2. As noted earlier, the game was designed to supplement the Foundations of Fitness & Wellness class that is part of the Living and Learning Center at the residence hall. Specifically, the game served as a substitute for the 50-minute weekly lab section associated with the class.
3. All of the players were first semester freshmen enrolled in the class. The players would be divided into teams of three or four students so they could compete with one another.
4. It was intended that the gameplay each week would echo wherever possible themes and issues (e.g., alcohol abuse, nutrition) students tackled during coursework without specifically referencing them. For example, during the week on nutrition students created recipes for Source Candy, an ultimate health bar. While providing tangential learning opportunities about course topics, *The Skeleton Chase* also provided opportunities to learn about the campus, develop critical thinking/problem-solving skills, and develop social skills related to team collaboration and cooperation.
5. University policies could not be violated. For example, we met with the university's campus police department to inform them of the game, including the fact that students might show up in odd places at odd hours and would be hacking into the fictional IU Security website. They were immensely helpful and supportive. The watch captain didn't even blink when we told him he might receive reports of someone asking about UFOs being spotted. Apparently this is a common occurrence!
6. Some of the most interesting university locations were not available 24/7.

For these reasons, *The Skeleton Chase* gameplay needed to be far more structured than that of a typical ARG. Puppetmasters (individuals working behind the scenes) of many ARGs launch the game and then monitor gameplay at arm's length, providing additional content as players make their way through the narrative. They anticipate players will become the phenomenon known as the "hive mind" – also known as collective intelligence or collaborative play, it occurs when many people working together are more intelligent than any individual – and increase the difficulty of the puzzles and challenges accordingly. However, because *The Skeleton Chase* was a supplement to a structured course and loosely tied to weekly class content, we could not be patient and allow our teams to proceed at their own individual paces. Rather, the teams needed to proceed as a whole, week to week, so their activity and progress could be monitored. In addition, we needed to make certain everyone was able to complete new mental puzzles and physical challenges by the end of each week. No one could be left behind.

As a result, it was necessary to establish a separate game "hot line" where players could buy hints and actual solutions to puzzles. They also were required to answer three questions at the end of each week to demonstrate they had in fact successfully completed that week's gameplay. This meant that puppetmasters had to be available for many hours each day. Some university locations such as the staged offices and certain recreation and sports venues were available only during regular work hours. As such, the puppetmasters did not have to be available at all hours of the day or night. Rather, one of the first elements of gameplay we adjusted was to limit the hotline help hours to 10 am to 8 pm. The starting hour was determined by the fact that the only students up before 10 am were usually in class. Students tend to sleep in whenever possible! The closing hour was determined by our geographical location. It gets dark early in Indiana in the fall. We couldn't require students to purchase flashlights to play. There was a safety factor to be considered as well. We're happy to report that no players were injured by playing the game.

Access to some locations was limited to regular working hours that were announced prior to the start of that week's play. Announcements were made at the start of play each week when student access to facilities was limited to regular working hours. The hotline remained open until 8 pm since some of the mental puzzles could still be solved using online resources. In the case of the Halloween event, the appearance of the actor was limited to a four hour period and the narrative was adjusted to account for the limited time. The resulting player traffic control required five puppetmasters on hand.

Unfortunately this necessary structure, the constant need to remind players they were playing a game, as well as Internal Review Board (human subjects) release forms and baseline testing, violated a key precept of alternate reality games – namely, TINAG (This Is Not a Game). ARG game designers work very hard to pull a veil down over the eyes of players during gameplay, making them forget at least temporarily that what they are experiencing isn't real.

At the end of this chapter, we discuss several lessons we learned from our pilot test of the ARG; and what we will adjust in the future. The ability to be flexible is an essential requirement for the designer of any ARG, even more so one that is successful as a physical intervention. In the next section, we report the results of our pilot study.

PILOT STUDY

A pilot study of *The Skeleton Chase* ARG was conducted to address two key questions:

1. Did *The Skeleton Chase* ARG influence the level of physical activity of gameplay participants?
2. How did participants react to the gameplay experience itself? Specifically, what game design elements influenced the attractiveness of the experience?

Participants

The pilot study involved 58 college freshmen (57% female) living in the FWLLC and enrolled in the Foundations of Fitness & Wellness semester-long course. For eight weeks, gameplay replaced a 50-minute lab period associated with course lectures. Students enrolled in non-gameplay lab sections met for structured physical activity sessions in the fitness center, receiving 10 class points per week for attendance. Each student received an ActiPed™ accelerometer, was given instructions regarding where to position the device on their shoe, and instructed to upload data (steps) at least once a week using the wireless access point installed in the FWLLC. Subjects wore the accelerometers for one week prior to the start of gameplay to allow for baseline data collection. During gameplay, participants could access and track their personal physical activity data stored on the ActiHealth™ web site.

Participants in *The Skeleton Chase* were randomly assigned to one of 18 three- or four-person teams based on baseline physical activity data. Due to roster changes immediately prior to the start of gameplay, members of one team were reassigned to other teams, resulting in 17 competing teams. Fifty-one students consented to having their data reported here.

Gameplay Procedures

Given that the game was played in a competition fashion, a point system was devised. Each team began gameplay with 1000 points. Finishing the week's narrative required completing puzzles, completing physical activities (e.g., finding real world objects/locations), and completing narrative elements. Weekly, a team could add (or lose) points based on: (1) correctly answering a set of three questions related to the week's activities, thus "confirming" completion; (2) whether *all* members of the team achieved 50,000 steps for the week, as recorded by the ActiPed™ accelerometers; and, (3) the purchase of hints and/or solutions related to that week's puzzles. Weekly bonus points were also awarded to the first three teams correctly answering the "confirming" questions. A web-based leader-board provided feedback on the weekly and overall relative performance of teams. In addition to the competitive nature of the game, the point system was necessary in order to award credit for "lab completion." Every week, small prizes (e.g., $5 gift certificates to local establishments) were awarded to members of that week's "winning" team. At the end of gameplay, members of the top three teams received monetary awards.

Gameplay began every week immediately after Wednesday's evening lecture session. Unless otherwise instructed, teams had until 8 pm on the following Tuesday to complete the week's narrative activities, accumulate steps, and upload step data to the ActiHealth™ system.

Measures

Regarding physical activity, via the ActiPed™ accelerometer, *baseline* (pre-game) step data was collected for one week, thus providing an indicator of students' current physical activity levels. Daily and weekly gameplay-related step data was collected on each student over the course of the game.

Perceptions of the *psychological attractiveness* of the gameplay experience and various design dimensions of *The Skeleton Chase* were measured via a survey instrument. (Further details regarding development and validation of the instrument are available from the authors, upon request). Prior research has examined factors that make games attractive to players (Malone, 1982; Yannakakis et al., 2006; Sinclair et al., 2007). The notion of *attractiveness* is related to the concept of flow, or the state of total engagement in an activity (Csikszentmihalyi, 1975). Sweetser and Wyeth (2005) offer eight elements – concentration, challenge, skill, control, clear goals, feedback, immersion, and social interaction – purported to influence the attractiveness of gameplay. For example, in the context of gameplay, if the skills of an individual are much higher than the challenge(s) at hand, the individual may quickly become bored. Conversely, if the challenge level is too high, the individual may feel a great deal of anxiety and frustration. Either scenario will likely make a game less attractive. Some elements (e.g., direct and immediate feedback) may be controllable by the game designer, while others (e.g., concentration) are characteristics of the player (Sinclair et al., 2007). Drawing from prior research (c.f., Jackson & Marsh, 1996; Sweetser & Wyeth, 2005), our instrument was designed to explore the relationship between these eight elements and the psychological attractiveness of *The Skeleton Chase*, thus providing some initial insight into its design strengths and weaknesses. The questionnaire items are reported in the Appendix.

PRELIMINARY RESULTS

Physical Activity

Descriptive statistics (n=49) of physical activity in terms of step counts (baseline, weekly, weekly average over 7 weeks), as recorded by the accelerometers, are presented in Table 1. Recall the goal for each student was 50,000 steps per week, with teams receiving game points if all members of the team met the goal in any given week. Overall, 57% of students met the individual weekly step goal 4 or more times during gameplay, while 25% of the teams met the team step goal 4 or more times.

To explore whether gameplay influenced students' level of physical activity, we ran paired

Table 1. Descriptive statistics: physical activity data

	Mean	Std. Deviation
Baseline (pre-game)	34,192.0	16,232.9
Week1	49,668.8	26,139.2
Week2	56,723.0	26,048.1
Week3	56,428.1	21,651.2
Week4	54,214.7	25,528.0
Week5	50,901.1	22,960.0
Week6	49,166.4	22,330.2
Week7	48,687.2	22,873.1
AvgSteps (7 weeks)	50,206.4	22,480.0

t-tests (see Table 2). The results indicate significant differences (p<.001) between baseline step data and *all* weeks of gameplay, and between baseline and the *overall average* steps/week of gameplay. Over the 7 weeks, on average, students accumulated 18,064 *more steps/week* (or approximately 9 miles!) than they had prior to the start of gameplay. It is worth noting that *The Skeleton Chase* varied, on a week-to-week basis, in terms of how much direct physical activity was involved in solving the week's narrative. In weeks 2–4, as examples, students participated in several extensive campus treasure hunts, moving from location to location, picking up clues, and solving puzzles. These activities involved significant walking as students traversed the campus. Other weeks (e.g., week 5) involved special events such as actually meeting game characters, partially explaining the variance in weekly physical activity shown in Table 1.

Reactions to Gameplay

Descriptive statistics (n=51) of the measures for psychological attractiveness and the eight gameplay dimensions are presented in Table 3.

To explore the question of what gameplay elements (e.g., concentration, skill, etc.) drive attractiveness, we ran a stepwise regression analysis using SPSS 16.0. The tests of this model are shown in Figure 5. The model explains 83% of the variance in psychological attractiveness. As shown, both concentration and immersion had a positive impact on attractiveness, with beta coefficients of .475 (p<.005) and .591 (p<.005),

Table 2. Paired samples T-test: physical activity

	Mean	t
Baseline - AvgSteps	-18,063.67	-8.121
Baseline - Week1	-15,476.8	-4.761
Baseline - Week2	-22,531.1	-7.316
Baseline - Week3	-22,236.1	-7.684
Baseline - Week4	-20,022.8	-6.993
Baseline - Week5	-16,709.1	-6.351
Baseline - Week6	-14,974.4	-5.814
Baseline - Week7	-14,495.2	-5.424

Table 3. Descriptive statistics: gameplay

Construct	Mean	Std. Deviation
Attractiveness	3.17	1.44
Concentration	4.10	1.13
Challenge	4.98	0.95
Skill	4.40	1.20
Control	3.83	1.43
Clear Goals	3.31	1.44
Feedback	4.33	1.37
Immersion	3.00	1.30
Social Interaction	4.02	1.18

respectively. For feedback, the coefficient of -.189 ($p<.05$) suggests that it had a negative impact on the attractiveness of *The Skeleton Chase*. While not significant predictors of the psychological attractiveness of the game, it is worth noting we found a significant difference between perceived challenge and skills (paired $t=3.54$, $p<.001$); students reported, on average .58 *more* challenge than skill level, suggesting the game's mental puzzles exceeded their perceived skill level.

Discussion

Preliminary results indicate that gameplay associated with *The Skeleton Chase* ARG had a positive influence on students' physical activity, as measured by step counts. In post-game interviews with students, they offered a number of comments regarding tangential learning (e.g., teamwork, time management) and the potential that gameplay may influence future physical activity behaviors, including:

"I liked getting to explore the campus and learning new things about it."

"I really liked getting out of the classroom."

"As my team was playing, we got to learn about facilities we could use like the pool."

"The game really did motivate me to walk – since I learned a lot about the campus, I know it is quicker and better for me to walk to class than take the bus."

"I liked working with my team and it was fun getting to know them."

"It took us awhile, but eventually our team figured out a strategy to play the game each week."

At the same time, students did raise concerns about wearing the Actiped™ accelerometers, and often claimed the device was under-reporting steps. Here, we learned a number of lessons! First, while the device is intended to be worn on a standard shoe or sneaker, we found that students wear all sorts of footwear (e.g., flip-flops, boots etc.) in all sorts of weather. Second, we found that students were concerned about what other, non-playing, students thought of them as they wore the device during gameplay. As examples, other students asked *"... did you steal your shoes?"* or *"... are you on house arrest?"* We never anticipated

Figure 5. Psychological attractiveness

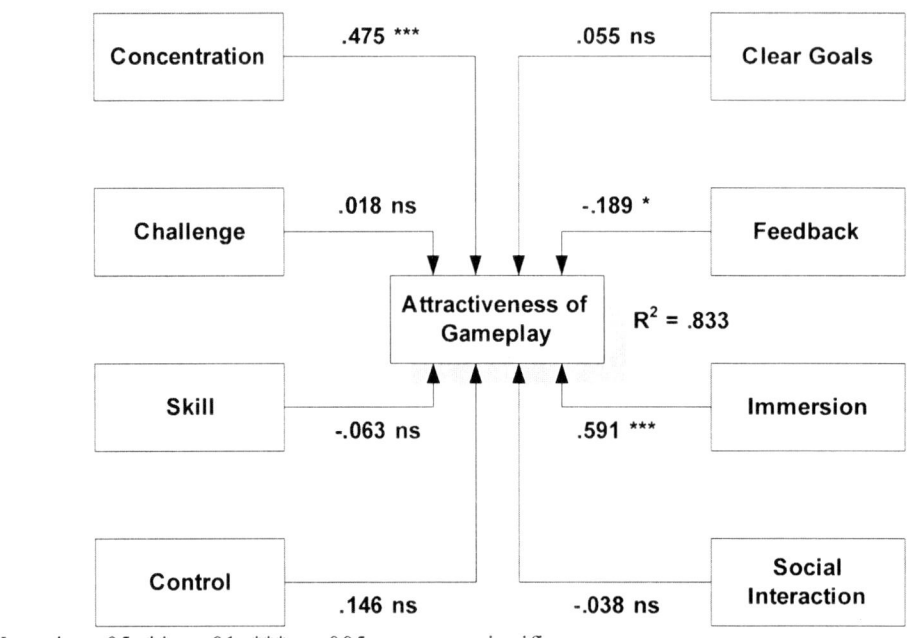

Note: * p<.05; ** p<.01; *** p<.005; ns = non-significant

these reactions, but fully appreciate the concerns of first semester freshmen. Overall, however, we are very encouraged regarding the impact of gameplay on physical activity.

Regarding reactions of students to *The Skeleton Chase* itself, as measured by psychological attractiveness, we are somewhat disappointed. However, as indicated by students' post-game comments, several implementation factors likely contributed to this finding, including:

"The time to play the game each week took too long compared to other lab [non-game] sections."

"I wish I could have picked my own team. We didn't know each other and our schedules were different."

Recall that students played *The Skeleton Chase* in lieu of their 50-minute lab section. Completing the mental puzzles and traversing the campus took many teams 3 or 4 hours per week, and required they coordinate their schedules. This was very different than what was required of individual students in the other lab sections, the basis of comparison for students playing *The Skeleton Chase*. While gameplay positively influenced physical activity and tangential learning occurred, many students complained that the game simply took too much time and that they had to work harder than students in other labs. As noted earlier, our results also suggest that students did not perceive their skills to be sufficient relative to the game's mental challenges (puzzles). Also, as first semester freshmen, we were forced to randomly assign students to teams, rather than let them choose who they wanted to work with. While the most successful teams learned to work with each other over time, others struggled with poor group communication and unequal effort among team members.

Overall, our results showed that immersion and concentration were the two key elements that positively contributed to the attractiveness of the game. Immersion reflects the degree to which a player experiences deep, but effortless involvement in a game. Immersive games (like ARGs) are intended to draw players into the game, and affect their senses through elements such as the narrative (Sweetser & Wyeth, 2005; Sweetser & Johnson, 2004). Student comments in this regard included:

"I enjoyed interacting with characters in the game, especially meeting Sam Clemens in the greenhouse."

"I really felt like I was using brain muscle and it was cool getting information from a bunch of different sources [email, text messages, web sites etc.]."

Depending on the motivation of a team and perhaps the ability to coordinate their schedules, we observed teams largely completed gameplay activities (mental and physical challenges) either at the beginning of the week *or* at the very end. While immersion positively influences attractiveness, the behavior of going "in and out" of gameplay witnessed here is likely an artifact of tying play to a weekly class schedule. We expect we would find a greater perception of immersion and, in turn, attractiveness in typical ARG play.

Similarly, concentration reflects workload and attention – to be enjoyable (or attractive), a game has to require concentration *and* the player must be able to concentrate on the game. Here, the more players felt concentration was present, the more attractive the game appeared. While *The Skeleton Chase* certainly required concentration (particularly to solve the mental puzzles), students' ability to concentrate was likely influenced by time pressures (weekly game completion, other courses etc.) and team dynamics (e.g., coordination, equal effort etc.). Again, as constraints are relaxed, attractiveness should increase as the ability to concentrate at one's own discretion is raised.

Finally, we found that feedback was negatively related to psychological attractiveness. This is contrary to expectations that appropriate and direct feedback would enhance the attractiveness of play. What was different about feedback as related to *The Skeleton Chase* ARG? First, feedback was largely received via communication with the puppetmaster manning the help hot line. To students, this feedback mechanism may have appeared to be outside of the game narrative. Second, the weekly and overall relative performance of teams was published at the end of each week's gameplay; while playing, teams did not know how they were actually doing on a relative basis.

FUTURE RESEARCH DIRECTIONS

While our initial results are promising, particularly with regard to impact on physical activity, tying *The Skeleton Chase* to a weekly class may have adversely affected student perceptions, particularly regarding immersion and concentration, both key drivers of psychological attractiveness. Tying play to a class also led to several implementation challenges. First, gameplay required a significant human resource commitment. Puppetmasters working long hours had to closely monitor gameplay and be available to test progress to confirm the completion of weekly activities. Second, competition between teams for prizes meant that the collective intelligence usually present in ARGs was not encouraged. Thus, puzzles and challenges were simplified to some degree. Finally, in some instances the availability of hints may have led to "lazy" gameplay as teams relied on buying hints, rather than their own capabilities.

To address the above issues, we will explore the impact of *The Skeleton Chase* again in spring 2009. Specifically, the game will not be associated with a class; rather, it will be played over eight weeks in a more typical ARG manner. Players will be recruited from the undergraduate student body at large, thus reflecting a wider range of ages and interests. While likely encompassing a range of casual to active to enthusiastic players, students will participate because they either want to play an ARG or because they are attracted by cash prizes. Collaboration will also be encouraged, leading to a better overall balance between skills and challenge. It will be up to students whether they play individually, or in teams of any size with members of their own choosing. Overall, as play constraints are relaxed, we anticipate that psychological attractiveness will increase.

While, as before, individual puzzles and physical challenges will be launched by characters from the game, we intend to reduce dependency on direct contact (live calls, text messages) with characters during gameplay, and replace them with automated computer and web interactions. In addition, gameplay will not need to be as closely monitored through direct contact with players on a week to week basis. No "catching up" each week will be necessary. Only as players complete content will new content be provided to advance the narrative. If it appears players are not progressing through the narrative in a timely enough fashion to complete it within the eight-week period, instead of overt hints from puppetmasters, clues will be added within the narrative itself in the form of new documents, messages from characters, etc.

Additional physical challenges called "sprints" will be launched at various times during gameplay to supplement the challenges driven by puzzles. These will feature tasks required by the narrative that must be completed in a short period of time ranging on average from 30–90 minutes. Sprints will also necessitate a more heightened level of physical activity (e.g., running instead of walking) for participants to be successful within the specific time limit.

Finally, the new session, while including as much of the previous session's gameplay as possible, will be re-structured to prevent previous players from being able to pass on their knowledge to new players, much in the same way a professor teaching similar course content from year to year alters exams and assignments to prevent cheating. New narrative material, as well as a new major character, will be added so that *The Skeleton Chase* ARG will play more like a sequel than a rerun.

CONCLUSION

In this chapter, we reported on *The Skeleton Chase*, a prototype alternate reality game intended to influence college students' physical activity and health outcomes. The game was designed in light of changing learning preferences and expectations. Our initial results are extremely promising relative to the impact on physical activity, as well as tangential learning such as teamwork and problem-solving. While much remains to be learned about the role of an ARG as an effective intervention, the lessons we have learned to date are serving as the impetus for further game design enhancements and testing. We hope our findings facilitate others interested in health intervention strategies and game design.

ACKNOWLEDGMENT

Support for this project was provided by a grant from the Robert Wood Johnson Foundation®. We would also like to acknowledge the contributions of graduate students Elizabeth Crosbie, Nicholas Cassidy, and Rickie Lee Marker-Hoffman. We would also like to thank Residential Programs and Services for their continued support.

REFERENCES

American College Health Association National College Health Assessment Spring 2006 Reference Group Data Report (Abridged) (2007). *Journal of American College Health, 55*(4), 195-206.

Anderson, I., Maitland, J., Sherwood, S., Barkhuus, L., Chalmers, M., & Hall, M. (2007). Shakra: Tracking and sharing daily activity levels with unaugmented mobile phones. *Mobile Networks and Applications, 12*, 185–199. doi:10.1007/s11036-007-0011-7

Bransford, J., Brown, A., & Cocking, R. (2000). *How people learn: Brain, mind, experience, and school*. Washington, DC: National Academy Press.

Bray, S. R., & Born, H. A. (2004). Transition to university and vigorous physical activity: Implications for health and psychological well being. *Journal of American College Health, 52*(4), 181–188. doi:10.3200/JACH.52.4.181-188

CDC. (2007). Department of Health and Human Services: *Overweight and Obesity*. Retrieved February 2, 2007: http://www.cdc.gov/nccdphp/dnpa/obesity/index.htm.

Csikszentmihalyi, M. (1975). *Beyond boredom and anxiety*. San Francisco, CA: Jossey-Bass.

Goran, M. I., & Reynolds, K. (2005). Interactive multimedia for promoting physical activity (IMPACT) in children. *Obesity Research, 13*, 762–771. doi:10.1038/oby.2005.86

Hivert, M. F., Langlois, M. F., Berard, P., Cuerrier, J. P., & Carpentier, A. C. (2007). Prevention of weight gain in young adults through a seminar-based intervention program. *International Journal of Obesity, 31*, 1262–1269. doi:10.1038/sj.ijo.0803572

Howe, N., & Strauss, W. (2000). *Millennials rising*. New York: Vintage Books.

Hoysniemi, J. (2006). International survey on the Dance Dance Revolution game. *Computers & Education, 4*(2), 1544–3574.

Jackson, S. A., & Marsh, H. W. (1996). Development and validation of a scale to measure optimal experience: The Flow State Scale. *Journal of Sport & Exercise Psychology, 18*, 17–35.

Kahn, E. B., Ramsey, L. T., Brownson, R. C., Heath, G. W., Howze, E. H., & Powell, K. E. (2002). The effectiveness of interventions to increase physical activity. *American Journal of Preventive Medicine, 22*, 73–107. doi:10.1016/S0749-3797(02)00434-8

Keating, X. D., Guan, J., Pinero, J. C., & Bridges, D. M. (2005). A meta-analysis of college students' physical activity behaviors. *Journal of American College Health, 54*(2), 116–125. doi:10.3200/JACH.54.2.116-126

Kim, J. Y., Allen, J. P., & Lee, E. (2008). Alternate reality gaming. *Communications of the ACM, 51*(2), 36–42. doi:10.1145/1314215.1340912

Leslie, E., Owen, N., Salmon, J., Bauman, A., Sallis, J. F., & Lo, S. K. (1999). Insufficiently active Australian college students: Perceived personal, social, and environmental influences. *Preventive Medicine, 28*(1), 20–27. doi:10.1006/pmed.1998.0375

Lieberman, D. A. (2001). Management of chronic pediatric diseases with interactive health games: theory and research findings. *The Journal of Ambulatory Care Management, 24*(1), 26–38.

Lin, J. J., Mamykina, L., Lindtner, S., Delajouz, G., & Strub, H. B. (2006). Fish'n'Steps: Encouraging physical activity with an interactive computer game. In P. Dourish & A. Friday (Eds.). *UbiComp 2006: Ubiquitous Computing,* (pp. 261-278). Berlin: Springer-Verlag

Lowry, R., Galuska, D. A., Fulton, J. E., Wechsler, H., Kann, L., & Collins, J. L. (2000). Physical activity, food choice, and weight management goals and practices among U.S. college students. *American Journal of Preventive Medicine, 18*(1), 18–27. doi:10.1016/S0749-3797(99)00107-5

Malone, T. W. (1980). *What makes things fun to learn? A study of intrinsically motivating computer games.* (Doctoral dissertation, Department of Psychology, Stanford University). Also available as technical report #CIS-7 (SSL-80-11), Xerox Palo Alto Research Center, Palo Alto, CA.

Mayo, M. J. (2007). Games for science and engineering education. *Communications of the ACM, 50*(7), 31–35. doi:10.1145/1272516.1272536

Must, A., & Anderson, S. E. (2003). Effects of obesity on morbidity in children and adolescents. *Nutrition in Clinical Care, 6*(1), 4–12.

National Center for Health Statistics. (2005). *National Health Interview Survey*. Retrieved December 3, 2007: http://www.cdc.gov/nchs/about/major/nhis/released200606.htm#7

Oblinger, D. (2003). Boomers, gen-xers, and millennials: Understanding the "new students.". *EDUCAUSE Review, 38*(4), 36–45.

Oblinger, D. (2004). The next generation of educational engagement. *Journal of Interactive Media in Education, 8,* 1–18.

Raines, C. (2002). *Connecting generations: The sourcebook for a new workplace*. Berkeley, CA: Fifth Street Design.

Sinclair, J., Hingston, P., & Martin, M. (2007). Considerations for the design of exergames. In *Proceedings of Graphite 2007: 5th International Conference on Computer Graphics and Interactive Techniques,* (pp. 289-295).

Singh, V., & Mathew, A. P. (2007). WalkMSU: An intervention to motivate physical activity in university students. In *Proceedings of Conference on Human Factors in Computing Systems CHI '07,* San Jose, CA, (pp. 2657-2662).

Sweetser, P., & Johnson, D. (2004). Player-centered game environments: Assessing player opinions, experiences, and issues. In *Entertainment Computing – ICEC 2004: Third International Conference,* (pp. 321-332). New York: Springer-Verlag.

Sweetser, P., & Wyeth, P. (2005). GameFlow: A model for evaluating player enjoyment in games. *ACM Computers in Education, 3*(3), Article 3A.

Szulborski, D. (2005). *This is not a game: A guide to Alternate Reality Gaming*. Morrisville, NC: Lulu.com.

Wang, Y., & Beydoun, M. A. (2007). The obesity epidemic in the United States – gender, age, socioeconomic, racial/ethnic, and geographic characteristics: A systematic review and meta-regression analysis. *Epidemiologic Reviews, 29*(1), 6–28. doi:10.1093/epirev/mxm007

Yannakakis, G. N., & Hallam, J. (2007). Towards optimizing entertainment in computer games. *Applied Artificial Intelligence, 21*(10), 933–971. doi:10.1080/08839510701527580

APPENDIX

MEASUREMENT ITEMS AND RELIABILITIES

All items were measured on a 7-point Likert scale from strongly disagree=1 to strongly agree=7.

Psychological Attractiveness (Reliability .887)

I was deeply engrossed in The Skeleton Chase game.

I was absorbed intensely in The Skeleton Chase game.

Playing The Skeleton Chase game was fun.

I thought The Skeleton Chase game was boring. (R)

Concentration (Reliability .824)

The Skeleton Chase game offered a lot of information from different sources (e.g., web sites, text messages, etc.).

Information provided from different sources was worth paying attention to.

The Skeleton Chase game quickly grabbed my attention.

While playing The Skeleton Chase, I was completely focused on the story as it unfolded.

Challenge (Reliability .761)

The overall outcome of The Skeleton Chase story was not easy to guess.

The solutions to the puzzles and challenges each week were not easy to guess.

The Skeleton Chase puzzles and challenges varied each week.

The Skeleton Chase story was revealed at an appropriate pace.

For me, The Skeleton Chase game felt challenging.

Skill (Reliability .811)

The Skeleton Chase's gameplay mechanics (e.g., use of the hot line) were easy to learn.

The Skeleton Chase game provided sufficient help (e.g., hints, solutions, weekly narrative summaries).

The Skeleton Chase puzzles and challenges matched my skills.

I loved the feeling of accomplishment I got solving puzzles and challenges, and want to capture that feeling again.

Control (Reliability .901)

I felt my actions mattered and were shaping the unfolding of the game.

I felt a sense of control over the actions and strategies used to play the game.

I felt I could play the game the way I wanted to play it.

Clear Goals (Reliability .877)

The Skeleton Chase game had clear goals.

I understood the goals of The Skeleton Chase game from the beginning.

I had a clear feeling that I knew what I was doing during gameplay.

Feedback (Reliability .848)

In my opinion, our team received sufficient feedback about our performance during gameplay.

I received sufficient feedback about my performance during gameplay.

I always knew my team's status relative to other teams (e.g., via the leader board).

In my opinion, my team had a clear understanding about how we were progressing during gameplay.

Immersion (Reliability .922)

As I was playing The Skeleton Chase game, I was less worried about everyday life.

I felt emotionally involved in the game.

I was really into pondering the puzzles and mental challenges of the game.

Time seemed to alter (slowing down or speeding up) when I was playing The Skeleton Chase game.

It was no effort to keep my mind on what was happening during gameplay.

Social Interaction (Reliability.852)

The Skeleton Chase game encouraged interaction within my team.

The Skeleton Chase game encouraged interaction between teams.

I discussed game events with my teammates even when we weren't playing.

I discussed game events with members of other teams.

    I discussed the game with people (friends, family) not playing the game.

Section 4
The Way Ahead:
The Future of Serious Games

Chapter 16
Establishing a Science of Game Based Learning

Alicia Sanchez
Defense Acquisition University, USA

Jan Cannon-Bowers
University of Central Florida, USA

Clint Bowers
University of Central Florida, USA

ABSTRACT

Using video games to train and educate is a notion that is gaining traction among gamers, parents, and serious educators alike. Unfortunately, to date there have been few rigorous studies to determine whether games can be effective learning tools. Given their inherent features, the authors feel certain that games can teach, and they are interested instead in addressing the question of how best to design games that will optimize learning. To accomplish this goal, the authors offer a simple framework for organizing variables and then discuss findings from psychology and education as a basis to formulate a research agenda for game-based training. In doing so, they hope to stimulate researchers to conduct appropriately controlled experiments that will begin to provide insight into how various features affect motivation and learning. In this way, a true science of educational games can be formed.

INTRODUCTION

While preliminary evidence from literature reviews supports the potential effectiveness of learning through electronic games (Garris, Ahlers, & Driskell, 2002; Gopher, Weil, & Bareket, 1994; Green & Bavelier, 2003; Knerr, Simutis & Johnson, 1979; Sims & Mayer, 2002; Vogel, et al., 2006), considerable theoretical and empirical work is still needed (Hays, 2005; O'Neil, Wainess, & Baker, 2005). Specifically of interest are the elements that differentiate games from other types of technologies used in learning and training settings; primarily simulations and virtual worlds.

The purpose of this chapter is to provide a foundational perspective from which a research agenda for studying games for education and training may be drawn. General conclusions from the science of learning serve as a basis to support the idea that games have the *potential* to teach and

Figure 1. Framework for organizing gaming features and learning

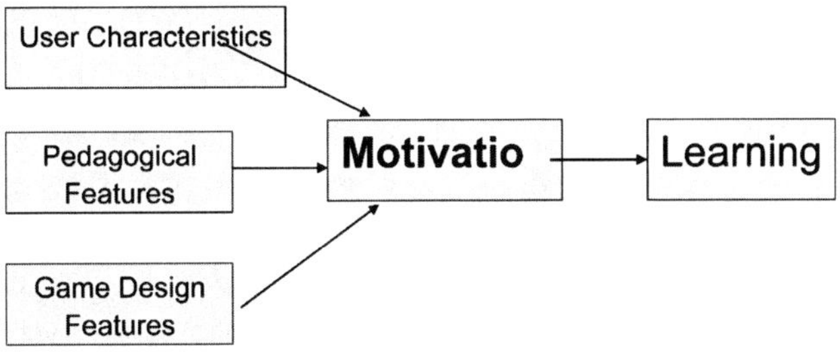

```
User Characteristics ──┐
                       ↓
Pedagogical Features → Motivation → Learning
                       ↑
Game Design Features ──┘
```

train effectively; however, the burden lies with the researchers seeking to determine *if*, *how*, and *why* games improve learning. Therefore, we must consider a range of characteristics that may contribute to the effectiveness of games as learning tools. Of course, not all games will have all of these characteristics. However, through this decomposition of potential research variables, it is hoped that their individual, mixed, and cumulative effects on learning will be parsed. Variables ranging from user, pedagogical, design, and implementation characteristics are discussed for use in future research campaigns geared towards rigorous evaluations of their effectiveness.

THE POTENTIAL OF GAMES AS EDUCATIONAL AND TRAINING TOOLS

Since empirical efforts to study educational games are few and far between, it is important to consider findings from psychology and education that have driven the education, simulation, and training industries for several decades. As such, we offer a fairly straightforward model of the learning process to serve as an organizing framework. As is evident from Figure 1, we posit that characteristics of the user, pedagogical features embedded in the game, and game design features can all affect the user's motivation to interact with the game, and in turn, influence learning. We also hypothesize that some of the features within these categories may exert a direct impact on learning.

An explanation of elements of this model, along with delineation of features that fall within these categories, is detailed in the following sections. We also include a discussion of implementation issues that could certainly have an impact on educational game effectiveness.

MOTIVATION AND LEARNING

Motivation has been implicated as a crucial antecedent to learning (e.g., Clark & Wittrock, 2000; Garris et al., 2002; Tannenbaum, Mathieu, Salas, & Cannon-Bowers, 1991). For example, it has been found that motivation to learn can affect learning outcomes. Moreover, learners who have higher motivation may be more likely to voluntarily play a learning game, and/or spend more time playing it. In both instances, we would expect higher learning games. Hence, any feature of a learning environment that enhances motivation should also, theoretically, enhance learning. Research is needed to further explicate the motivational aspects of games. For example, we hypothesize two types of motivation that could have an impact on learning, *motivation to play* and *motivation to learn,* which may exert differential influences on learning, and be determined by different features of the game or environment.

Motivation to play. It is quite possible that what is traditionally measured as a person's willingness to participate in an activity could be separate from and might actually negatively impact a person's acquisition of knowledge. Motivation to play does not inherently indicate that learning will take place within an activity, but does indicate a person's willingness to participate in an activity for some purpose other than learning. An interesting question becomes, then, can we exploit the natural motivation to play a game for the purposes of improving learning?

Motivation to learn. Conversely, motivation to learn does indicate a person's willingness to participate in an activity because they see some benefit in possessing knowledge, and they view their participation in an activity as a way to further their understanding. In this case, further research can help to explicate how a game might be designed so that it triggers this type of motivation in a user. For example, narrative elements in a game could extol the virtues of obtaining a particular type of knowledge that becomes real for the learner.

USER CHARACTERISTICS

Turning again to Figure 1, the first class of variables that we believe will affect motivation and learning is user characteristics. That is, the individual learner or game player brings their own unique experiences, attitudes, and beliefs to any learning situation. Considerations must be made in order to understand how these differences might impact motivational and learning outcomes, and how individuals might be accommodated or situated into the use of games for learning and training. Characteristics of individual users such as perceived relevance, preferences, prior knowledge, previous experiences using games, and expectations must be considered cumulatively and in isolation in order to effectively parse out their potential impacts on motivation and learning outcomes.

Relevance

A factor related to motivation is the relevance to a learner of the learning outcomes. Belanich, Orvis, and Mullin (2004) found that content relevant to game play was learned significantly more than content irrelevant to game play. In this form, relevance refers to a person's determination that the activity they are going to participate in will in some way be beneficial to them later. For example, explaining the relevance of Algebra to an eighth grader may be much more difficult than explaining the need for a soldier to learn how to identify improvised explosive devices. Kanfer and McCombs (2000) suggest that a learner's interest in the topic and perceived utility of the information could affect their choices and persistence to learn that material. Indeed, a learner's judgment of relevance of learning content is likely to affect motivation and hence learning. Research is needed to better understand these findings with respect to games. For example, it may be possible to use game elements to enhance perceived relevance in users.

Preferences

While the theory of learning styles has received considerable attention and adoption, to date no research has concluded that individuals are predisposed to learning in one format or mode over another (Mayer & Massa, 2003). Instead, it is likely that individuals hold preferences related to how they receive information, and that those preferences can impact their learning (Clark & Feldon, 2005). This becomes relevant as it's easy to believe that some people will hold strong negative feelings about the use of games in education and training (Sandford, Ulicsak, Facer, and Rudd, 2005), while others will be delighted with the notion. Research is needed to better understand what might drive a user to prefer game-based learning.

Prior Knowledge

The prior knowledge that any individual user holds in a subject area is certain to impact learning outcomes when information is presented via any format. Constructivist (Bruner, 1966) theory suggests that individuals assimilate new information into existing frameworks of prior knowledge. The extent to which a person has the ability to form relationships between targeted learning content in games and their existing knowledge may positively or negatively impact their demonstration of acquired knowledge. Determining baseline levels of knowledge and/or employing design characteristics that may be adaptive of a user's prior knowledge, such as scaffolding and/or leveling (discussed later), may help researchers determine how much learning has occurred as a result of the use of a game as an educational or training tool.

Experience

Another variable that might impact learning outcomes is a user's experience with the technology that is being employed. We often overestimate the number of people who play games; for example a study by Orvis, Orvis, Belanich, and Mullin (2008) recently found that while only 17% of Army Cadets polled indicated no video game play, 44% indicated only limited experience playing video games (much below what many would assume). In addition, an avid game player might be more comfortable with a gaming interface, and thus be bored with lengthy directions. Conversely, an individual who never plays games might find some the logistics of game play to be overwhelming. Expectations regarding game dynamics such as graphics and controls may also play a role in learning outcomes varying on the level of experience a user has with playing games. More research in this area is needed.

Expectations

Findings from research have indicated that expectations can moderate a training experience (Tannenbaum et al., 1991). Within educational and training activities, it's possible that using one term instead of another might evoke expectations that do not align with their definition of the activity. For example, if the term *game* is used to describe an activity and the activity shares no characteristics with an individual's conception of a game, they may become disappointed, disenchanted, and frustrated with that activity. These negative perceptions of the activity may have an impact on the learning. Conversely, when a person thinks that they are going into a serious learning session, finding a game might be disconcerting. Attention to this matter is important and could pay off in terms of motivational and learning gains.

PEDAGOGY

Years of empirical research in the fields of education, psychology, and training have provided some heuristics that could be generalized to the use of pedagogy in a variety of technological tools such as games. In this section, several dominant constructs determined to be relevant to the use of games will be examined and recommended for future research initiatives.

Experiential Learning

Experiential learning (or learning through experience) has been cited as a fundamental human process (Kolb, 1984). To accomplish experiential learning in a typical learning situation (e.g., classroom) typically requires some type of technology. Applications provide users the opportunity to have experiences that they might not be able to have otherwise for many reasons including expense, risk levels, or the infrequency of similar situations. Certainly, games can be constructed so that they mimic real world experience. A driving question becomes, then, what features of the environment must be represented in the game world to ensure that appropriate learning will occur?

Situated Learning/ Anchored Instruction

Well-designed activities can also provide *anchored instruction* and *situated learning*: two constructs that are based on the pedagogical principle that in order for learning to be effective, it must be presented in a meaningful context to the learner (Bransford, Sherwood, Hasselbring, Kinzer, & Williams, 1990). Learning experiences within simulated activities can be varied and reinforced by providing opportunities for learners to catalog instances in a way that enables them to recall those experiences when necessary. This repertoire of instances or templates provides a greater breadth of situations for them to use as the basis for future decisions. Anchored instruction is also expected to induce novice information organization to be more similar to the information organization of experts (Bransford, Brown, & Cocking, 2000). This type of knowledge acquisition, sometimes called *learning for understanding*, can also lead to facilitated transfer of the knowledge by supporting task-relevant knowledge structures that can be easily accessed and utilized. Once again, games provide a perfect backdrop for situated instruction. Research is needed to determine how best to exploit this possibility in games.

Active Learning

Educators and instructional theorists have also converged on the conclusion that *active participation* by learners is a key element of good instruction (Chi, 2000). Evidence suggests that courses emphasizing interactive education and active involvement in learning activities show better results in students regardless of the quality of the instructor. Mayer (2001), however, distinguishes between behavioral activity and cognitive activity in learning, citing cognitive activity to be crucial for learning. According to Mayer, the appearance of active participation (i.e., behavioral activity like button pushing or steering) may not be important as long as cognitive participation is occurring. Activities such as games may require several inputs within a few seconds of one another, forcing a user to make decisions regarding their actions constantly. Hence, games are inherently active environments and, as such, are hypothesized to enhance motivation and learning. Specific hypothesis testing in this area is needed.

Scaffolding/Leveling

The use of scaffolding in the form of leveling within games could also have positive learning outcomes. Evolved from Vygotsky's theory related to the 'zone of proximal development' (1978), Wood, Bruner, and Ross (1976) developed the concept of scaffolding as a mechanism for facilitating the completion of a task that a learner might not be able to accomplish themselves. Originally, scaffolding was used as a method of instruction by teachers who 'tutored' a learner through a task in order to increase their level or understanding so that they could move on to more complex learning objectives. Within games, levels often mimic the scaffolding approach by incorporating mini-tasks or challenges designed to increase a player's skill in order to prepare them to move on to more complex levels of play. Further research on the mechanisms of leveling and their relationship to the notion of scaffolding is needed.

Self-Efficacy

Self-efficacy can be defined as a learner's belief that they have the necessary competence required to accomplish a task (Bandura, 1977; Bandura, 1986; Gist, Schwoerer, & Rosen, 1989; Gist, Stevens, & Bavetta, 1991). In general, studies converge on the conclusion that high self-efficacy learners perform better in training than low self-efficacy learners. Games often do a good job at maintaining a player's self-efficacy by providing constant opportunities for success. And people seem to like doing things they're good at and avoid doing things they are not.

Self-efficacy has also been theorized as a consequence of instruction as well as a cause. Moreover, self-efficacy has been linked to other important motivational variables such as goal setting and self-regulation (Kanfer & McCombs, 2000; Pintrich & Zusho, 2002). Hence, instructional interventions that build self-efficacy would be expected to be more successful than those that do not. There is a vast opportunity within game design to investigate how best to develop and maintain self-efficacy in users.

Goal Orientation/Goal Setting

Goal setting involves establishing a standard or objective for performance. In learning systems, goals help to focus learners on the task and help them to select or construct strategies for goal accomplishment; hence, they serve to direct attention (Locke & Latham, 1990). Goal commitment (i.e., the degree to which the learner is committed to the learning goal) is a determinant of how much the goal affects performance (Locke, Shaw, & Saari, 1981). In addition, goal setting has been linked to self-regulatory processes (Schunk & Ertmer, 1999), which are associated with effective learning. Research has also shown that to be most effective in enhancing performance, goals must be specific, difficult but achievable, and proximal (Locke & Latham, 1990).

Goal setting provides another excellent target for learning games. Unlike traditional learning environments (classroom and virtual), information within games is usually essential to the player's success. In other words, within a game, if a player is given a key, they knows they are going to need use that key at some point in the game. Essentially, most activities are goal driven, with smaller (proximal) accomplishments building gradually to larger and larger (distal) ones. Given this inherent feature, it should be rather natural for educational game developers to help learners set meaningful goals for learning within the context of the story.

Metacognition & Self-Regulation

Game players often have excellent metacognitive and self-regulation skills. They often know where they stand in relation to all aspects of the game, they are keenly aware of their strengths and weaknesses, and they are able to develop strategies for remediating deficiencies in and acquiring knowledge. These processes are important to human learning and can enhance learning outcomes. Metacognition (Bransford et al., 2000) is broadly defined as having insight into one's own learning process. It includes a learner's knowledge about learning, knowledge of their own strengths and weaknesses, and the demands of the learning task at hand. Metacogntion also includes self-regulation, which is described as the processes by which students accomplish learning goals (Schunk & Zimmerman, 2003). Self-regulated learning emphasizes metacognitive, motivational, and behavioral processes including planning, goal setting, self-monitoring, and self-efficacy (Kanfer & Ackerman, 1989). Research is needed to better understand how to trigger and manage metacognition and self-regulation using gaming elements.

Engagement

Engagement refers to the degree to which the learner is motivated by tasks, and interacts and takes part socially in the task environment. A person's engagement could certainly be causally related to the amount of time a person spends interacting with a game. Determining what characteristics of an activity and a user lead to engagement could provide insight into what characteristics could be used in further initiatives.

With the advent of virtual environments, researchers have explored variables associated with engagement or immersion in learning. The notion of immersion is hypothesized to be a psychological state resulting from a participant's intense feelings of "presence" in a virtual experience (Gerhard, Moore, & Hobbs, 2004). A related topic is the idea of "flow." Csikszentmihalyi (1990) describes *flow* as an experience where an individual becomes so engaged in an activity that time becomes distorted, self-consciousness is forgotten, and external rewards disappear. Instead, people engage in complex, goal-directed behavior because it is inherently motivating. From a pedagogical standpoint, it is not clear whether or how this notion of flow may affect the learning process. However, from a strictly time-on-task perspective (i.e., the amount of time engaging instructional content), it would follow that intense engagement should benefit learning.

Feedback

In game activities, feedback is often immediate and meaningful. If a player tries to jump over a ravine that is too wide for them, they die or are seriously injured. If they miss a turn in a racing game, they crash. Feedback is information provided to learners regarding their performance as well as their progress in achieving learning objectives. A tremendous amount of work has been done regarding how to provide feedback to enhance learning and transfer. Feedback is the central mechanism by which learners can regulate their own performance and understand how to improve. According to Bransford et al. (2000), opportunities for feedback should be frequent and/or continuous. It has also been argued that feedback should help learners to understand how to change their performance in order to improve (i.e., simply providing learners with knowledge of results is insufficient) (Bransford et al., 2000). Two types of feedback have been linked to learning: process and outcome (Smith-Jentsch, Zeisig, Acton, & McPherson, 1998).

Process feedback often provides information regarding the performance in terms of strategy and guidance (e.g., you focused on identifying and engaging targets, your position on the field can help you with spatial awareness of where those targets might be), while outcome feedback often involves performance based data only (e.g., you killed 19 of 20 targets). Obviously, both types of feedback have the potential to affect motivation and learning; research is needed to provide specific information on how best to incorporate feedback into a learning game.

DESIGN

The design characteristics that conventionally constitute games are driven by theories of play, entertainment, challenge, and fun. Many of these characteristics often serve as a delineator between games and simulations, but many of these characteristics also overlap into educational applications. These characteristics are not exclusive of entertainment games, but are more typical of the potential for educational games that are gleaned from entertainment games.

Game Type

When considering the genre of a game, the most important distinctions lie in the objectives being targeted for learning. Could we teach reading with a first person shooter? Maybe, so long as it was impossible for a learner to successfully play the game without accomplishing some reading. Could an arcade style game teach business negotiation? Could a driving game be used to teach bomb detection? Maybe, but would it be a great idea? We really don't know. Hence, game genre must be considered a variable in researching learning outcomes and may factor into user characteristics like preferences, motivation, and expectations. What is needed is a taxonomy of game genres, along with their essential features, that can be used to match against learning objectives.

Game Quality

Anyone who has ever played a bad game—one that was so complicated that it took longer to figure out *how* to play it than to actually play the game itself; or so ill conceived that it left the users feeling like they had been robbed of a segment of their lives—knows that these are not enjoyable experiences. There are bad games. And people don't want to play them. Within the games industry, a commercial quality game for a newer console ranges from $6-10 million dollars, unfortunately, as few of 16% of those are likely to be profitable (Campbell, 2005). If a game is bad, it may seriously reduce learning outcomes. Unfortunately, there isn't a validated scale to measure how good a game is. In order to consider the quality of the game in total (including graphics, genre, theme, play, user design, etc.), perhaps we should take a tip from our industry counterparts and increase our use of play-testing, or form advisory groups of people who play games.

Story

The use of story within games can have powerful impact on the emotion and engagement of a player. It can also serve as a distraction in some cases. The use of story, scenarios, backstory, and themes can impact learning, transfer, and a user's perceptions of a game experience. The quality of a story, and its alignment with the targeted learning objectives is sure to have an impact on the quality of the game, and a player's belief of the value of the learning experience. Research that takes advantage of work done in areas like scenario design (Cannon-Bowers et al, 1998) is needed in this regard.

Rewards

Rewards within a simulation or game can be demonstrated through scoring, gaining new tools or possessions based on accomplishments, and player ranking or player leveling. These rewards or social status indicators could influence a player's motivation to continue playing the game. Decades of research into motivation, and especially intrinsic motivation, generally show that externalizing the reward for engaging in a behavior can reduce a person's intrinsic motivation to engage in it (Deci, Koestner, & Ryan, 1999; Ryan & Deci, 2000). However, researchers have also found that under certain conditions, externalized rewards and competition can increase intrinsic motivation (Reeve & Deci, 1996). In particular, when an activity is seen as challenging, allows the user to gain feelings of competence, and is not perceived as being controlling (i.e., forced on the user), intrinsic motivation can be enhanced.

In addition, when the user places importance on doing well, competition generates effective involvement in the activity and increases personal meaningfulness (Epstein & Harackiewicz, 1992). It should be noted that in at least one study (with children), a gender difference was found with respect to public acknowledgment of achievement as a reward. In that case, boys were more likely to respond to such public recognition than girls (Nemeth, 1999). Certainly, further research in this area as it relates to games is needed.

Embodiment/ Personalization

A player's avatar can influence their feelings of investment within a game. As players' embodiments or avatars are often seen as true reflections of themselves, the ability to personalize an avatar might increase a user's sense of emotional attachment with the success or failure of that avatar within a game (Gee, 2003). User "embodiment" has become a topic of interest in recent years as the development of collaborative computer systems and virtual worlds has increased. In a learning system context, Moreno and Mayer (2004) found that personalizing the interaction with the student in a virtual environment improved learning and retention of science content. Baylor's (2000) work on intelligent agents/mentors similarly indicates a measurable impact of personalization and presentation style on learning outcomes. Hence, it appears that increasing a student's level of embodiment and personalization can increase engagement and learning. In addition, avatars provide a basis for conversation and social interaction (Slater, Sedagic, Usoh, & Schroeder, 2000), which increase the sense of engagement in the game. The details of this type of personalization and its affect on learning are needed.

Authenticity

Authenticity is the degree to which the synthetic experience causes learners to engage in cognitive processes that are similar to those in the real world. Authenticity can refer to the reality of the graphics, the probability that this situation could occur in the real world, or the adherence to situational rules. Researchers have also concluded that the *authenticity* or *fidelity* of the learning environment is critical to efficient high-level learning, as these allow learners to more easily transfer what they have learned to real-world tasks (Hays & Singer, 1989; Honebein, Duffy, & Fishman, 1993; Jonassen, 2000; Petraglia, 1998; Savery & Duffy, 1996). Findings generally indicate that instruction, however, need not be carried out in tasks that completely mimic the real world, as long as the experience elicits cognitive processes that are similar to those that are actually required. Simulations and related technologies have also been identified as ideal tools for fostering understanding of complex and abstract phenomena because they allow learners to make connections between abstract ideas and concrete (real-world) examples (Dede, Salzman, Loftin, & Sprague, 1999). Here again, games provide a perfect backdrop for authentic learning environments; more work is needed to translate these findings into design guidance.

Fidelity

Fidelity is the degree to which the synthetic experience needs to be a faithful representation of the real phenomenon or task. There are two types of fidelity that might impact a user's experience: physical fidelity and cognitive fidelity. Within physical fidelity, a synthetic experience might or might not be an accurate representation of the real world object or environment (i.e., a cockpit simulation might need to have exact replication of the switches and buttons). Cognitive fidelity, however, is related to the similarity between the cognitive processes required (i.e., do the thought processes of a task or learning experience within a simulated environment or a game represent thought processes that would be required in the real world). Both types of fidelity occur within educational games, but it is unclear how best to formulate meaningful guidelines that game developers can follow in building educational games.

Immersion/Emotional Intensity

Immersion and emotional intensity can be likened to the degree to which a learner feels they are a part of the synthetic experience. It is well known that training researchers have wrestled with the question of transfer of learning for decades (Baldwin & Ford, 1988). Suffice it to say that many variables may affect transfer, and that inducing learners to apply what they have learned is not as straightforward as it may appear. One factor that may affect transfer is the intensity of the learning experience, so that game play can drive behavioral change long after the game is over. While evidence in adult populations is lacking, a study of diabetic children given a diabetes management game had 77% fewer follow-up visits (emergency and urgent care) than a control group. More work along these lines is needed.

Collaborative Learning

Games can serve as excellent collaborative learning environments. The image of gamers isolating themselves from contact with others no longer applies thanks to the Internet; instead, game players form elaborate social networks (Beck & Wade, 2004). Massively multiplayer online games (MMOGs) in particular often involve teams of players working towards a variety of specific objectives. Even outside of game play, some players play a game for years and become members in communities (often virtually) that surround those games. Within these communities, players share tips and strategies about game play, as well as personal information, and often become ingrained in the culture surrounding the game.

The notion that collaboration can aid learning is beginning to gain traction with instructional researchers (e.g., CTGV, 2000; Clark & Wittrock, 2000). For example, Nelson (1999) outlines an approach for collaborative learning that includes specific guidelines for implementing various methods, including instructor-implemented methods, learner-implemented methods, instructor-and learner-implemented methods, and interactive-methods. According to Nelson, collaborative learning is effective because it takes advantage of learners' natural collaborative processes, while fostering exchange in rich social contexts and allowing for multiple perspectives. See Bowers, Smith, and Cannon-Bowers (this volume) for more detail on how MMOGs might be used as collaborative learning environments.

IMPLEMENTATION

Thus far, we have discussed potential learners or game players, the pedagogy behind the use of games for learning and training, and how the design characteristics may impact learning; but equally important to yielding learning outcomes is how those games are used. The implementation options for games can often be as complex as the creation. Games certainly have several uses, and depending on the targeted outcomes can impact learning in different ways. Decisions regarding whether a game will serve as a standalone teaching experience or supplement materials and concepts presented within a traditional or e-learning setting, the timing of these implementations, and the use of games as evaluations, will be discussed in the next sections. In all cases, further research is needed to get a better handle on how to optimize learning outcomes through appropriate implementations.

Supplemental

Games that are implemented to supplement existing courses or training programs may serve as opportunities to practice the skills being taught or as application tools to add relevance and understanding for conceptual knowledge being presented. Alternatively, games can be used for the purpose of increasing the depth of a learner's knowledge on a particular topic or learning objective. These types of games should have targeted learning opportunities weaved into the game play dynamics and should be closely aligned with curriculum objectives.

Teaching

While the use of games as tools that add practice opportunities, relevance, and application to conceptual knowledge provided via more traditional classroom experiences are the most commonly accepted concepts, it is possible to imagine games serving as standalone educational opportunities. These types of games would most likely avoid discovery or incidental learning opportunities

(Mayer, 2004) and would serve as guided learning opportunities in which conceptual and practical knowledge are provided and exercised.

Practice Opportunities

Games that are implemented to provide practice opportunities for learners can often provide varied situations that are authentic to the learning objectives or can serve as repetitions that can add to a learner's automation of the task. These types of implementations should be designed to allow users to apply as much of the knowledge they have gained through traditional learning in order to enhance their expertise. While no new learning may occur during this type of implementation, as a student should already possess all the knowledges and skills to complete a practice opportunity, it is possible that this type of game implementation may lead to increased retention, increased depth of knowledge, and increased ability to transfer the knowledge to the real world.

Performance Evaluation

Games may additionally serve as predictors of preparedness or future performance. The use of games as evaluations or diagnoses of performance may serve to fine tune skills, provide opportunities to experience varied situations, or provide insight into a person's ability to transfer their skills into real world applications. Performance evaluations might be additionally used to determine an individual's level of proficiency prior to traditional learning formats and may serve as useful tools in placement based on performance.

Timing

When a game is implemented into a course, it may serve to hinder learning outcomes. For example, instituting a game too early in a learning experience, when students may not possess the knowledge required to process the implications of the game, may lead to frustration, confusion, negative learning, and/or poor transfer. Using a supplemental game late in a course may not have a significant impact on learning since a learner may have reached a ceiling in their learning, or it may increase the depth of processing or the ability for a learner to transfer knowledge. Research focused on timing will assist in determining the effect of timing on learning outcomes.

FUTURE DIRECTIONS AND CONCLUSIONS

It is hoped that the definition of variables associated with games for education and training, and the identification of research questions that surround these characteristics, will contribute to the evolving use of technology within education and training. Developing a science of game-based learning may only serve to further blur the lines between the technologies that are commonly associated (or confused with) games (e.g., virtual worlds, simulations, MMOGs, collaborative learning environments), in that a dissection of the characteristics that contribute to learning should serve as the driving force behind the development of education and training tools. It is hoped that through carefully planned research and adoption of the characteristics that emerge as contributors to learning outcomes, other technologies will begin to leverage these characteristics. If this happens, it may be that labels such as games or simulations become superfluous, and that a true science of instructional technology that feeds multiple media can emerge.

Video games are quickly becoming the most popular form of entertainment in the United States and throughout the world. Given such popularity, it is not surprising that instructional experts have become interested in using video games as serious instructional platforms. Our hope is that such enthusiasm be followed quickly by interest on the part of the training R&D community. This paper only begins to delineate all of the issues that need to be addressed in order to reach the potential of educational games. The use of the characteristics outlined here should serve as a catalyst for stimulating rigorous empirical research studies involving games.

REFERENCES

Baldwin, T. T., & Ford, J. K. (1988). Transfer of training: A review and directions for future research. *Personnel Psychology, 41*, 63–105. doi:10.1111/j.1744-6570.1988.tb00632.x

Bandura, A. (1977). Self-efficacy: Toward a unifying theory of behavioral change. *Psychological Review, 84*, 191–215. doi:10.1037/0033-295X.84.2.191

Bandura, A. (1986). *Social foundations of thought and action: A social cognitive theory*. Englewood Cliffs, NJ: Prentice-Hall.

Baylor, A. L. (2000). Beyond butlers: Intelligent agents as mentors. *Journal of Educational Computing Research, 22*, 373–382. doi:10.2190/1EBD-G126-TFCY-A3K6

Beck, J. C., & Wade, M. (2004). *Got Game: How the Gamer Generation is Reshaping Business Forever*. Boston: Harvard Business School Press.

Belanich, J., Orvis, K. L., & Mullin, L. N. (2004). *Training Game Design Characteristics that Promote Instruction and Motivation*. The Interservice/Industry Training, Simulation & Education Conference, Orlando, FL.

Berlyne, D. E. (1960). *Conflict, arousal, curiosity*. New York: McGraw-Hill.

Bransford, J. D., Brown, A. L., & Cocking, R. R. (Eds.). (2000). *How people learn: Brain, mind, experience, and school*. Washington, DC: National Academy Press.

Bransford, J. D., Sherwood, R. D., Hasselbring, T. S., Kinzer, C. K., & Williams, S. M. (1990). Anchored instruction: Why we need it and how technology can help. In D. Nix & R. J. Spiro (Eds.), *Cognition, education, and multimedia: Exploring ideas in high technology* (pp. 115-141). Hillsdale, NJ: Lawrence Erlbaum Associates.

Bruner, J. (1966). *Toward a Theory of Education*. Cambridge, MA: Harvard University Press.

Campbell, C. (2005). Only 80 games a year will succeed. *Business Week Online*.

Cannon-Bowers, J. A., Burns, J. J., Salas, E., & Pruitt, J. S. (1998). Advanced technology in decision-making training: The case of shipboard embedded training. In J. A. Cannon-Bowers & E. Salas (Eds.), *Making decisions under stress: Implications for individual and team training* (pp. 365-374). Washington, DC: APA Press.

Chi, M. T. H. (2000). Self-explaining: The dual processes of generating inference and repairing mental models. In R. Glaser (Ed), *Advances in instructional psychology: Educational design and cognitive science, Vol. 5.* (pp. 161-238). Mahwah, NJ: Lawrence Erlbaum Associates.

Clark, R., & Wittrock, M. C. (2000). Psychological principles in training. In S. Tobias & J. D. Fletcher (Eds.), *Training and retraining: A handbook for business, industry, government, and the military* (pp. 51-84). New York: Macmillan.

Clark, R. E., & Feldon, D. F. (2005). Five common but questionable principles of multimedia learning. In R. E. Mayer (Ed.), *Cambridge handbook of multimedia learning*. Cambridge, MA: Cambridge University Press.

Cognition & Technology Group at Vanderbilt (CTGV). (2000). Adventures in anchored instruction: Lessons from beyond the ivory tower. In R. Glaser (Ed), *Advances in instructional psychology: Educational design and cognitive science,* (Vol. 5, pp. 35-99). Mahwah, NJ: Lawrence Erlbaum Associates.

Csikszentmihalyi, M. (1990). *Flow: The psychology of optical experience*. New York: Harper Perennial.

Deci, E. L., Koestner, R., & Ryan, R. M. (1999). A meta-analytic review of experiments examining the effects of extrinsic rewards on intrinsic motivation. *Psychological Bulletin, 125*, 627–668. doi:10.1037/0033-2909.125.6.627

Dede, C., Salzman, M., Loftin, R. B., & Sprague, D. (1999). Multisensory immersion as a modeling environment for learning complex scientific concepts. In W. Feurzeig & N. Roberts (Eds.), *Modeling and Simulation in Science and Mathematics Education*. New York: Springer-Verlag.

Epstein, J. A., & Harackiewicz, J. M. (1992). Winning is not enough: The effects of competition and achievement motivation on intrinsic interest. *Personality and Social Psychology Bulletin, 18*, 128–138. doi:10.1177/0146167292182003

Garris, R., Ahlers, R., & Driskell, J. E. (2002). Games, motivation, and learning: A research and practice model. *Simulation & Gaming, 33*(4), 441–467. doi:10.1177/1046878102238607

Gee, J. (2003). *What Video Games Have to Teach Us About Learning and Literacy*. New York: Palgrave Macmillan.

Gerhard, M., Moore, D., & Hobbs, D. (2004). Embodiment and copresence in collaborative interfaces. *International Journal of Human-Computer Studies, 61*, 453–480. doi:10.1016/j.ijhcs.2003.12.014

Gist, M. E., Schwoerer, C., & Rosen, B. (1989). Effects of alternative training methods on self-efficacy and performance in computer software training. *The Journal of Applied Psychology, 74*, 884–891. doi:10.1037/0021-9010.74.6.884

Gist, M. E., Stevens, C. K., & Bavetta, A. G. (1991). Effects of self-efficacy and post-training intervention on the acquisition and maintenance of complex interpersonal skills. *Personnel Psychology, 44*, 837–861.

Gopher, D., Weil, M., & Bareket, T. (1994). Transfer of skill from a computer game trainer to flight. *Human Factors, 36*, 387–405.

Green, C. S., & Bavelier, D. (2003). Action video game modifies visual selective attention. *Nature, 423*, 534–537. doi:10.1038/nature01647

Hays, R. T. (2005). *The effectiveness of instructional games: a literature review and discussion* (Technical Report 2005-004). Orlando, FL: Naval Air Warfare Center Training Systems Division.

Hays, R. T., & Singer, M. J. (1989). *Simulation fidelity in training system design*. New York: Springer-Verlag.

Honebein, P. C., Duffy, T. M., & Fishman, B. J. (1993). Constructivism and the design of learning environments: Context and authentic activities for learning. In T.M. Duffy, J. Lowyck, & D.H. Jonassen (Eds.), *Designing environments for constructive learning* (pp. 87-108). New York: Springer-Verlag.

Jonassen, D. H. (2000). Revisiting activity theory as a framework for designing student-centered learning environments. In D. H. Jonassen & S. M. Land (Eds.). *Theoretical foundations of learning environments* (pp. 89-121). Mahwah, NJ: Lawrence Erlbaum Associates.

Kanfer, R., & Ackerman, P. L. (1989). Motivation and cognitive abilities: An integrative/aptitude-treatment interaction approach to skill acquisition. *The Journal of Applied Psychology, 74*, 657–690. doi:10.1037/0021-9010.74.4.657

Kanfer, R., & McCombs, B. L. (2000). Motivation: Applying current theory to critical issues in training. In S. Tobias & J. D. Fletcher (Eds.), *Training and retraining: A handbook for business, industry, government, and the military* (pp. 85-108). New York: Macmillan.

Knerr, B. W., Simutis, Z. M., & Johnson, R. M. (1979). *Computer-based simulations for maintenance training: Current ARI research,* (Defense Technical Information Center No. ADA 139371, Technical Report 544). Alexandria, VA: U.S. Army Research Institute for the Behavioral and Social Sciences.

Kolb, D. A. (1984). *Experiential learning: Experience as the source of learning and development.* Englewood Cliffs, NJ: Prentice Hall.

Locke, E. A., & Latham, G. P. (1990). Work motivation: The high performance cycle. In U. Kleinbeck & H. Quast, (Eds.), *Work motivation* (pp. 3-25). Hillsdale, NJ: Lawrence Erlbaum Associates.

Locke, E. A., Shaw, K. N., & Saari, L. M. (1981). Goal setting and task performance: 1969-1980. *Psychological Bulletin, 90,* 125–152. doi:10.1037/0033-2909.90.1.125

Mayer, R. E. (2001). *Multimedia learning.* Cambridge, UK: Cambridge University Press.

Mayer, R. E. (2004). Should there be a three strikes rule against pure discovery? The case for guided methods of instruction. *The American Psychologist, 59*(1), 14–19. doi:10.1037/0003-066X.59.1.14

Mayer, R. E., & Massa, L. J. (2003). Three facets of visual and verbal learners: Cognitive ability, cognitive style, and learning preference. *Journal of Educational Psychology, 95*(4), 833–846. doi:10.1037/0022-0663.95.4.833

Moreno, R., & Mayer, R. E. (2004). Personalized messages that promote science learning in virtual environments. *Journal of Educational Psychology, 96,* 165–173. doi:10.1037/0022-0663.96.1.165

Nelson, L. M. (1999). Collaborative problem solving. In C. M. Reigeluth (Ed.), *Instructional-design theories and models: A new paradigm of instructional theory,* (Vol. II, pp. 241-267). Mahwah, NJ: Lawrence Erlbaum Associates.

Nemeth, E. (1999). Gender differences in reaction to public achievement feedback. *Educational Studies, 25,* 297–310. doi:10.1080/03055699997819

O'Neil, H. F., Wainess, R., & Baker, E. L. (2005). Classification of learning outcomes: Evidence from the computer games literature. *Curriculum Journal, 16*(4), 455–474. doi:10.1080/09585170500384529

Orvis, K. A., Orvis, K. I., Belanich, J., & Mullin, L. N. (2008). The influence of trainee gaming experiences on affective and motivational learner outcomes of videogame-based training environments. In H. F. O'Neil & R. S. Perez (Eds.), *Computer games and team and individual learning.* New York: Elsevier.

Petraglia, J. (1998). *Reality by design: The rhetoric and technology of authenticity in education.* Mahwah, NJ: Lawrence Erlbaum Associates.

Pintrich, P. R., & Zusho, A. (2002). The development of academic self-regulation: The role of cognitive and motivational factors. In A. Wigfield & J. Eccles (Eds.), *Development of achievement motivation,* (pp. 249-284). San Diego, CA: Academic Press.

Reeve, J., & Deci, E. L. (1996). Elements of the competitive situation that affect intrinsic motivation. *Personality and Social Psychology Bulletin, 22,* 24–33. doi:10.1177/0146167296221003

Ryan, R. M., & Deci, E. L. (2000). Self-determination theory and the facilitation of intrinsic motivation, social development, and well-being. *The American Psychologist, 55,* 68–78. doi:10.1037/0003-066X.55.1.68

Sandford, R., Ulicsak, M., Facer, K., & Rudd, T. (2006). *Teaching with games: Using commercial off-the-shelf computer games in formal education.* Retrieved from http://www.futurelab.org.uk/resources/documents/project_reports/teaching_with_games/TWG_report.pdf

Savery, J. R., & Duffy, T. M. (1996). Problem based learning: An instructional model and its constructivist framework. In B. G. Wilson (Ed.), *Constructivist learning environments: Case studies in instructional design,* (pp. 135-148). Hillsdale, NJ: Educational Technology Publications.

Schunk, D. H., & Ertmer, P. A. (1999). Self-Regulatory processes during computer skill acquisition: Goal and self-evaluative influences. *Journal of Educational Psychology, 91,* 251–260. doi:10.1037/0022-0663.91.2.251

Schunk, D. H., & Zimmerman, B. J. (2003). Self-regulation in learning. In W.M. Reynolds, & G.E. Miller, (Eds.), *Handbook of psychology: Educational psychology,* (Vol. 7, pp. 59-78). New York: John Wiley.

Sims, V. K., & Mayer, R. E. (2002). Domain specificity of spatial expertise: The case of video game players. *Applied Cognitive Psychology, 16,* 97–115. doi:10.1002/acp.759

Slater, M., Sadagic, A., Usoh, M., & Schroeder, R. (2000). Small group behaviour in a virtual and real environment: A comparative study. *Presence (Cambridge, Mass.), 9,* 37–15. doi:10.1162/105474600566600

Smith-Jentsch, K. A., Zeisig, R. L., Acton, B., & McPherson, J. A. (1998). Team dimensional training: A strategy for guided team self-correction. In J.A. Cannon-Bowers & E. Salas (Eds.), *Making decisions under stress: Guidelines for individual and team training,* (pp. 299-312). Washington, DC: American Psychological Association.

Tannenbaum, S. I., Mathieu, J. E., Salas, E., & Cannon-Bowers, J. A. (1991). Meeting trainees' expectations: The influence of training fulfillment on the development of commitment, self-efficacy, and motivation. *The Journal of Applied Psychology, 76,* 759–769. doi:10.1037/0021-9010.76.6.759

Vogel, J., Vogel, D., Cannon-Bowers, J., Bowers, C., Muse, K., & Wright, K. (in press). Computer and interactive simulations for learning: A meta-analysis. *Journal of Educational Computing Research.*

White, R. W. (1959). Motivation reconsidered: The concept of competence. *Psychological Review, 66,* 297–333. doi:10.1037/h0040934

Wood, D., Bruner, J. S., & Ross, G. (1976). The role of tutoring in problem solving. *Journal of Child Psychology and Psychiatry, and Allied Disciplines, 17*(2), 89–100. doi:10.1111/j.1469-7610.1976.tb00381.x

Chapter 17
The Way Ahead for Serious Games

Jan Cannon-Bowers
University of Central Florida, USA

INTRODUCTION

Despite the fact that people have been playing games since before recorded history, the field of *Serious Games*—as a scientifically valid and viable area of investigation and application—is really in its infancy. In fact, the rise in availability and popularity of video games is a relatively recent phenomenon and actual applications of video game technology to serious pursuits are relatively rare. That said, I believe that Serious Games are coming into their own, and predict that the next few years will witness an explosion of new games, design features and guidelines, success stories, and scientific findings regarding their effectiveness. How quickly and fruitfully this happens depends, in part, on how well the community of researchers, designers, developers, evaluators, and end users can come together to systematically conceive of, and deploy games for serious purposes. Haphazard attempts—i.e., those that do not build on the findings of others' experiences—will retard the speed at which viable games are consistently produced. Likewise, rigorous evaluation of those games that are developed cannot be neglected or future developers will likely fall prey to the same problems and pitfalls as their predecessors.

This vision of a unified field of Serious Games may be far off, but I think that there are several precursors (or enablers) that the field can work on now as a means to effectively evolve. These include: working towards common definitions and a common language for describing Serious Games; adopting an overarching framework in which the various types

Table 1. Serious Game Labels (adapted from Sawyer & Smith, 2008)

Alternative Labels for Serious Games	
Educational Games	Digital Game-Based Learning
Learning Games	Immersive Learning
Training Games	Simulations
Games for Health	Social Impact Games
Health Games	Persuasive Games
Simulation/Simulators	Games for Change
Virtual Reality	Synthetic Learning Environments
Virtual Worlds	Game-Based "X"
Alternative Purpose Games	Games for Good

of Serious Games can be organized as a means to establish commonalities and differences in various applications; and identifying challenges and barriers to development and implementation of Serious Games early so that these can be confronted and dealt with before they can impede progress. In the following sections, I address these issues and outline what I believe are the challenges in developing Serious Games and ultimately in creating a viable field of study.

DEFINITIONS

Clearly, a unified field of Serious Games is not possible until an agreement can be reached about how Serious Games are defined. According to Sawyer and Smith (2008), there is a tendency to equate Serious Games with those devoted to learning or training, but they strongly disagree with this position. Instead, they list a variety of labels that have been used to describe different types of Serious Games (I've added a few to the original list); see Table 1.

The problem with this group of labels is that many are used to refer to the same type of game (i.e., that have a common goal). For example, some people use the term "educational game" while others may use "immersive learning environment," but mean the same thing. Alternatively, lumping together all games that have some kind of learning as their objective and calling them "learning games" is probably not meaningful since there may be very different applications within this space.

As an alternative to this loose collection of labels, Smith and Sawyer (2008) contend that what is needed in the field is a better way to define and discuss the various kinds of games that have been produced or that can be conceived.

Addressing first the definition issue, my preferred definition would be: *Serious Games are games that are not exclusively designed for entertainment purposes* (A. Smith, personal communication, July 2009). This definition is broad enough to cover the scope of possible applications (i.e., any purpose besides purely entertainment) and also allows for the fact that some games may have a serious as well as entertainment component (hence the phrase "not exclusively"). However, since the definition is so broad, it does not reflect or indicate the types of games that are part of the Serious Game space. For this reason, Sawyer and Smith (2008) provided a taxonomy of games (based on their usage); this is described next.

USES OF GAMES: A TAXONOMY

The taxonomy provided by Sawyer and Smith (2008) addresses the definitional issues discussed above and demonstrates the breadth and diversity of Serious Game applications being conceived. According to these authors, a taxonomy is necessary for several reasons, including: creating a shared mental model for those involved in Serious

Games regarding what they are and what they could be; erasing various myths about Serious Games; providing a snapshot of the current state of the Serious Games industry; determining where R&D is needed; and creating a foundation for future efforts by organizing the space. I also believe that such a taxonomy can be a viable communication device so that potential sponsors and users get a sense of the scope and breadth of possible applications.

The taxonomy includes several categories of Serious Games that have been developed, proposed, or can be conceived. These include: Games for health, Advergames, Games for training, Games for education, Games for science and research, Games in production, and Games as work. Several of these are self-explanatory; but a few require further explanation. For example, *Advergaming* is the use of gaming in advertising. *Games in production* refer to games that are used to actually produce a product or outcome. An example here is when a game is used to forecast the likelihood of various outcomes as input to an actual planning process. Related to this, *Games as work* refers to situations where the game itself is the medium with which the employee interacts to get his/her job done. To date, there are few, if any, examples of this type of game in actual practice, but the idea that it can be done is certainly plausible.

Sawyer and Smith (2008) then cross the categories of serious games listed above with several broad sectors that could make use of the games. These sectors include: Government, Defense, Healthcare, Marketing & Communications, Education, Corporate, and Industry. The resulting matrix (categories X sectors) displays a host of specific game applications. For example, in Defense, the Games for health category might manifest itself as games for rehabilitation and wellness for military members while in Industry, Games for health could be represented by an occupational safety game.

While some of the cells in the taxonomy are less obvious than others, the overarching framework has value as a means to organize the space of Serious Games. It can certainly help guide researchers and developers to understand how their efforts fit into a larger landscape. But ultimately, the value of the taxonomy may be in defining the boundaries of generalizability of the results, guidelines, and lesson learned from development efforts. For example, it may be found that Games for training share common features and/or development processes, but these features may not necessarily be applicable to Advergaming efforts. Eventually, a taxonomy like this one, if adopted by a majority of the field, can provide a common language for researchers, developers, sponsors and consumers.

CHALLENGES IN SERIOUS GAMES

Given that we can address the definitional and classification issues noted above, there are a number of challenges that must be addressed to ensure that the potential of Serious Games is realized. Indeed, the future of Serious Games depends, I believe, on the ability of researchers to show how they can be applied successfully. This is no small task, especially since, as is obvious from the taxonomy shown above, the term "successfully" is relative--it will be defined in terms of the purpose or goals of the game. But, achieving goals such as improved health and well being, increased learning, or improved sales are all complex outcomes that are difficult to demonstrate no matter what the intervention. Moreover, a thorough understanding of measurement theory is necessary so that it is possible to reveal true changes that can be attributed to the game.

Beyond measurement issues, I believe that building truly effective games—those where the game is clearly a causal factor in achieving the desired outcome—hinges on the ability of the developer to make accurate decisions during the game design process. And these decisions must be based on strong theoretical justification and empirical results where possible. Hence, I see the greatest challenge to the future of Serious Games as conducting high quality programs of research that produce usable and informative results. Beyond this, it is imperative that developers devote time and effort into identifying and addressing implementation issues. As with other technology-based interventions, even well designed games may fail if barriers to implementation are not addressed. Finally, there are several challenges related to the Serious Game development process itself that need to be addressed if they are to be successfully developed and implemented. The following sections offer a bit more detail about these challenges.

RESEARCH NEEDS IN SERIOUS GAMES

Even though many gamers—and those who observe them—are convinced of the value of game play, I do not think that games will be accepted as a viable medium or serious pursuit until their benefits are demonstrated empirically. The chapter by Sanchez et al. (this volume) presented some ideas for the content of such study. In addition, among the requirements for R&D that I see in this area, the following are most pressing:

- Employ rigorous experimental designs that conform to accepted standards of proof (e.g., in educational research)
- Construct compelling theories and scientific justification for why the game will work for its intended purpose
- Isolate game features that contribute to goal attainment so that these can be applied in other games with similar objectives
- Figure out what works by testing the propositions in a variety of settings and with different users

- Conduct studies that lead to specific design guidelines by testing various possibilities and comparing the results
- Produce generalizable results by specifying the probable boundaries of the findings and their applicability to other tasks/situations
- Gain an understanding of assessment and measurement issues so that appropriate outcomes are adequately addressed
- Provide for large scale testing so that statistical conclusions can be drawn

If headway can be made in at least some of these areas, I think that we will begin to see a *science* of Serious Games emerge. At least one incumbent value of this will be to attract grant funding and corporate interest in supporting Serious Game design efforts. Furthermore, the likelihood that the goals of the game will be reached effectively and efficiently will be enhanced.

IMPLEMENTATION ISSUES IN SERIOUS GAMES

In my mind there are several pitfalls that can plague any effort to design and deliver an intervention aimed at achieving specific goals for an individual or an organization, and developing Serious Games is no exception. First off, it is incumbent on the developer to work with sponsors/users to ensure that requirements are carefully and fully specified. This is a well-defined process in that can be found in, for example, training needs analysis or user-centered design. Serious game developers should not attempt to reinvent the wheel here (and likely fall prey to many of the same avoidable problems that others have before them). Entire volumes and texts exist to guide this process.

Other tenets that anyone who has developed a successful intervention or program can recite include: engage with users early and often; try to anticipate and mitigate barriers to implementation; and carefully manage the expectations of sponsors, users, and others involved in implementation.

Again, guidance for these types of development efforts is readily available. Moreover, modern design engineering techniques that emphasize rapid prototyping and iterative design, "build a little, test a little" strategies will help ensure that an effective end product is produced.

With respect to Serious Games, in particular, there may be push back on the part of user groups to accept a *game* as a serious intervention. In sectors like healthcare, the military and even education, it is likely that sponsors and users will need to be convinced of the value of games to achieve desired outcomes. This also highlights the need for well designed scientific studies as discussed above; in my experience, nothing is as convincing as hard data to show that something works and silence critics.

DEVELOPMENT CHALLENGES

As several authors in this volume have commented, one of the greatest challenges in developing Serious Games is managing a diverse, multi-disciplinary team of professionals, each with their own ideas, approaches, language, culture, and motives. At a minimum, a well-rounded team would include: project manager, subject matter experts, programmers, artists, other creatives (e.g., story tellers), human performance experts, measurement experts, and gamers. Now, several of these perspectives can be represented in a single individual, but combining too many roles will likely lead to overload. Among the issues here are even seemingly mundane factors such as acceptable work hours (not to over-generalize, but most gamers I know are usually not happy starting a meeting as early as the military folks I have worked with). While this sounds like a minor issue, it may be more appropriately viewed as a symptom of two very different cultures that we hope will mesh so that work can get done. My feeling is that a good team leader will manage these differences deliberately, and seek compromises that keep everyone satisfied.

A related consequence of working in a multi-disciplinary team is that there is typically a natural tension among the values and desires of the team members. Hence, it is essential that project managers actively balance the serious goals of the game with the more entertaining (or fun) aspects. All too often in the educational game area I come across the attitude that "instructional features suck the fun out of games," or the flip side of that, "if they are having too much fun, they can't be learning." Obviously, I think that both of these are dead wrong. The entire premise of Serious Games is that it is possible to achieve serious objectives within the context of an engaging game environment. Indeed, really successful Serious Games seamlessly bridge the gap between the *serious* and the *frivolous*, the *desired outcomes* and the *experience*, the *science* and the *art*.

On the more technical side, an issue that needs attention in Serious Game development is reuse—of content, engines, assets, and approaches. This imperative is perhaps the only way that costs can be contained, since it is simply not feasible for Serious Game developers to start from scratch every time. In addition, an iterative design process—as advocated above—where frequent user testing is built into the development schedule can help avoid costly errors.

The final issue that I will address in this area is funding sources. As stated, I am convinced that well conceived, systematic research programs are essential to the future of the filed. Without funding for such programs, it will be difficult for Serious Game developers to move the field forward. At this point, it is not clear where such funding will exist. Again, it may be that a few more well-publicized success stories (most likely in the educational sector) will catalyze funding agencies to create such programs. This is particularly essential to support efforts to "scale up"—i.e., those programs that can demonstrate successful large-scale implementation of a Serious Game with important, measurable outcomes.

THE FUTURE

My personal decision to become involved in Serious Game development was based on many hours of watching Navy personnel learn from low-fidelity simulations, most often while thoroughly enjoying the process. And the highlight of my more recent experiences in conceiving and developing Serious Games is always when we watch users encounter them and *play*. Good games, I believe, have the potential to transform the way we think about education by providing the learner with authentic, engaging experiences that enhance their learning and retention. Likewise, in other areas—games for change, games for peace, health games, and others shown above—the power of games to effect positive change in the human condition is immense. But the future I see is not one that will happen arbitrarily. Rather, it will only be realized if we can collectively build a field of Serious Games that is self-reflective and evolves (and improves itself) continuously, based on well conceived and executed research and experiences. I remain optimistic that this is vision can be achieved.

REFERENCES

Sanchez, A., Cannon-Bowers, J. A., & Bowers, C. (2010). Establishing a Science of Game-Based Learning. In C. Bowers & J. A. Cannon-Bowers, (Eds.), *Serious Game Design and Development: Technologies for Training and Learning.* Hershey, PA: IGI Global.

Sawyer, B., & Smith, P. A. (2008, February,). *Serious Games Taxonomy*. Paper presented at the Game Developers Conference, San Francisco, CA.

Compilation of References

110th Congress (2008). *Higher Education Opportunity Act Conference Report to accompany H.R. 4137* (House Report 110-803). Retrieved June 22, 2009, from http://www.congress.gov/cgi-bin/cpquery/?sel=DOC&&item=&r_n=hr803.110&&&sid=cp110pOPWa&&refer=&&&db_id=cp110&&hd_count=& Allen, B., Otto, R., & Hoffman, B. (2000). Case-based learning: contexts and communities of practice. In S. Tobias & J. Fletcher (Eds.), *Training and Retraining* (pp. 453-472). New York: Macmillan Reference.

AAMC Task Force On the Clinical Skills Education of Medical Students. (2005). *Recommendations for clinical skills curricula for undergraduate medical education*. Washington, DC.

AAMC. (2006). *Help wanted: More U.S. doctors*. Retrieved June 1, 2009, from the AAMC Web site, http://www.aamc.org/workforce/helpwanted.pdf

Abbott, H. P. (2002). *The Cambridge introduction to narrative*. Cambridge, UK: Cambridge University Press.

Accreditation Council for Graduate Medical Education (ACGME) Outcomes Project. (2006). *Educating physicians for the 21st century*. Chicago, IL: Author.

Adler, M. D., & Johnson, K. B. (2000). Quantifying the literature of computer-aided instruction in medical education. *Academic Medicine, 75*(10), 1025–1028. doi:10.1097/00001888-200010000-00021

AGCME Outcomes Project. (2000). *Toolbox of assessment methods*. Chicago, IL: Author.

Ahl, D. H. (1983). Editorial. *Video & Arcade Games, 1*(1), 4. (See also http://www.atarimagazines.com/cva/v1n1/editorial.php) Barker, T., & Bernhard, B. (n.d.). Wii leages mix virtual exercise with real results. *Examiner*. Retrieved on November 3, 2008 from http://www.examiner.com/a-1669072~Wii_leagues_mix_virtual_exercise_with_real_results.html

Akinbami, L. J., Mooreman, J. E., Garbe, P. L., & Sondik, E. J. (2009). Status of childhood asthma in the United States, 1980-2007. *Pediatrics, 123*(Suppl. 3), S131–S145. doi:10.1542/peds.2008-2233C

Aldrich, C. (2005). *Learning by doing: A comprehensive guide to simulations, computer games and pedagogy in e-Learning and other educational experiences*. San Francisco, CA: Pfeiffer Books.

Alessi, S. M., & Trollip, S. R. (1991). *Computer-Based Instruction: Methods and Development*. Englewood Cliffs, NJ: Prentice-Hall.

American College Health Association National College Health Assessment Spring 2006 Reference Group Data Report (Abridged) (2007). *Journal of American College Health, 55*(4), 195-206.

Ammerman, A. S., Lindquist, C. H., Lohr, K. N., & Hersey, J. (2002). The efficacy of behavioral interventions to modify dietary fat and vegetable intake: A review of the evidence. *Preventive Medicine, 35*(1), 25–41. doi:10.1006/pmed.2002.1028

An integrated mobile wireless system for capturing physiological data streams during a cognitive-motor task: Applications for aging. (2007, November 11-12). In *IEEE Engineering in Medicine and Biology Workshop, IEEE Dallas University of Texas at Dallas*, (pp. 67-70).

Analyzing Motoric and Physiological Data in Describing Upper Extremity Movement in the Aged (with G.N. Pradhan, N. Engineer, B. Prabhakaran). (2008, July 16-19). In *International Conference on Pervasive Technologies Related to Assistive Environments (PETRA 2008), ACM International Conference Proceeding Series,* Athens, Greece, (Vol. 282, pp.8).

Anand, P. G., & Ross, S. M. (1987). Using computer-assisted instruction to personalize arithmetic materials for elementary school children. *Journal of Educational Psychology, 79*(1), 72–78. doi:10.1037/0022-0663.79.1.72

Anderson, I., Maitland, J., Sherwood, S., Barkhuus, L., Chalmers, M., & Hall, M. (2007). Shakra: Tracking and sharing daily activity levels with unaugmented mobile phones. *Mobile Networks and Applications, 12,* 185–199. doi:10.1007/s11036-007-0011-7

Anderson, R. C., & Anderson, K. E. (1990). Success and failure attributions in smoking cessation among men and women. *AAOHN, 38*(4), 180–185.

Anonymous. (2009). China moves to solve graduate unemployment issue. *People's Daily Online.* Retrieved January 16, 2009 from http://english.peopledaily.com.cn/90001/90781/90879/6571767.html

Arthur, L. J. (1992). Quick & dirty. *Computerworld, 26*(50), 109–112.

Asher, J. (1965). The Strategy of the Total Physical Response: an Application to Learning Russian. *International Review of Applied Linguistics, 3*(4), 291–300. doi:10.1515/iral.1965.3.4.291

Asher, J. (1966). The Learning Strategy of The Total Physical Response: A Review. *Modern Language Journal, 50*(2), 79–84. doi:10.2307/323182

Asher, J. (1969). The Total Physical Response Approach to the Second Language Learning. *Modern Language Journal, 53*(1), 3–17. doi:10.2307/322091

Asher, J. (1977). Children Learning Another Language: A Developmental Hypothesis. *Child Development, 58*(1-2), 24–32.

Asher, J., & Adamski, J. C. (2003). *Learning another language through actions.* Los Gatos, CA: Sky Oaks Productions.

Asher, J., Kusudo, J., & De la Torre, R. (1974). Learning a Second Language through Commands: The Second Field Test. *Modern Language Journal, 58*(1/2), 24–32. doi:10.2307/323986

Associated Press. (2006). A video game that seeks to make peace, not war; 'PeaceMaker' simulates Israeli-Palestinian conflict in all its complexity. *MSNBC.* Retrieved from http://www.msnbc.msn.com/id/12423759

Association of American Medical Colleges (AAMC). (2006). *Questions and answers about the AAMC's new physician workforce position.* Retrieved June 1, 2009, from the AAMC Web site, http://www.aamc.org/workforce/workforceqa.pdf

Asynchronous Learning Networks Magazine, 1(1). (1997, March).

Avedon, E. M., & Sutton-Smith, B. (Eds.). (1971). *The Study of Games.* New York: John Wiley & Sons, Inc.

Backer, J. A. (1999). *Multi-user domain object oriented as a high school procedure for foreign language acquisition.* Unpublished Ph.D. thesis, Nova Southeastern University.

Baddeley, A. D. (1986). *Working Memory.* Oxford, UK: Clarendon Press.

Baddeley, A. D. (2000). The episodic buffer: A new component of working memory. *Trends in Cognitive Sciences, 4*(11), 417–423. doi:10.1016/S1364-6613(00)01538-2

Baker, A., Navarro, E. O., & van der Hoek, A. (2005). An experimental card game for teaching software engineering processes. *Journal of Systems and Software, 75*(1-2), 3–16. doi:10.1016/j.jss.2004.02.033

Bal, M. (1997). *Narratology: Introduction to the theory of narrative* (2nd ed.). Toronto, Canada: University of Toronto Press.

Baldwin, T. T., & Ford, J. K. (1988). Transfer of training: A review and directions for future research. *Personnel*

Psychology, 41, 63–105. doi:10.1111/j.1744-6570.1988. tb00632.x

Ballagas, R., & Walz, S. P. (2007). REXplorer: Using Player-Centered Iterative Design Techniques for Pervasive Game Design. In C. Magerkurth & C. Röcker (Eds.), *Pervasive Gaming Applications - A Reader for Pervasive Gaming Research Vol. 2*. Aachen, Germany: Shaker Verlag.

Bandura, A. (1977). Self-efficacy: Toward a unifying theory of behavioral change. *Psychological Review, 84*, 191–215. doi:10.1037/0033-295X.84.2.191

Bandura, A. (1977). Self-efficacy: Toward a unifying theory of behavioral change. *Psychological Review, 84*, 191–215. doi:10.1037/0033-295X.84.2.191

Bandura, A. (1986). *Social foundations of thought and action: A social cognitive theory*. Englewood Cliffs, NJ: Prentice-Hall.

Bandura, A. (1991). Social Cognitive Theory of Self Regulation. Special Issue: Theories of cognitive self-regulation. *Organizational Behavior and Human Decision Processes, 50*, 248–287. doi:10.1016/0749-5978(91)90022-L

Bandura, A. (1997). *Self-efficacy. The Exercise of Control*. New York, NY: W.H. Freeman.

Bandura, A., & Schunk, D. H. (1981). Cultivating competence, self-efficacy, and intrinsic interest through proximal self-motivation. *Journal of Personality and Social Psychology, 41*, 586–598. doi:10.1037/0022-3514.41.3.586

Barab, S., Thomas, M., Dodge, T., Carteaux, R., & Tuzun, H. (2005). Quest Atlantis: A game without guns. *Educational Technology Research and Development, 50*(1), 86–107. doi:10.1007/BF02504859

Baranowski, T., Buday, R., Thompson, D. I., & Baranowski, J. (2008). Playing for real. Video game and stories for health-related behavior change. *American Journal of Preventive Medicine, 34*, 74–82. doi:10.1016/j.amepre.2007.09.027

Barksdale, J., & Hundt, R. (2005). *Digital future initiative: Challenges and opportunities for public service media in the digital age*. Retrieved on June 22, 2009 from http://www.newamerica.net/publications/policy/digital_future_initiative_report

Baron, R. A., & Byrne, D. (1987). *Social Psychology: Understanding Human Interaction*. (5th ed.). Newton, MA: Allyn and Bacon.

Barry, M. J., Fowler, F. J. Jr, Mulley, A. G. Jr, Henderson, J. V. Jr, & Wennberg, J. E. (1995). Patient reactions to a program designed to facilitate patient participation in treatment decisions for benign prostatic hyperplasia. *Medical Care, 33*, 771–782. doi:10.1097/00005650-199508000-00003

Bartholomew, L. K., Gold, R. S., Parcel, G. S., Czyzewski, D. I., Sockrider, M. M., & Fernandez, M. (2000b). Watch, Discover, Think, and Act: Evaluation of computer assisted instruction to improve asthma self-management in inner-city children. *Patient Education and Counseling, 39*(2-3), 269–280. doi:10.1016/S0738-3991(99)00046-4

Bartholomew, L. K., Parcel, G. S., Kok, G., & Gottlieb, N. H. (2001). *Intervention Mapping: Designing theory- and evidence-based health promotion programs*. Mountain View, CA: Mayfield Publishing Company.

Bartholomew, L. K., Parcel, G. S., Kok, G., & Gottlieb, N. H. (2006). *Planning Health Promotion Programs. An intervention mapping approach*. Hoboken, NJ: John Wiley & Sons, Inc.

Bartholomew, L. K., Shegog, R., Parcel, G. S., Gold, R. S., Fernandez, M. E., & Czyzewski, D. I. (2000a). Watch, Discover, Think, and Act: A model for patient education program development. *Patient Education and Counseling, 39*(2-3), 253–268. doi:10.1016/S0738-3991(99)00045-2

Bartholomew, L.K., Parcel, G.S., & Kok, G. (1998). Intervention mapping: A process for developing theory- and evidence-based health education programs. *Health Education & Behavior, Oct 25*(5), 545-63.

Bartlett, F. C. (1958). *Thinking: An Experimental and Social Study*. London: Allen and Unwin.

Bateman, C. (Ed.). (2007). *Game writing: narrative skills for videogames*. Boston: Charles River Media.

Baylor, A. L. (2000). Beyond butlers: Intelligent agents as mentors. *Journal of Educational Computing Research, 22*, 373–382. doi:10.2190/1EBD-G126-TFCY-A3K6

Baylor, A. L. (2001). Agent-based learning environments for investigating teaching and learning. *Journal of Educational Computing Research, 26*, 249–270.

Beauvois, M. H. (1992). Computer-assisted classroom discussion in the foreign language classroom: conversations in slow motion. *Foreign Language Annals, 25*(1), 455–464. doi:10.1111/j.1944-9720.1992.tb01128.x

Beauvois, M. H., & Eledge, J. (1996). Personality types and megabytes: Student attitudes toward computer-mediated communication (CMC) in the language classroom. *CALICO Journal, 13*(2), 27–45.

Beck, I., & McKeown, M. (1991). Conditions of vocabulary acquisition. In R. Barr, M. Kamil, P. Mosenthan & P. D. Pearson (Eds.), *Handbook of Reading Research* (Vol. 2, pp. 789-814). New York: Longman.

Beck, J. C., & Wade, M. (2004). *Got Game: How the Gamer Generation is Reshaping Business Forever*. Boston: Harvard Business School Press.

Begg, M., Dewhurst, D., & Macleod, H. (2005). Game-Informed Learning: Applying computer game processes to higher education. *Innovate: Journal of Online Education, 1*(6). Retrieved from http://innovateonline.info/?view=article&id=176

Belanich, J., Orvis, K. L., & Mullin, L. N. (2004). *Training Game Design Characteristics that Promote Instruction and Motivation*. The Interservice/Industry Training, Simulation & Education Conference, Orlando, FL.

Bell-Gredler, M. E. (1986). *Learning and Instruction: Theory into Practice*. New York: Macmillan.

Benevolo, L. (1967). *The origins of modern town planning*. London, United Kingdom: Routledge and Kegan Paul.

Berlyne, D. E. (1960). *Conflict, arousal, curiosity*. New York: McGraw-Hill.

Berlyne, J. (1983). Conceptualizing student motivation. *Educational Psychologist, 18*, 200–215.

Bernanke, B. S. (2006). *Before Leadership South Carolina, Greenville South Carolina* [PDF]. Retrieved from http://www.federalreserve.gov/newsevents/speech/Bernanke20060831a.htm

Betrus, A. (2001). *The many hats of an instructional designer card game – overview*. Retrieved November 10, 2008 from http://www2.potsdam.edu/betrusak/idcardgame/idcardgameoverview.html

Blickensderfer, E. L., Cannon-Bowers, J. A., & Salas, E. (1998). Cross training and team performance. In J.A. Cannon-Bowers & E. Salas (Eds.), *Making decisions under stress: Implications for individual and team training* (pp. 299-312). Washington, DC: American Psychological Association.

Block, G., Miller, M., Harnack, L., Kayman, S., Mandel, S., & Cristofar, S. (2000). An interactive CD-ROM for nutrition screening and counseling. *American Journal of Public Health, 90*, 781–785. doi:10.2105/AJPH.90.5.781

Bogost, I. (2007). *Persuasive Games. The expressive power of videogames*. Cambridge, MA: Massachusetts Institute of Technology.

Bonk, C. J., & Dennen, V. P. (2005, March). *Massive Multiplayer Online Gaming: A Research Framework for Military Training and Education*. Office of the Undersecretary of Defense For Personnel and Readiness. Technical Report 2005-1.

Booker, C. (2005). *The seven basic plots: Why we tell stories*. London: Continuum International Publishing Group.

Bosworth, B., & Triplett, J. (2006). *Is the 21st Century Productivity Expansion Still in Services? And What Should Be Done About It?* Paper from the 2006 Summer Institute sponsored by the National Bureau of Economic Research and the Conference on Research in Income and Wealth, Cambridge, MA. Retrieved from http://www.nber.org/confer/2006/si2006/prcr/bosworth.pdf

Bower, G. H. (1975). Cognitive Psychology: An introduction. In W. K. Estes (Ed.), *Handbook of learning and cognitive processes, Volume 1, Introduction to concepts and issues*, (pp. 25-80). Hillsdale, NJ: Erlbaum.

Bradley, B. P., Gossip, M., Brewin, C. R., Phillips, G., & Green, L. (1992). Attributions and relapse in opiate addicts. *Journal of Consulting and Clinical Psychology*, *60*(3), 470–472. doi:10.1037/0022-006X.60.3.470

Bradley, P. (2006). The history of simulation in medical education and possible future directions. *Medical Education*, *40*, 254–262. doi:10.1111/j.1365-2929.2006.02394.x

Bransford, J. D., Brown, A. L., & Cocking, R. R. (Eds.). (1999). *How people learn: Brain, mind, experience, and school*. Washington, DC: National Academy Press.

Bransford, J. D., Sherwood, R. D., Hasselbring, T. S., Kinzer, C. K., & Williams, S. M. (1990). Anchored instruction: Why we need it and how technology can help. In D. Nix & R. J. Spiro (Eds.), *Cognition, education, and multimedia: Exploring ideas in high technology* (pp. 115-141). Hillsdale, NJ: Lawrence Erlbaum Associates.

Bransford, J., Brown, A., & Cocking, R. (2000). *How people learn: Brain, mind, experience, and school*. Washington, DC: National Academy Press.

Bray, S. R., & Born, H. A. (2004). Transition to university and vigorous physical activity: Implications for health and psychological well being. *Journal of American College Health*, *52*(4), 181–188. doi:10.3200/JACH.52.4.181-188

Brehmer, B. (1992). Dynamic decision making: Human control of complex systems. *Acta Psychologica*, *81*(3), 211–241. doi:10.1016/0001-6918(92)90019-A

Brehmer, B., & Dörner, D. (1993). Experiments with computer-simulated microworlds: Escaping both the narrow straits of the laboratory and the deep blue sea of the field study. *Computers in Human Behavior*, *9*(2-3), 171–184. doi:10.1016/0747-5632(93)90005-D

Bremner, J., Randall, R., Scott, T., Bronen, R., Seibyl, J., & Southwick, S. (1995). MRI-based measurement of hippocampal volume in patients with combat-related posttraumatic stress disorder. *The American Journal of Psychiatry*, *152*, 973–981.

Briskman, L. (1981). Creative product and creative process in science and art. In D. Dutton and M. Krausz (Eds.), *The Concept of Creativity in Science and Art*. Boston: Martinus Nijhoff.

Brooke, J. (1986). *SUS- A quick and dirty usability scale*. Earley, UK: Readhatch.

Brookhaven National Laboratory. (1981). *Video Games – Did They Begin at Brookhaven?* Retrieved November 3, 2008 from http://osti.gov/accomplishments/videogame.html

Brown, A. L. (1988). Motivation to learn and understand: On taking charge of one's own learning. *Cognition and Instruction*, *5*(4), 311–321. doi:10.1207/s1532690xci0504_4

Brown, J. S., & Thomas, D. (2006). You Play World of Warcraft? You're Hired! Why multiplayer games may be the best kind of job training. *Wired (San Francisco, Calif.)*, *14*(4).

Brubaker, B. H. (1988). An attributional analysis of weight outcomes. *Nursing Research*, *37*(5), 282–287. doi:10.1097/00006199-198809000-00005

Bruner, J. (1966). *Toward a Theory of Education*. Cambridge, MA: Harvard University Press.

Bruner, J. (1990). *Acts of meaning*. Cambridge, MA: Harvard University Press.

Bueno de Mesquita, B. (2006). Game theory, political economy, and the evolving study of war and peace. *The American Political Science Review*, *100*(4), 637–642. doi:10.1017/S0003055406062526

Buscaglia, T. (2007). Game law: Scrum deals – good, bad or ugly! *Gamasutra: The Art & Science of Making Games*. Retrieved November 17, 2008 from http://www.gamasutra.com/view/feature/1662/game_law_scrum_deals__good_bad_.php

Cameron, J., & Pierce, W. D. (1994). Reinforcement, reward, and intrinsic motivation: A meta-analysis. *Review of Educational Research*, *64*(3), 363–423.

Campbell, C. (2005). Only 80 games a year will succeed. *Business Week Online*.

Campbell, J. (1949). *The hero with a thousand faces*. Princeton, NJ: Princeton University Press.

Cannon-Bowers, J. A., & Salas, E. (1998a). *Making decisions under stress: Implications for individual and team training*. Washington, DC: American Psychological Association.

Cannon-Bowers, J. A., & Salas, E. (1998b). Team performance and training in complex environments: Recent findings from applied research. *Current Directions in Psychological Science*, 7(3), 83–87. doi:10.1111/1467-8721.ep10773005

Cannon-Bowers, J. A., Burns, J. J., Salas, E., & Pruitt, J. S. (1998). Advanced technology in decision-making training: The case of shipboard embedded training. In J. A. Cannon-Bowers & E. Salas (Eds.), *Making decisions under stress: Implications for individual and team training* (pp. 365-374). Washington, DC: APA Press.

Cannon-Bowers, J., & Bowers, C. (2007). *Learning and Technology-Based Solutions for PTSD Prevention: An Example of Future Medical Simulation*. Paper presented at the Interservice/Industry Training, Simulation & Education Conference (IITSEC), Orlando, FL.

Carroll, J. M., Stein, C., Byron, M., & Dutram, K. (1996). Using Interactive Multimedia to deliver nutrition education to Maine's WIC clients. *Journal of Nutrition Education*, 28, 19–25. doi:10.1016/S0022-3182(96)70011-0

Carson, L. (2006). *WV Games for Health*. Retrieved on November 17, 2008 from statewide-ddr-wv.ppt

Carson, L. (2008, November 3). Video games promoted in fight against childhood obesity. *Canada.com*. Retrieved on November 17, 2008 from http://www.canada.com/topics/news/story.html?id=3dfaf40b-0c53-4a89-9b31-306deecbcf5e

Cartier, J. L., & Stewart, J. (2000). Teaching the nature of inquiry: Further development in a high school genetics curriculum. *Science and Education*, 9, 247–267. doi:10.1023/A:1008779126718

CDC. (2007). Department of Health and Human Services: *Overweight and Obesity*. Retrieved February 2, 2007: http://www.cdc.gov/nccdphp/dnpa/obesity/index.htm.

Centers for Disease Control and Prevention. (2009). Retrieved June 22, 2009 from http://www.cdc.gov/HomeandRecreationalSafety/Playground-Injuries/index.html.

Centre for Pervasive Healthcare (CfPH). (2007). *HealthyHome: Supportive technology for pregnant women with diabetes*. Retrieved from http://www.healthyhome.dk/

Champion, V. L., & Huster, G. (1995). Effect of interventions on stage of mammography adoption. *Journal of Behavioral Medicine*, 18, 169–187. doi:10.1007/BF01857868

Chapelle, C. A. (1996). *Computer applications in applied linguistics*. Cambridge, UK: Cambridge University Press.

Chatham, R. E. (2007). Games for training. *Communications of the ACM*, 50(7), 36–43. doi:10.1145/1272516.1272537

Chatham, R., & Braddock, J. (2001). Training Superiority & Training Surprise, Final Report, *Defense Science Board*. Retrieved from http://www.acq.osd.mil/dsb/reports/trainingsuperiority.pdf

Chaudhuri, A., Khan, S. A., Lakshmiratan, A., Py, A., & Shah, L. (2003). Trust and trustworthiness in a sequential bargaining game. *Journal of Behavioral Decision Making*, 16, 331–340. doi:10.1002/bdm.449

Chenova, J. (2004). Flow in games and everything else. *Communications of the ACM*, 50(4), 31–34.

Cheung-Blunden, V., & Blunden, B. (2008). Paving the road to war with group membership, appraisal antecedents, and anger. *Aggressive Behavior*, 34(3), 175–189. doi:10.1002/ab.20234

Chi, M. T. H. (2000). Self-explaining: The dual processes of generating inference and repairing mental models. In R. Glaser (Ed), *Advances in instructional psychology: Educational design and cognitive science, Vol. 5*. (pp. 161-238). Mahwah, NJ: Lawrence Erlbaum Associates.

Chi, M. T. H., Glaser, R., & Farr, M. J. (Eds.). (1988). *The nature of expertise*. Hillsdale, NJ: Lawrence Erlbaum Associates.

Chwif, L., & Barretto, M. R. P. (2003). Simulation models as an aid for teaching and learning process in operations management. In S. Chick, P. J. Sánchez, D. Ferrin, & D. J. Morrice (Eds.), *2003 Winter Simulation Conference* (pp. 1994-2000).

Clark, J. E., Lanphear, A. K., & Riddick, C. C. (1987). The effects of video game playing on the response selection of elderly adults. *Journal of Gerontology, 42*(1), 82–85.

Clark, N. M. (1989). Asthma self-management education. Research and implications for clinical practice. *Chest, 95*(5), 1110–1113. doi:10.1378/chest.95.5.1110

Clark, N. M., & Starr-Schneidkraut, N. J. (1993). Management of asthma by patients and families. *American Journal of Respiratory and Critical Care Medicine, 149*, s54–s66.

Clark, N. M., & Zimmerman, B. J. (1990). A social cognitive view of self-regulated learning about health. *Health Education Research, 5*(3), 371–379. doi:10.1093/her/5.3.371

Clark, R. E., & Feldon, D. F. (2005). Five common but questionable principles of multimedia learning. In R. E. Mayer (Ed.), *Cambridge handbook of multimedia learning*. Cambridge, MA: Cambridge University Press.

Clark, R., & Wittrock, M. C. (2000). Psychological principles in training. In S. Tobias & J. D. Fletcher (Eds.), *Training and retraining: A handbook for business, industry, government, and the military* (pp. 51-84). New York: Macmillan.

Clifton, T. (2006). *Training Developer Forum Use of Business Simulations to Grow General Manager Skills*. PowerPoint, Palm Desert, CA.

Cognition & Technology Group at Vanderbilt (CTGV). (2000). Adventures in anchored instruction: Lessons from beyond the ivory tower. In R. Glaser (Ed), *Advances in instructional psychology: Educational design and cognitive science,* (Vol. 5, pp. 35-99). Mahwah, NJ: Lawrence Erlbaum Associates.

Cognition and Technology Group at Vanderbilt (CGTV). (1992). Anchored instruction in science and mathematics: Theoretical basis, developmental projects, and initial research findings. In R. A. Duschl & R. J. Hamilton (Eds.), *Philosophy of science, cognitive psychology, and educational theory and practice. SUNY series in science education* (pp. 244-273). Albany, NY: State University of New York Press.

Cognition and Technology Group at Vanderbilt (CGTV). (1997). *The Jasper Project: Lessons in curriculum, instruction, assessment, and professional development*. Mahwah, NJ: Lawrence Erlbaum Associates.

Cognition and Technology Group at Vanderbilt (CTGV). (2000). Adventures in anchored instruction: Lessons from beyond the ivory tower. In R. Glaser (Ed.), *Advances in instructional psychology: Educational design and cognitive science,* (Vol. 5, pp. 35-99). Mahwah, NJ: Lawrence Erlbaum Associates.

Cohen, A. (2006). Video-Game Makers Discover New, Older, Market. *NPR*. Retrieved on November 17, 2008 from http://www.npr.org/templates/story/story.php?storyId=6589941.

Collada (2009). *Collada* [computer software]. www.collada.org.

Cornett, S. (2004). The usability of massively multiplayer online roleplaying games: Designing of new users. *Computer Human Interaction*, 703-710.

Council on Graduate Medical Education (COGME). (2005). *Physician Workforce Policy Guidelines for the U.S. for 2000 – 2020*. Rockville, MD: U.S. Department of Health and Human Services.

Craik, F. I., & Lockhart, R. S. (1972). Levels of processing: A framework for memory research. *Journal of Verbal Learning and Verbal Behavior, 11*(6), 671–684. doi:10.1016/S0022-5371(72)80001-X

Crawford, C. (2002). *The Art of Interactive Design: A Euphonious and Illuminating Guide to Building Successful Software*. San Francisco, CA: No Starch Press.

Crawford, C. (2005). *On interactive storytelling.* Berkeley, CA: New Riders Games.

Creer, T. L. (1990a). Strategies for judgment and decision-making in the management of childhood asthma. *Pediatric Asthma Allergy & Immunology, 4*(4), 253–264. doi:10.1089/pai.1990.4.253

Creer, T. L. (1991). The application of behavioral procedures to childhood asthma: Current and future perspectives. *Patient Education and Counseling, 17,* 9–22. doi:10.1016/0738-3991(91)90047-9

Creer, T. L., Wigal, J. K., Kotses, H., & Lewis, P. (1990b). A critique of 19 self-management programs for childhood asthma: Part II. Comments regarding the scientific merit of the programs. *Pediatric Asthma Allergy & Immunology, 4*(1), 41–55. doi:10.1089/pai.1990.4.41

Csikszentmihalyi, M. (1975). *Beyond boredom and anxiety.* San Francisco, CA: Jossey-Bass.

Csikszentmihalyi, M. (1990). *Flow: The psychology of optical experience.* New York: Harper Perennial.

Cullen, K. W., Bartholomew, L. K., Parcel, G. S., & Kok, G. (1998). Intervention Mapping: Use of theory and data in the development of a fruit and vegetable nutrition program for Girl Scouts. *Journal of Nutrition Education, 30*(4), 188–195. doi:10.1016/S0022-3182(98)70318-8

Da Vinci, L. (1498). *Trattato della Pittura.* Milan, Italy.

Davis, S., Jacobs, R., Moar, M., & Watkins, M. (2007). *Exploring the Subjective City.* In F. von Borries, S.P. Walz, and M. Böttger (Eds.), *Space Time Play: Computer Games, Architecture and Urbanism: The Next Level.* Basel, Germany: Birkhäuser Publishing.

de Freitas, S. I. (2006). Using games and simulations for supporting learning. *Learning, Media and Technology, 31*(4), 343–358. doi:10.1080/17439880601021967

Deardorff, V. W. (1986). Computerized health education: A comparison with traditional formats. *Health Education & Behavior, 13,* 61–72. doi:10.1177/109019818601300107

Deci, E. L., Koestner, R., & Ryan, R. M. (1999). A meta-analytic review of experiments examining the effects of extrinsic rewards on intrinsic motivation. *Psychological Bulletin, 125,* 627–668. doi:10.1037/0033-2909.125.6.627

Dede, C., Salzman, M., Loftin, R. B., & Sprague, D. (1999). Multisensory immersion as a modeling environment for learning complex scientific concepts. In W. Feurzeig & N. Roberts (Eds.), *Modeling and Simulation in Science and Mathematics Education.* New York: Springer-Verlag.

deHaan, J., & Diamond, J. (2007, August 04-05). The experience of telepresence with a foreign language video game and video. *Sandbox Symposium 2007,* San Diego, CA, (pp. 39–46).

DeMarle, M. (2007). Nonlinear game narrative. In C. Bateman (Ed.), *Game writing: Narrative skills for videogames* (pp. 71-84). Boston: Charles River Media.

Department of Defense Task Force on Mental Health. (2007). *An achievable vision: Report of the Department of Defense Task Force on Mental Health.* Falls Church, VA: Defense Health Board.

Desurvire, H. W. (2008). Evaluating User Experience and Other Lies in Evaluating Games. *Computer Human Interaction Conference,* (p. 6). Florence, Italy.

Desurvire, H., & Wiberg, C. (2008). Evaluating User Experience and Other Lies in Evaluating. *Computer Human Interaction Conference,* Florence, Italy.

DiClemente, C. C., & Prochaska, J. O. (1982). Self-change and therapy change of smoking behavior: A comparison of processes of change in cessation and maintenance. *Addictive Behaviors,* 133–142. doi:10.1016/0306-4603(82)90038-7

DiClemente, C. C., & Prochaska, J. O. (1985). Process and stages of self-change: Coping and competence in smoking behavioral change. In S. Shiffman & T.A Wills, (Eds.), *Coping and Substance Abuse.* New York: Academic Press.

DiClemente, C. C., Prochaska, J. O., Fairhurst, S. K., Velicer, W. F., Velasquez, M., & Rossi, J. S. (1991). The process of smoking cessation: An analysis of precontemplation, contemplation, and preparation stages of change. *Journal of Consulting and Clinical Psychology*, *59*, 259–304. doi:10.1037/0022-006X.59.2.295

DiFonzo, N., Hantula, D. A., & Bordia, P. (1998). Microworlds for experimental research: Having your (control and collection) cake, and realism too. *Behavior Research Methods, Instruments, & Computers*, *30*(2), 278–286.

Dille, F., & Platten, J. Z. (2007). *The ultimate guide to videogame writing and design*. New York: Lone Eagle Publishing.

Diller, D., Roberts, B., Blankenship, S., & Nielsen, D. (2004). DARWARS Ambush! – Authoring lessons learned in a training game. In *Proceedings of the 2004 Interservice/Industry Training, Simulation and Education Conference (I/ITSEC)*, Orlando, FL.

Dinse, H. R. (2001). The Aging Brain. In N. Elsner & G.W. Kreutzberg (Eds.), *The Neurosciences at the turn of the century: Vol. 1. Proceedings of the 4th Meeting of the German Neuroscience Society*, (pp. 356-363). Thieme: 28th Göttingen Neurobiology Conference.

Dinse, H. R. (2005). Treating the aging brain: cortical reorganization and behavior. *Acta Neurochirurgica*, *93*(Suppl.), 79–84. doi:10.1007/3-211-27577-0_12

Donchin, E. (1995). Video games as research tools: The Space Fortress game. *Behavior Research Methods, Instruments, & Computers*, *27*(2), 217–223.

Donovan, M., Bransford, J., & Pellegrino, J. (Eds.). (1999). *How People Learn*. Washington, DC: National Academy Press.

Donyaee, M., Seffah, A., & Rilling, J. (2007). Benchmarking Usability of Early Designs Using Predictive Metrics. *IEEE*, (pp. 2514-2519).

Duncan, T. E., Duncan, S. C., Beauchamp, N., Wells, J., & Ary, D. V. (2000). Development and evaluation of an interactive CD-ROM refusal skills program to prevent youth substance use: "refuse to use. *Journal of Behavioral Medicine*, *23*, 59–72. doi:10.1023/A:1005420304147

Dunn, R. S., & Dunn, K. J. (1993). *Teaching Secondary Students Through Their Individual Learning Styles: Practical Approaches for Grades 7-12*. Boston: Allyn and Bacon.

Eagleton, T. (1996). *Literary theory: An introduction*, (2nd ed.). Minneapolis, MN: The University of Minnesota Press.

Eck, R. V. (2006). Digital Game-Based Learning: It's not just the digital natives who are restless. *EDUCAUSE Review*, *41*(2), 16–30.

Ediger, M. (2001). *Reading: Intrinsic versus Extrinsic Motivation*. An opinion paper supported by ERIC.

Edwards, W. (1962). Dynamic decision theory and probabilistic information processing. *Human Factors*, *4*, 59–73.

Eiser, R. J., van der Pligt, J., Raw, M., & Sutton, S. R. (1985). Trying to stop smoking: Effects of perceived addiction, attributions for failure, and expectancy of success. *Journal of Behavioral Medicine*, *8*(4), 321–341. doi:10.1007/BF00848367

Ellison, K. (2007, May 21). Video Games vs. the Aging Brain. *Discover Magazine*. Retrieved on November 5, 2008 from http://discovermagazine.com/2007/may/the-elastic-brain/article_print

Elton, M. When will the information explosion reach older Americans? *The American Behavioral Scientist*, *31*(5), 564–575. doi:10.1177/000276488031005005

Entertainment Software Association. (2007). *The 2007 essential facts about computer and video games industry*. Washington, DC: Entertainment Software Association.

Entertainment Software Association. (2008 August). *Statistic of the Month: Essential Facts—Sales, Demographic, and Usage Data*. Retrieved November 4, 2008, from http://www.theesa.com/newsroom/esa_newsletter/august2008/index.html

Epstein, J. A., & Harackiewicz, J. M. (1992). Winning is not enough: The effects of competition and achievement motivation on intrinsic interest. *Person-*

ality and Social Psychology Bulletin, 18, 128–138. doi:10.1177/0146167292182003

Ericsson, K. A. (1996). The acquisition of expert performance: an introduction to some of the issues. In K.A. Ericsson, (Ed.), *The road to excellence: the acquisition of expert performance in the arts and sciences, sports, and games.* Mahwah, NJ: Lawrence Erlbaum Associates.

Fall, A. J., Henry, R. L., & Hazell, T. (1998). The use of an interactive computer program for the education of parents of asthmatic children. *Journal of Paediatrics and Child Health, 34*(2), 127–130. doi:10.1046/j.1440-1754.1998.00178.x

Falsetti, J., & Schweitzer, E. (1995). SchMOOze University: A MOO for ESL/EFL students. In M. Warschauer (Ed.), *Virtual connections: On-line activities and projects for networking language learners* (pp. 231-232). Honolulu, HI: University of Hawai'i, Second Language Teaching and Curriculum Center.

Farrell, C. M., Klimack, W. K., & Jaquet, C. R. (2003). Employing interactive multimedia instruction in military science education at the U.S. Military Academy. In *Proceedings of the Interservice/Industry Training, Simulation, and Education Conference* (I/ITSEC); Orlando, FL, Dec. 1-4, 2003.

FAS. (2009a). Retrieved June 22, 2009, from http://www.fas.org/programs/ltp/games/medulla.html

FAS. (2009b). Retrieved June 22, 2009 from http://vworld.fas.org/wiki/FAS:Introduction.

Federation of American Scientists. (FAS). (2008). *A small survey taken by FAS asked students whether they were interested in biology before and after being given conventional instruction in the subject. A majority were less interested after completing the course.*

Fehl, G., & Rodriguez-Lores, J. (Eds.). (1997). *Die Stadt wird in der Landschaft sein und die Landschaft in der Stadt - Bandstadt und Bandstruktur als Leitbilder im modernen Städtebau* (transl. by MK). Basel, Germany: Birkhäuser Publishing. Games for Health/The Serious Games Initiative (n.d.). *Games for Health – About*. Retrieved from http://www.gamesforhealth.org/about2.html

Feldner, T. M., Monson, C. M., & Friedman, M. J. (2007). A Critical Analysis of Approaches to Targeted PTSD Prevention Current Status and Theoretically Derived Future Directions. *Behavior Modification, 31*(1), 80–116. doi:10.1177/0145445506295057

Filbeck, G., & Smith, L. (1996). Learning styles, teaching strategies, and predictors of success for students in coporate finance. *Financial Practice and Education, 48*, 1299–1300.

Finkelstein, J. A., Brown, R. W., Schneider, L. C., Weiss, S. T., Quintana, J. M., & Goldmann, D. A. (1995). Quality of care for preschool children with asthma: the role of social factors and practice setting. *Pediatrics, 95*, 389–394.

Finnegan, J. R., & Viswanath, K. (2008). Communication theory and health behavior change. In K. Glanz, B.K. Rimer, & K. Viswanath, (Eds.), *Health Behavior and Health Education. Theory, Research, and Practice,* (4th ed). San Francisco, CA: Jossey-Bass, John Wiley & Sons, Inc.

Fitts, P. M., & Posner, M. I. (1967). *Learning and skilled performance in human performance.* Belmont, CA: Brock-Cole.

Flanagan, M., & Nissenbaum, H. (2007). A game design methodology to incorporate social activist themes. In *CHI '07: Proceedings of the SIGCHI conference on Human factors in computing systems* (pp. 181-190). New York: ACM Press.

Fletcher, J. D. (2009). Education and Training Technology in the Military. *Science, 323*(5910), 72–75. doi:10.1126/science.1167778

Foss, B. A., & Eikaas, T. I. (2006). Game play in engineering education: Concept and experimental results. *International Journal of Engineering Education, 22*(5), 1043–1052.

Frasca, G. (2003). Ludologists love stories, too: notes from a debate that never took place. In M. Copier & J. Raessens (Eds.) *Proceedings of the Levelup 2003 Conference.* Retrieved 11 Feb 2009 from www.ludology.org/articles/Frasca_LevelUp2003.pdf

Fullerton, T., Swain, C., & Hoffman, S. (2008). *Game design workshop: A playcentric approach to creating innovative games,* (2nd ed.). Burlington, MA: Morgan Kaufmann Publishers.

Gabler, K., Gray, K., Kucic, M., & Shodhan, S. (2005). How to prototype a game in under 7 days: Tips and tricks from 4 grad students who made over 50 games in 1 semester. *Gamasutra: The Art & Science of Making Games.* Retrieved November 2, 2008 from http://www.gamasutra.com/features/20051026/gabler_01.shtml

Gagne, R. M., Briggs, L. J., & Wager, W. W. (1988). *Principles of Instructional Design.* New York: Holt, Rinehart, and Winston, Inc. Games for health, available at http://www.gamesforhealth.org/

Gallese, V. (2001). The "Shared Manifold" Hypothesis. From Mirror Neurons to Empathy. *Journal of Consciousness Studies, 5*(7), 33–50.

Garris, R., Ahlers, R., & Driskell, J. E. (2002). Games, motivation, and learning: A research and practice model. *Simulation & Gaming, 33*(4), 441–467. doi:10.1177/1046878102238607

Gartzke, E., & Gleditsch, K. S. (2006). Identity and conflict: Ties that bind and differences that divide. *European Journal of International Relations, 12*(1), 53–87. doi:10.1177/1354066106061330

Gee, J. P. (2003). *What video games have to teach us about learning and literacy.* New York: Palgrave Macmillan.

Gee, J. P. (2004). *Situated language and learning: a critique of traditional schooling.* New York: Routledge.

Gee, J. P. (2007). Good video games + Good Learning. In *Collected essays on video games, learning, and literacy.* New York: Peter Lang Publishing Inc.

Genette, G. (1980). *Narrative discourse* (J. E. Lewin, Trans.). Ithaca, NY: Cornell UP.

Gentner, D. (1983). Structure-mapping: A theoretical framework for analogy. *Cognitive Science, 7,* 155–170.

Gerbert, B., Berg-Smith, S., Mancusco, M., Caspers, N., McPhee, S., & Null, D. (2003). Using innovative video doctor technology in primary care to deliver brief smoking and alcohol intervention. *Health Promotion Practice, 4*(3), 249–261. doi:10.1177/1524839903004003009

Gerhard, M., Moore, D., & Hobbs, D. (2004). Embodiment and copresence in collaborative interfaces. *International Journal of Human-Computer Studies, 61,* 453–480. doi:10.1016/j.ijhcs.2003.12.014

Gibbs, V., & Auerbach, A. (2001, July). Learning Curves for New Procedures—the Case of Laparoscopic Cholecystectomy. In *University of California at San Francisco (UCSF) –Stanford University, Evidence-Based Practice Center, Making Health Care Safer: A Critical, Analysis of Patient Safety Practices, AHRQ* (Publication 01-E058). Retrieved June 22, 2001 from www.ahcpr.gov/clinic/ptsafety/pdf/front.pdf.

Gilutz, H., Bar-On, D., Billing, E., Rehnquist, N., & Cristal, N. (1991). The relationship between causal attribution and rehabilitation in patients after their first myocardial infarction. A cross cultural study. *European Heart Journal, 12,* 883–888.

Girgensohn, A., & Lee, A. (2002, November). Making websites be social places for social interaction. In *Proceedings of Computer Supported Collaborative Work '02,* New Orleans, LA, 136–145.

Gist, M. E., Schwoerer, C., & Rosen, B. (1989). Effects of alternative training methods on self-efficacy and performance in computer software training. *The Journal of Applied Psychology, 74,* 884–891. doi:10.1037/0021-9010.74.6.884

Gist, M. E., Stevens, C. K., & Bavetta, A. G. (1991). Effects of self-efficacy and post-training intervention on the acquisition and maintenance of complex interpersonal skills. *Personnel Psychology, 44,* 837–861.

Glanz, K., & Rimer, K. (2005). *Theory at a glance. A guide for health promotion practice,* (2nd ed.). Washington, DC: NIH Pub. No. 05-3896.

Glanz, K., Patterson, R. E., Kristal, A. R., DiClemente, C. C., Heimendinger, J., & Linnan, L. A. (1994). Stages of change in adopting healthy diets: Fat, fiber, and correlates of nutrient intake. *Health Education Quarterly, 21,* 499–519.

Glanz, K., Rimer, B. K., & Viswanath, K. (2008). The scope of health behavior and health education. In Glanz, K., Rimer, B.K., & Viswanath, K., (Eds.), *Health Behavior and Health Education: Theory, Research, and Practice*, (4th ed.). San Francisco, CA: Jossey-Bass, John Wiley & Sons, Inc.

Glaser, R. (1989). Expertise in learning: How do we think about instructional processes now that we have discovered knowledge structure? In D. Klahr & D. Kotosfky (Eds.), *Complex information processing: The impact of Herbert A. Simon* (pp. 269-282). Hillsdale, NJ: Lawrence Erlbaum Associates.

Gochman, D. S. (1982). Labels, systems, and motives: Some perspectives on future research. *Health Education Quarterly*, *9*, 167–174.

Gochman, D. S. (1997). Health Behavior Research: Definitions and Diversity. In D.S. Gochman (Ed.), *Handbook of Health Behavior Research*, (Vol I: Personal and Social Determinants). New York: Plenum Press.

Goldstein, J., Cajko, L., Oosterbroek, M., Michielsen, M., van Houten, O., & Salverda, F. (1997). Video games and the elderly. *Social Behavior and Personality*, *25*(4), 345–352. doi:10.2224/sbp.1997.25.4.345

Gonzalez, C., Czlonka, L., Saner, L., & Eisenberg, L. (under review). *Learning to stand in the other's shoes: A computer video game experience of the Israeli-Palestinian conflict*. Manuscript under review.

Gonzalez, C., Lerch, J. F., & Lebiere, C. (2003). Instance-based learning in dynamic decision making. *Cognitive Science*, *27*(4), 591–635.

Gonzalez, C., Martin, M., & Hansberger, J. T. (2006). Feedforward effects on predictions in a dynamic battle scenario. In *Proceedings of the Human Factors and Ergonomics Society 50th Annual Meeting* (pp. 265-269). Santa Monica, CA: Human Factors and Ergonomics Society.

Gonzalez, C., Vanyukov, P., & Martin, M. K. (2005). The use of microworlds to study dynamic decision making. *Computers in Human Behavior*, *21*(2), 273–286. doi:10.1016/j.chb.2004.02.014

Gopher, D., Weil, M., & Bareket, T. (1994). Transfer of skill from a computer game trainer to flight. *Human Factors*, *36*, 387–405.

Goran, M. I., & Reynolds, K. (2005). Interactive multimedia for promoting physical activity (IMPACT) in children. *Obesity Research*, *13*, 762–771. doi:10.1038/oby.2005.86

Gorriz, C. M., & Medina, C. (2000). Engaging girls with computers through software games. *Communications of the ACM*, *43*, 42–49. doi:10.1145/323830.323843

Green, C. S., & Bavelier, D. (2003). Action video game modifies visual selective attention. *Nature*, *423*, 534–537. doi:10.1038/nature01647

Green, L. W., & Kreuter, M. W. (2005). *Health Promotion Planning: An Educational and Environmental Approach*, (4th Ed.). New York: McGraw-Hill.

Green, M. C. (2004). Transportation into narrative worlds: The role of prior knowledge and perceived realism. *Discourse Processes*, *38*(2), 247–266. doi:10.1207/s15326950dp3802_5

Green, M. C. (2006). Narratives and cancer communication. *The Journal of Communication*, *56*(s1), S163–S183. doi:10.1111/j.1460-2466.2006.00288.x

Greitzer, F. L., Hershman, R. L., & Kelly, R. T. (1981). The Air Defense Game: A microcomputer program for research in human performance. *Behavior Research Methods and Instrumentation*, *13*(1), 57–59.

Griffiths, M. (2002). The educational benefits of videogames. *Education and Health*, *20*(3), 47–51.

Grove, R. J. (1993). Attributional correlates of cessation self-efficacy among smokers. *Addictive Behaviors*, *18*, 311–320. doi:10.1016/0306-4603(93)90032-5

Hader, R., Saver, C., & Steltzer, T. (2006). No time to lose. *Nursing Management*, *37*(7), 23–29, 48.

Hadley, O. (2001). *Teaching Language in Context*. Florence, KY: Heinle and Heinle.

Hagman, J., & Rose, A. (1983). Retention of military skills: A review. *Human Factors*, *25*(2), 199–213.

Halfon, N., & Newacheck, P. W. (1993). Childhood asthma and poverty: differential impacts and utilization of health services. *Pediatrics, 91*(1), 56–61.

Halverson, R. (2005). What can K-12 school leaders learn from video games and gaming. *Innovate: Journal of Online Education, 1*(6).

Hanna, J. L. (1978). African dance: Some implications for dance therapy. *American Journal of Dance Therapy, 2*(1), 3–15. doi:10.1007/BF02579589

Hanna, J. L. (1995). African dance: some implications for dance therapy. *Journal of Alternative and Complementary Medicine (New York, N.Y.), 1*(4), 323–331. doi:10.1089/acm.1995.1.323

Harter, S. (1981). A new self-report scale of intrinsic versus extrinsic orientation in the classroom: Motivational and informational components. *Developmental Psychology, 17*(3), 300–312. doi:10.1037/0012-1649.17.3.300

Hays, R. T. (2005). *The effectiveness of instructional games: a literature review and discussion* (Technical Report 2005-004). Orlando, FL: Naval Air Warfare Center Training Systems Division.

Hays, R. T., & Singer, M. J. (1989). *Simulation fidelity in training system design.* New York: Springer-Verlag.

Heffner, S. (Ed.). (2007). *Virtual Reality Applications in the U.S. Healthcare Market.* New York: Kalorama Information.

Herron, R. E., & Sutton-Smith, B. (Eds.). (1971). *Child's Play.* New York: John Wiley & Sons, Inc.

Herz, J. C. (1997). *Joystick Nation: How computer games ate our quarters, won our hearts, and rewired our minds.* Boca Raton, FL: Little, Brown, and Company.

Hill, R. W., Jr., Belanich, J., Lane, H. C., Core, M., Dixon, M., Forbell, E., et al. (2006*). Pedagogically Structured Game-Based Training: Development of the Elect BiLAT Simulation.* National Technical Information Service. Retrieved from http://handle.dtic.mil/100.2/ADA461575

Hivert, M. F., Langlois, M. F., Berard, P., Cuerrier, J. P., & Carpentier, A. C. (2007). Prevention of weight gain in young adults through a seminar-based intervention program. *International Journal of Obesity, 31,* 1262–1269. doi:10.1038/sj.ijo.0803572

Hobbs, F. B., & Damon, B. L. (1996). *65+ in the United States: U.S. Bureau of the Census, current population reports, special studies, P23-190.* Retrieved June 1, 2009, from the U.S. Census Bureau Web site, http://www.census.gov/prod/1/pop/p23-190/p23-190.pdf

Hoge, C. W., Castro, C. A., Messer, S. C., McGurk, D., Cotting, D. I., & Koffman, R. L. (2004). Combat Duty in Iraq and Afghanistan, Mental Health Problems and Barriers to Care. *The New England Journal of Medicine, 351*(1), 13–22. doi:10.1056/NEJMoa040603

Holcomb, J. B., Caruso, J., McMullin, N. R., Wade, C. E., Pearse, L., Oetjen-Gerdes, L., et al. (2006). *Causes of death in U.S. special operations forces in the global war on terrorism: 2001-2004.* MacDill AFB, Tampa, FL: U.S. Special Operations Command.

Holzinger, A., & Searle, G. (2008). Investigating Usability Metrics for the Design and Development of Applications for the Elderly. In *11th International Conference on Computers Helping People with Special Needs* (pp. 98-105). Linz, Austria: Springer.

Homer, C. J., Szilagyi, P., Rodewald, L., Bloom, S. R., Greenspan, P., & Yazdgerdi, S. (1996). Does quality of care effect rates of hospitalization for childhood asthma? *Pediatrics, 98*(1), 18–23.

Honebein, P. C., Duffy, T. M., & Fishman, B. J. (1993). Constructivism and the design of learning environments: Context and authentic activities for learning. In T.M. Duffy, J. Lowyck, & D.H. Jonassen (Eds.), *Designing environments for constructive learning* (pp. 87-108). New York: Springer-Verlag.

Hoot, J. L., & Hayship, B. (1983). Microcomputers and the Elderly: New directions for self-sufficiency and life-long learning. *Educational Gerontology, 9,* 493–499. doi:10.1080/0380127830090513

Hornbaek, K., & Lai-Chong Law, E. (2008). Meta-Analysis of Correlations Among Usability Measures. *Computer Human Interaction* (pp. 617-626). San Jose, CA: ACM.

Hornung, R. L., Lennon, P. A., Garrett, J. M., DeVellis, R. F., Weinberg, P. D., & Stretcher, V. J. (2000). Interactive computer technology for skin cancer prevention targeting children. *American Journal of Preventive Medicine*, *90*, 781–785.

Hospers, H. J., Kok, G., & Strecher, V. J. (1990). Attributions for previous failures and subsequent outcomes in a weight reduction program. *Health Education Quarterly*, *17*(4), 409–415.

Howe, N., & Strauss, W. (2000). *Millennials rising*. New York: Vintage Books.

Hoysniemi, J. (2006). International survey on the Dance Dance Revolution game. *Computers & Education*, *4*(2), 1544–3574.

Hsu, S., Lee, F., & Wu, M. (2005). Designing action games for appealing to buyers. *Cyberpsychology & Behavior*, *8*, 585–591. doi:10.1089/cpb.2005.8.585

Hubbard, P. (1991). Evaluating computer games as language learning tools. *Simulation & Gaming*, *22*(2), 220–223. doi:10.1177/1046878191222006

Hudson, J. M., & Bruckman, A. (2002). IRC Francais: The creation of an internet-based SLA community. *Computer Assisted Language Learning*, *15*(2), 109–134. doi:10.1076/call.15.2.109.8197

Huss, K., Rand, C. S., Butz, A. M., Eggleston, P. A., Murigande, C., & Thompson, L. (1994). Home environmental risk factors in urban minority asthmatic children. *Annals of Allergy*, *72*, 173–177.

Hussain, T. S., & Ferguson, W. (2005). Efficient development of large-scale military training environments using a multi-player game. In *2005 Fall Simulation Interoperability Workshop*, (pp. 421-431).

Hussain, T. S., Weil, S. A., Brunye, T., Sidman, J., Ferguson, W., & Alexander, A. L. (2008). Eliciting and evaluating teamwork within a multi-player game-based training environment. In H.F. O'Neil & R.S. Perez (Eds.), *Computer Games and Team and Individual Learning* (pp. 77-104). Amsterdam, The Netherlands: Elsevier.

IEAA. (2008). Retrieved Feb 12, 2008, from The Interavtive Entertainmnt Association of Australia http://www.ieaa.com.au/

ImpactGames. (2008). *PeaceMaker [game]*. Retrieved from http://www.peacemakergame.com

Institute for Creative Technologies. (2009). *Post-Traumatic Stress Disorder Assessment and Treatment: Project Description*. Retrieved on June 2, 2009, from http://ict.usc.edu/projects/post_traumatic_stress_disorder_assessment_and_treatment_ptsd/C45

Integration of Motion Capture and EMG data for Classifying the Human Motions (with Balakrishnan Prabhakaran, Gaurav Pradhan, Navzer Engineer). (2007, April 17-20). In *Proceedings of the IEEE 23rd International Conference on Data Engineering Workshops* (pp. 56-63).

International Playing-Card Society. (2006). Playing-card games. *The International Playing-Card Society*. Retrieved November 15, 2008 from http://www.cs.man.ac.uk/~daf/i-p-c-s.org/faq/games.php

Invetstment-U. (2008). Retrieved April 25, 2008, from Investment U http://www.investmentu.com/

ISFE. (2005, June). Retrieved April 20, 2008, from Interactive Software Federation of Europe http://www.isfe-eu.org/

Isselsteijn, W., Nap, H. H., deKort, Y., & Poels, K. (2007). Digital game design for elderly users. In *Proceedings of the 2007 Conference on Future Play*, Toronto, Canada (pp. 17-22). New York: ACM.

Issenberg, S. B., McGaghie, W. C., Petrusa, E. R., Lee, G. D., & Scalese, R. J. (2005). Features and uses of high-fidelity medical simulations that lead to effective learning: A BEME systematic review. *Medical Teacher*, *27*(1), 10–28. doi:10.1080/01421590500046924

Jackson, S. A., & Marsh, H. W. (1996). Development and validation of a scale to measure optimal experience: The Flow State Scale. *Journal of Sport & Exercise Psychology*, *18*, 17–35.

Jacobs, S. (2007). The Basics of Narrative. In C. Bateman (Ed.), *Game writing: narrative skills for videogames* (pp. 25-42). Boston: Charles River Media.

James, C. V., & Rosenbaum, S. (2009). Paying for quality care: Implications for racial and ethnic health disparities in pediatric asthma. *Pediatrics, 123*(Suppl. 3), S205–S210. doi:10.1542/peds.2008-2233L

Janoff-Bulman, R. (1979). Characterological versus behavioral self-blame: Inquiries into depression and rape. *Journal of Personality and Social Psychology, 37*(10), 1798–1809. doi:10.1037/0022-3514.37.10.1798

Janoff-Bulman, R., & Lang-Gunn, L. (1988). Coping with disease, crime, and accidents: The role of self-blame attributions. In L.Y. Abramson (Ed.), *Social Cognition and Clinical Psychology: A synthesis* (pp. 116-147). New York: Guilford Press.

Jaycox, K., & Hicks, K. (1976). *Elders, Students, and Computer—A New Team* (ISEAC No. 7). Urbana, IL: University of Illinois, Illinois Series on Educational Applications of Computers.

Jenkins, H. (2006). Game design as narrative architecture. In K. Salen & E. Zimmerman (Eds.), *The game design reader,* (pp. 670-689). Cambridge, MA: The MIT Press.

Johnson, S. (2006). *Everything Bad is Good for You: How Popular Culture is Making Us Smarter*. London: Penguin.

Johnson, W. L., Wang, N., & Wu, S. (2007). Experience with serious games for learning foreign languages and cultures. In *Proceedings of the SimTecT Conference,* Australia.

Jonassen, D. H. (2000). Revisiting activity theory as a framework for designing student-centered learning environments. In D. H. Jonassen & S. M. Land (Eds.). *Theoretical foundations of learning environments* (pp. 89-121). Mahwah, NJ: Lawrence Erlbaum Associates.

Jones, D. (2007). *Mounting pressures facing the U.S. workforce and the increasing need for adult education and literacy*. National Commission on Adult Literacy. Retrieved June 22, 2009 from http://www.nationalcommissiononadultliteracy.org/content/nchemspresentation.pdf

Jones, T., & Richey, R. (2000). Rapid prototyping methodology in action: A developmental study. *Educational Technology Research and Development, 48*(2), 63–80. doi:10.1007/BF02313401

Juul, J. (2001). Games telling stories? *Game Studies, 1*(1).

Kahn, E. B., Ramsey, L. T., Brownson, R. C., Heath, G. W., Howze, E. H., & Powell, K. E. (2002). The effectiveness of interventions to increase physical activity. *American Journal of Preventive Medicine, 22*, 73–107. doi:10.1016/S0749-3797(02)00434-8

Kahn, G. (1993). Computer-based patient education: A progress report. *Patient Informatics, 10*, 93–99.

Kanfer, R., & Ackerman, P. L. (1989). Motivation and cognitive abilities: An integrative/aptitude-treatment interaction approach to skill acquisition. *The Journal of Applied Psychology, 74*, 657–690. doi:10.1037/0021-9010.74.4.657

Kanfer, R., & Ackerman, P. L. (1989). Motivation and cognitive abilities: an integrative/aptitude-treatment interaction approach to skill acquisition. *The Journal of Applied Psychology, 74*, 657–690. doi:10.1037/0021-9010.74.4.657

Kanfer, R., & McCombs, B. L. (2000). Motivation: Applying current theory to critical issues in training. In S. Tobias & J. D. Fletcher (Eds.), *Training and retraining: A handbook for business, industry, government, and the military* (pp. 85-108). New York: Macmillan.

Kanfer, R., & McCombs, B. L. (2000). Motivation: Applying current theory to critical issues in training. In S. Tobias & J. D. Fletcher (Eds.), *Training and retraining: A handbook for business, industry, government, and the military* (pp. 85-108). New York: Macmillan.

Kasl, S. V., & Cobb, S. (1966). Health behavior, illness behavior, and sick-role behavior: I. Health and illness behavior. *Archives of Environmental Health, 12*, 246–266.

Kawashima, R. (2003). *Train Your Brain*. Teaneck, NJ: Kumon Publishing Co.

Keating, X. D., Guan, J., Pinero, J. C., & Bridges, D. M. (2005). A meta-analysis of college students' physical activity behaviors. *Journal of American College Health*, *54*(2), 116–125. doi:10.3200/JACH.54.2.116-126

Keith, C. (2007). Scrum and long term project planning for video games. *Gamasutra: The Art & Science of Making Games*. Retrieved November 14, 2008 from http://www.gamasutra.com/view/feature/3142/scrum_and_long_term_project_.php

Kelle, A. (2008). Experiential learning in an arms control simulation. *Political Science & Politics*, *41*, 379–385.

Keller, J. M. (1987b). Development and use of the ARCS model of instructional design. *Journal of Instructional Development*, *10*(3), 2–10. doi:10.1007/BF02905780

Keller, J. M., & Kopp, T. (1987). Application of the ARCS model of motivational design. In C.M. Reigeluth (Ed.), *Instructional Theories in Action: Lessons Illustrating Selected Theories and Models*. Hillsdale, NJ: Lawrence Erlbaum Associates.

Keller, J. M., & Suzuki, K. (1987a). Use of ARCS motivation model courseware design. In D.H. Jonassen (Ed.), *Instructional Designs for Microcomputer Courseware*. Hillsdale, NJ: Lawrence Erlbaum Associates.

Kelly, H. (2005). Games, cookies, and the future of education. *Issues in Science and Technology*, (21): 33–40.

Kelly, H., Howell, K., Glinert, E., Holding, L., Swain, C., & Burrowbridge, A. (2007). How to build serious games. *Communications of the ACM*, *50*(7), 44–50. doi:10.1145/1272516.1272538

Kelly, H., Howell, K., Glinert, E., Holding, L., Swain, C., & Burrowbridge, A. (2007). How to Build Serious Games. *Communications of the ACM*, *50*(7), 45. doi:10.1145/1272516.1272538

Kelman, H. C. (2008). A social-psychological approach to conflict analysis and resolution. In D. Sandole, S. Byrne, I. Sandole-Staroste, & J. Senehi (Eds.), *Handbook on conflict analysis and resolution* (pp. 170-183). London: Routledge [Taylor & Francis].

Kim, J. Y., Allen, J. P., & Lee, E. (2008). Alternate reality gaming. *Communications of the ACM*, *51*(2), 36–42. doi:10.1145/1314215.1340912

Kindt, T., & Müller, H.-H. (Eds.). (2003). *What is narratology? Questions and answers regarding the status of a theory*. Berlin: Walter de Gruyter.

Kirreimuir, J., & McFarlane, A. (2006). Literature review in games and learning. [FutureLabs Report 8]. Accessed February 2009, available from http://www.futurelab.org.uk/resources/documents/lit_reviews/Games_Review.pdf

Kirsch, I., Braun, H., & Yamamoto, K. (2007). *America's Perfect Storm: Three Forces Changing Our Nation's Future*. Princeton, NJ: Educational Testing Service. Retrieved June 22, 2009 from http://www.ets.org/Media/Education_Topics/pdf/AmericasPerfectStorm.pdf

Klingelhofer, E. L., & Gershwin, M. E. (1988). Asthma self-management programs: Premises, not promises. *The Journal of Asthma*, *25*(2), 89–101. doi:10.3109/02770908809071359

Knerr, B. W., Simutis, Z. M., & Johnson, R. M. (1979). *Computer-based simulations for maintenance training: Current ARI research*, (Defense Technical Information Center No. ADA 139371, Technical Report 544). Alexandria, VA: U.S. Army Research Institute for the Behavioral and Social Sciences.

Kohn, L., Corrigan, J., & Donaldson, M. (Eds.). (1999). *To Err is Human: Building a Safer Health System*. Committee on Quality of Health Care in America, Institute of Medicine. Washington, DC: National Academy Press.

Kolb, D. A. (1984). *Experiential learning: Experience as the source of learning and development*. Englewood Cliffs, NJ: Prentice Hall.

Korea, I. T. *Times*. (n.d.). Government likes serious games, May 18, 2009. Retrieved June 1, 2009 from the Korea IT Times Web site, http://www.koreaittimes.com/story/1403/government-likes-serious-games

Kotses, H., Stout, M. A., Wigal, J. K., Carlson, B., Creer, T., & Lewis, P. (1991). Individualized asthma self-management: A beginning. *The Journal of Asthma*, *28*(4), 287–289. doi:10.3109/02770909109073386

Krashen, S. D., & Terrell, T. D. (1983). *The natural approach: Language acquisition in the classroom.* London: Prentice Hall Europe.

Krawczyk, M., & Novak, J. (2006). *Game development essentials: game story & character development.* Clifton Park, NY: Thomson.

Kreuter, M., Farrell, D., Olevitch, L., & Brennan, L. (2000). *Tailoring Health Messages. Customizing Communication with Computer Technology.* Mahwah, NJ: Lawrence Erlbaum Associates.

Krishna, S., Balas, E. A., Spencer, D. C., Griffin, J. Z., & Boren, S. A. (1997). Clinical trials of interactive computerized patient education: implications for family practice. *The Journal of Family Practice*, *45*, 25–33.

Krolikowska, K., Kronenberg, J., Maliszewska, K., Sendzimir, J., Macnuszewski, P., & Dunaksi, A. (2007). Role-playing simulations as a communication tool in community dialogue: Karkonosze Mountains case study. *Simulation & Gaming*, *38*(2), 195–210. doi:10.1177/1046878107300661

Krupnick, J. L., & Green, B. L. (2008). Psychoeducation to Prevent PTSD: A Paucity of Evidence. *Psychiatry*, *71*(4), 329–331.

Krzywinska, T. (2006). Blood Scythes, Festivals, Quests, and Backstories: World Creation and Rhetorics of Myth in World of Warcraft. *Games and Culture*, 383–396. doi:10.1177/1555412006292618

Lake, A., & Townshend, T. (2006). Obesogenic environments: exploring the built and food environments . *The Journal of the Royal Society for the Promotion of Health*, (June): 2006.

Larimer, M. E., Palmer, R. S., & Marlatt, G. A. (1999). Relapse prevention: An overview of Marlatt's cognitive-behavioral model. *Alcohol Research & Health*, *23*(2), 151–160.

Lee, A., Danis, C., Miller, T., & Jung, Y. (2001). Fostering Social Interactions in Online Spaces. In *Proceedings of INTERACT 2001: IFIP TC.13 International Conference on Human-Computer Interaction* (pp. 59–66). Amsterdam, The Netherlands: IOS Press.

Legler, J. (2002). The effectiveness of interventions to promote mammography among women with historically lower rates of screening. *Cancer Epidemiology, Biomarkers & Prevention*, *11*(1), 59–71.

Lepper, M. R., & Cordova, D. I. (1992). A desire to be taught: instructional consequences of intrinsic motivation. *Motivation and Emotion*, *16*(3), 187–208. doi:10.1007/BF00991651

Lepper, M. R., & Greene, D. (1979). *The Hidden Costs of Reward.* Morristown, NJ: Lawrence Erlbaum Associates.

Lepper, M. R., & Malone, T. W. (1987). Intrinsic motivation and instructional effectiveness in computer-based education. In R.E. Snow & M.J. Farr (Eds.), *Aptitude, Learning, and Instruciton, III. Conative and Affective Process Analysis.* Hillsdale, NJ: Lawrence Erlbaum Associates.

Lepper, M. R., & Malone, T. W. (1987). Intrinsic motivation and instructional effectiveness in computer-based education. In R.E. Snow and M.J. Farr (Eds), *Aptitude, Learning, and Instruction, III. Conative and Affective Process Analysis.* Hillsdale, NJ: Lawrence Erlbaum Associates.

Lepper, M. R., Corpus, J. H., & Iyengar, S. S. (2005). Intrinsic and extrinsic motivation orientations in the classroom: Age differences and academic correlates. *Journal of Educational Psychology*, *91*(2), 184–196. doi:10.1037/0022-0663.97.2.184

Leslie, E., Owen, N., Salmon, J., Bauman, A., Sallis, J. F., & Lo, S. K. (1999). Insufficiently active Australian college students: Perceived personal, social, and environmental influences. *Preventive Medicine*, *28*(1), 20–27. doi:10.1006/pmed.1998.0375

Levant, R. F. (2008). *Evidence-Based Practice in Psychology.* Paper presented at the Culturally Informed Evidence-Based Practices (CIEBP) National Conference, Bethesda, MD.

Lewis, F. M., & Daltroy, L. H. (1990). How causal explanations influence health behavior: Attribution theory. In K. Glanz, F.M. Lewis, & B.K. Rimer, (eds.), *Health behavior and health education*. San Francisco, CA: Jossey-Bass Publishers.

Lieberman, D. A. (1997). Interactive video games for health promotion: Effects on knowledge, self-efficacy, social support, and health. In R.L. Street, W.R. Gold, & T. Manning (Eds), *Health Promotion and Interactive Technology*. Mahwah, NJ: Lawrence Erlbaum Associates.

Lieberman, D. A. (2001). Management of chronic pediatric diseases with interactive health games: theory and research findings. *The Journal of Ambulatory Care Management*, 24(1), 26–38.

Lin, J. J., Mamykina, L., Lindtner, S., Delajoux, G., & Strub, H. B. (2006). Fish'n'Steps: Encouraging physical activity with an interactive computer game. In P. Dourish & A. Friday (Eds.). *UbiComp 2006: Ubiquitous Computing*, (pp. 261-278). Berlin: Springer-Verlag

Lindstromberg, S., & Boers, F. (2005). From Movement to Metaphor with Manner-of-Movement Verbs. *Applied Linguistics*, 26(2), 241–261. doi:10.1093/applin/ami002

Linn, D. (2008). Fallout 3 (PC). *1UP.com*. Retrieved February 3, 2009, from http://www.1up.com/do/reviewPage?cId=3170949

Lintern, G., & Kennedy, R. S. (1984). Video game as a covariate for carrier landing research. *Perceptual and Motor Skills*, 58(1), 167–172. doi:10.2466/PMS.58.1.167-172

Locke, E. A., & Latham, G. P. (1990). Work motivation: The high performance cycle. In U. Kleinbeck & H. Quast, (Eds.), *Work motivation* (pp. 3-25). Hillsdale, NJ: Lawrence Erlbaum Associates.

Locke, E. A., Shaw, K. N., & Saari, L. M. (1981). Goal setting and task performance: 1969-1980. *Psychological Bulletin*, 90, 125–152. doi:10.1037/0033-2909.90.1.125

Lowood, H. (2006). Storyline, Dance/Music, or PvP?: Game movies and community players in World of Warcraft. *Games and Culture*, 362–384. doi:10.1177/1555412006292617

Lowry, R., Galuska, D. A., Fulton, J. E., Wechsler, H., Kann, L., & Collins, J. L. (2000). Physical activity, food choice, and weight management goals and practices among U.S. college students. *American Journal of Preventive Medicine*, 18(1), 18–27. doi:10.1016/S0749-3797(99)00107-5

Lugo, L. (2007). International obligations and the morality of war. *Society*, 44(6), 109–112. doi:10.1007/s12115-007-9021-0

Lundy, H. (2002). *Using dance/movement therapy techniques to augment the effectiveness of therapeutic holding with children*. Master of Arts thesis, College of Nursing and Health Professions, MCP Hahnemann University. Archived at http://hdl.handle.net/1860/1100

Mak, H., Johnson, P., Abbey, H., & Talamo, R. C. (1982). Prevalence of asthma and health utilization of asthmatic children in an inner city. *The Journal of Allergy and Clinical Immunology*, 70, 367–372. doi:10.1016/0091-6749(82)90026-4

Malaby, T. M. (2007). Beyond play: A new approach to games. *Games and Culture*, 2(2), 95–113. doi:10.1177/1555412007299434

Malone, T. W. (1980). *What makes things fun to learn? A study of intrinsically motivating computer games*. (Doctoral dissertation, Department of Psychology, Stanford University). Also available as technical report #CIS-7 (SSL-80-11), Xerox Palo Alto Research Center, Palo Alto, CA.

Malone, T. W. (1981a). Towards a theory of intrinsically motivating instruction. *Cognitive Science*, 5, 333–369.

Malone, T. W. (1981b). What makes computer games fun? *BYTE Publications*.

Marcus, B. H., Emmons, K. M., Simkin-Silverman, L. R., Linnan, L. A., Taylor, E. R., & Bock, B. C. (1998). Evaluations of motivationally-tailored vs. standard self-help physical activity interventions at the workplace. *American Journal of Health Promotion*, 12, 246–253.

Mateas, M. (2002). *Interactive drama, art, and artificial intelligence*. Unpublished PhD Dissertation: Carnegie Mellon University, Pittsburgh, PA.

Mayer, R. E. (2001). *Multimedia learning.* Cambridge, UK: Cambridge University Press.

Mayer, R. E. (2001). *Multimedia learning.* Cambridge, UK: Cambridge University Press.

Mayer, R. E. (2004). Should there be a three strikes rule against pure discovery? The case for guided methods of instruction. *The American Psychologist, 59*(1), 14–19. doi:10.1037/0003-066X.59.1.14

Mayer, R. E., & Massa, L. J. (2003). Three facets of visual and verbal learners: Cognitive ability, cognitive style, and learning preference. *Journal of Educational Psychology, 95*(4), 833–846. doi:10.1037/0022-0663.95.4.833

Mayo, M. J. (2007). Games for science and engineering education. *Communications of the ACM, 50*(7), 30–35. doi:10.1145/1272516.1272536

McCurry, J. (2006). Video games for the elderly: an answer to dementia or a marketing tool? *The Guardian*, March 7. Retrieved on November 3, 2008 from http://www.guardian.co.uk/technology/2006/mar/07/nintendods.games

McDonald, C., Cannon-Bowers, J., Whatley, D., & Dunne, J. (2009). *The Pulse!! collaboration: Academe & industry, building trust.* Panel tutorial presented at the Medicine Meets Virtual Reality annual conference. Long Beach, CA: January 19-22, 2009.

McGuire, W. (1972). Social psychology. In P.C. Dodwell (Ed.), *New Horizons in Psychology*, (pp. 219-242). Middlesex, UK: Penguin Books.

McPherson, A., Glazerbrook, C., & Smyth, A. (2001). Double click for health: the role of multimedia in asthma education. *Archives of Disease in Childhood, 85*, 447–449. doi:10.1136/adc.85.6.447

Meadows, M. S. (2003). *Pause & effect: The art of interactive narrative.* Indianapolis, IN: New Riders.

Mental Health Advisory Team (MHAT IV) (2006). *Operation Iraqi Freedom 05-07 Final Report.* Office of the Surgeon Multinational Force-Iraq and Office of the Surgeon General Unites States Army Medical Command.

Metacritic.com. (2009). *Fallout 3.* Retrieved February 3, 2009, from http://www.metacritic.com/games/platforms/xbox360/fallout3

Meyer, C., Ganascia, J.-G., & Zucker, J.-D. (1997). Learning Strategies in Games by Anticipation. *International Conference on Artificial Intelligence*, Nagoya, Japan. Merzenich, M. (n.d.). *On the Brain.* Retrieved on November 3, 2008 from http://merzenich.positscience.com/. See also Posit Science: Brain Fitness Program http://www.positscience.com/products/brain_fitness_program/

Milheim, W. D., & Martin, B. L. (1991). Theoretical bases for the use of learner control: three different perspectives. *Journal of Computer-Based Instruction, 18*(3), 99–105.

Miller, G. A. (1956). The magical number seven, plus or minus two: Some limits on our capacity for processing information. *Psychological Review, 63*, 81–97. doi:10.1037/h0043158

Mintz, A., Geva, N., Redd, S. B., & Carnes, A. (1997). The effect of dynamic and static choice sets on political decision making: An analysis using the decision board platform. *The American Political Science Review, 91*(3), 553–566. doi:10.2307/2952074

Mokdad, A. H., Marks, J. S., Stroup, D. F., & Gerberding, J. L. (2000). Actual causes of death in the United States. *Journal of the American Medical Association, 10*(291), 1238–1245.

Montet, V. (2006). Video games aim to spice up old people's lives *Yahoo News*, October 1. Originally appearing at http://fullcoverage.yahoo.com/s/afp/afplifestyleusitgameshealthelderly. Retrieved on November 3, 2008 from http://www.nadin.ws/archives/477

Moreno, R., & Mayer, R. E. (2004). Personalized messages that promote science learning in virtual environments. *Journal of Educational Psychology, 96*, 165–173. doi:10.1037/0022-0663.96.1.165

Morris, L. A., & Kranouse, D. E. (1982). Informing patients about drug side effects. *Journal of Behavioral Medicine, 5*, 363–373. doi:10.1007/BF00846163

Morrison, J. E., & Meliza, L. L. (1999). *Foundations of the after action review process* (Special Report 42). Alexandria, VA: U.S. Army Research Institute for the Behavioral and Social Sciences.

Morrison, W., & Labonte, M. (2008, November). *China's currency: A summary of the economic issues* (Congressional Research Service Report RS21625). Retrieved June 22, 2009 from http://www.fas.org/sgp/crs/row/RS21625.pdf.

Murray, J. H. (1998). *Hamlet on the Holodeck*. Cambridge, MA: The MIT Press.

Musgrove, M. (2005). Video games give world peace a chance. *The Washington Post*. Retrieved from http://www.washingtonpost.com/wp-dyn/content/article/2005/10/15/AR2005101500218.html

Must, A., & Anderson, S. E. (2003). Effects of obesity on morbidity in children and adolescents. *Nutrition in Clinical Care*, *6*(1), 4–12.

Nadin, M. (1991). *Mind—Anticipation and Chaos*. Stuttgart, Germany: Belser Presse.

Nadin, M. (2003). *Anticipation—The end is where we start from*. Basel, Switzerland: Lars Mueller Verlag.

Nadin, M. (2004). *Project Seneludens*. Richardson, TX: University of Texas at Dallas, anté – Institute for Research in Anticipatory Systems.

Nadin, M. (2005). Anticipating Extreme Events – the need for faster-than-real-time models. In *Extreme Events in Nature and Society*, (Frontiers Collection). New York: Springer Verlag.

Nadin, M. (2006). *Anticipation, Games, and Brain Plasticity*. Syllabus for the graduate course, Special Topics in Art &Technology. Retrieved on November 19, 2008 from http://www.utdallas.edu/bbs/news_events/2005/ATEC7390.001.html and http://ante.utdallas.edu/agbp/

Nash, H., & Snowling, M. (2006). Teaching new words to children with poor existing vocabulary knowledge: A controlled evaluation of the definition and context methods. *International Journal of Language & Communication Disorders*, *41*(3), 335–354. doi:10.1080/13682820600602295

National Center for Health Statistics. (2005). *National Health Interview Survey*. Retrieved December 3, 2007: http://www.cdc.gov/nchs/about/major/nhis/released200606.htm#7

National Commission on Adult Literacy. (2008). *Reach higher America: Overcoming crisis in the U.S. workforce*. Retrieved June 22, 2009 from http://www.nationalcommissiononadultliteracy.org/ReachHigherAmerica/ReachHigher.pdf

National Institute on Media and the Family. *Sixth Annual Video and Computer Report Card*. (2001). Retrieved September 12, 2002, from http://www.mediaandthefamily.org/research/vgrc/2001-2.shtml

National League for Nursing. (2009). *NLN annual nursing data review*. Retrieved June 1, 2009, from http://www.nln.org/newsreleases/annual_survey_031609.htm

Nelson, L. M. (1999). Collaborative problem solving. In C. M. Reigeluth (Ed.), *Instructional-design theories and models: A new paradigm of instructional theory*, (Vol. II, pp. 241-267). Mahwah, NJ: Lawrence Erlbaum Associates.

Nemeth, E. (1999). Gender differences in reaction to public achievement feedback. *Educational Studies*, *25*, 297–310. doi:10.1080/03055699997819

Neuman, S. B. (2005). Readiness for reading and writing – What do we mean? *Early Childhood Today 3*, *20*(2), 8.

Neuman, S. B. (2006). Building vocabulary to build literacy skills: How to help children build a rich vocabulary day by day. *Early Childhood Today 3*, *20*(3), 12-14.

NHLBI. (1995). *Global initiative for asthma: global strategy for asthma management and prevention*. NHLBI/WHO Workshop Report. 1995; Pub. No. 95-3659.

Nicolelis, M. (2005). *Neural Control of Artificial Limb Movement*. Paper presented at the symposium: Reprogramming the Human Brain, April 2005, Dallas, TX.

O'Connell, D. O., & Velicer, W. F. (1988). A decisional balance measure and the stages of change model for weight loss. *The International Journal of the Addictions*, *23*, 729–740.

O'Neil, H. F., & Perez, R. S. (2008). *Computer Games and Team and Individual Learning*. Amsterdam, The Netherlands: Elsevier.

O'Neil, H. F., Wainess, R., & Baker, E. L. (2005). Classification of learning outcomes: Evidence from the computer games literature. *Curriculum Journal, 16*(4), 455–474. doi:10.1080/09585170500384529

O'Neil, H. F., Wainess, R., & Baker, E. L. (2005). Classification of learning outcomes: evidence from the computer games literature. *Curriculum Journal, 16*(4). doi:10.1080/09585170500384529

Oblinger, D. (2003). Boomers, gen-xers, and millennials: Understanding the "new students." *EDUCAUSE Review, 38*(4), 36–45.

Oblinger, D. (2004). The next generation of educational engagement. *Journal of Interactive Media in Education, 8*, 1–18.

Odenweller, C. M., Hsu, C. T., & DiCarlo, S. E. (1998). Educational card games for understanding gastrointestinal physiology. *The American Journal of Physiology, 275*(6 Pt 2).

Omodei, M. M., & Wearing, A. J. (1995). The Fire Chief microworld generating program: An illustration of computer-simulated microworlds as an experimental paradigm for studying complex decision-making behavior. *Behavior Research Methods, Instruments, & Computers, 27*, 303–316.

Onega, S., & Landa, J. A. G. (Eds.). (1996). *Narratology: An introduction*. London: Longman.

Orvis, K. A., Orvis, K. I., Belanich, J., & Mullin, L. N. (2008). The influence of trainee gaming experiences on affective and motivational learner outcomes of videogame-based training environments. In H. F. O'Neil & R. S. Perez (Eds.), *Computer games and team and individual learning*. New York: Elsevier.

Owen, N., Fotheringham, M. J., & Marcus, B. H. (2002). Communication technology and health behavior change. In Glanz, K., Rimer, B.K., & Lewis, F.M. (Eds.), *Health Behavior and health Education. Theory, research, and Practice,* (3rd ed.). San Francisco, CA: Jossey-Bass, John Wiley & Sons, Inc.

Paivio, A. (1986). *Mental Representations: A Dual Coding Approach*. Oxford, UK: Oxford University Press.

Palmgreen, P., & Donohew, L. (2002). Effective mass media for drug abuse prevention campaigns. In Z. Sloboda & W. Bukoski (eds), *Effective Strategies for Drug Abuse Prevention*. New York: Plenum Press.

Papaloukas, S., & Xenos, M. (2008). Usability and education of games through combined assessment methods. *PETRA*.

Papert, S. (1993). *Mindstorms: Children, computers, and powerful ideas*. New York: Basic Books.

Papert, S. (1993). *The children's machine: Rethinking school in the age of the computer*. New York: Basic Books.

Papert, S. (1993). *The Children's Machine: Rethinking School in the Age of the Computer*. New York: Basic Books.

Papert, S. (June, 1998). Does Easy Do It? Children, Games, and Learning. *Game Developer Magazine*, 88.

Parker, L. E., & Lepper, M. R. (1992). Effects of fantasy contexts on children's learning and motivation: Making learning more fun. *Journal of Personality and Social Psychology, 62*(4), 625–633. doi:10.1037/0022-3514.62.4.625

Parks, C. D., & Hulbert, L. G. (1995). High and low trusters' responses to fear in a payoff matrix. *The Journal of Conflict Resolution, 39*, 718–730. doi:10.1177/0022002795039004006

Parks, C. D., Henager, R. F., & Scamahorn, D. S. (1996). Trust and reactions to messages of intent in social dilemmas. *The Journal of Conflict Resolution, 40*, 134–151. doi:10.1177/0022002796040001007

Peterson, C., Seligman, M. E. P., & Vaillant, G. E. (1988). Pessimistic explanatory style is a risk factor for physical illness: A thirty-five year longitudinal study. *Journal of Personality and Social Psychology, 55*, 23–27. doi:10.1037/0022-3514.55.1.23

Petraglia, J. (1998). *Reality by design: The rhetoric and technology of authenticity in education*. Mahwah, NJ: Lawrence Erlbaum Associates.

Piaget, J. (1963). *The Origins of Intelligence in Children*, (M. Cook, Trans.). New York: Norton.

Pillay, H., Brownlee, J., & Wilss, L. (1999). Cognition and recreational computer games: Implications for educational technology. *Journal on Research in Computing Education*, *32*(1), 203–216.

Pinelle, D., Wong, N., & Stach, T. (2008). Heuristic Evaluation for Games: Usability Principles for Video Game Design. *Computer Human Interaction Proceedings* (pp. 1453-1462). Florence, Italy: ACM.

Pintrich, P. R., & Zusho, A. (2002). The development of academic self-regulation: The role of cognitive and motivational factors. In A. Wigfield & J. Eccles (Eds.), *Development of achievement motivation* (pp. 249-284). San Diego, CA: Academic Press.

Pliske, R., McCloskey, M., & Klein, G. (2001). Decision Skills Training: Facilitating Learning From Experience. In E. Salas & G. Klein (Eds.), *Linking Expertise and Naturalistic Decision Making* (pp. 37-42). Retrieved June 22, 2009 from http://books.google.com/books?hl=en&lr=&id=lAo5kOgrdRoC&oi=fnd&pg=PA37&dq=Skills+Training:+Facilitating+Learning+From+Experience&ots=4mpqoAVod-&sig=sobMZht5N5lsPU1hTFL-Jf5ssL8

Pollock, J. C., & Sullivan, H. J. (1990). Practice mode and learner control in computer-based instruction. *Contemporary Educational Psychology*, *15*, 251–260. doi:10.1016/0361-476X(90)90022-S

Prensky, M. (2001). *Digital Game-Based Learning*. New York: McGraw-Hill.

Prince, G. (2003a). *Dictionary of narratology*. Lincoln, NE: University of Nebraska Press.

Prince, G. (2003b). Surveying narratology. In T. Kindt & H.-H. Müller (Eds.), *What is narratology? Questions and answers regarding the status of a theory*, (pp. 1-16). Berlin: Walter de Gruyter.

Prochaska, J. J., Zabinski, M. F., Calfas, K. J., Sallis, J. F., & Patrick, K. (2000). PACE+: Interactive communication technology for behavior change in clinical settings. *American Journal of Preventive Medicine*, *19*, 127–131. doi:10.1016/S0749-3797(00)00187-2

Prochaska, J. O., & DiClemente, C. C. (1992). Stages of change in the modification of problem behaviors. In M. Hersen, R.M. Eisler, P.M. Miller (Eds.), *Progress in Behavior Modification* (pp 184-218). Sycamore, IL: Sycamore Publishing Company.

Prochaska, J. O., DiClemente, C. C., Velicer, W. F., & Rossi, J. S. (1993). Standardized, individualized, interactive and personalized self-help programs for smoking cessation. *Health Psychology*, *12*, 399–405. doi:10.1037/0278-6133.12.5.399

Prochaska, J. O., Redding, C. A., Harlow, L. L., Rossi, J. S., & Velicer, W. F. (1994). The Transtheoretical Model of Change and HIV prevention: A review. *Health Education Quarterly*, *21*, 471–486.

Prochaska, J. O., Velicer, W. F., DiClemente, C. C., & Fava, J. L. (1988). Measuring processes of change: Applications to the cessation of smoking. *Journal of Consulting and Clinical Psychology*, *56*, 520–528. doi:10.1037/0022-006X.56.4.520

Propp, V. I. (1968). *Morphology of the folktale* (2nd ed.). Austin, TX: University of Texas Press.

Quellmalz, E. S., & Pellegrino, J. W. (2009). Technology and Testing. *Science*, *323*(5910), 75–79. doi:10.1126/science.1168046

Rachelefsky, G. S. (1987). Review of asthma self-management programs. *The Journal of Allergy and Clinical Immunology*, *80*, 506–511. doi:10.1016/0091-6749(87)90087-X

Raghavan, K., Satoris, M. L., & Glaser, R. (1997). The impact of model-centered instrbction on student learning: The area and volume units. *Journal of Computers in Mathematics and Science Teaching*, *16*, 363–404.

Raghavan, K., Satoris, M. L., & Glaser, R. (1998). The impact of MARS curriculum: The mass unit.

Science Education, 82, 53–91. doi:10.1002/(SICI)1098-237X(199801)82:1<53::AID-SCE4>3.0.CO;2-#

Raines, C. (2002). *Connecting generations: The sourcebook for a new workplace.* Berkeley, CA: Fifth Street Design.

Rakowski, W. A., Dube, C. A., & Goldstein, M. G. (1996b). Considerations for extending the transtheoretical model of behavior change to screening mammography. *Health Education Research, 11*, 77–96. doi:10.1093/her/11.1.77

Rakowski, W. A., Ehrich, B., Dube, C. A., Pearlman, D. N., Goldstein, M. G., & Peterson, K. K. (1996a). Screening mammography and constructs from the Transtheoretical Model: Associations using two definitions of the stages-of- adoption. *Annals of Behavioral Medicine, 18*(2), 91–100. doi:10.1007/BF02909581

Randel, J. M., Morris, B. A., Wetzel, C. D., & Whitehill, B. (1992). The effectiveness of games for educational purposes: A review of recent research. *Simulation & Gaming, 23*(23), 261–276. doi:10.1177/1046878192233001

Rankin, Y., Gold, R., & Gooch, B. (2006a). Evaluating interactive gaming as a language learning tool. In *Conference Proceedings of SIGGRAPH 2006,* Boston.

Rankin, Y., Gold, R., & Gooch, B. (2006b). 3D role-playing games as language learning tools. In *Conference Proceedings of EuroGraphics 2006, 25,*Vienna, Austria.

Rasch, R. F. R. (2006). *The nurse educator shortage: Opportunities for education and career advancement.* Unpublished paper. Nashville, TN: Vanderbilt University School of Nursing.

Rathbun, G. A., Saito, R. S., & Goodrum, D. A. (1997). Reconceiving isd: Three perspectives on rapid prototyping as a paradigm shift. In O. Abel, N.J. Maushak & K.E. Wright (Eds.), *19th Annual Proceedings of Selected Research and Development Presentations at the 1997 National Convention of the Association for Educational Communications and Technology* (pp. 291-296). Ames, IA: Iowa State University.

Reed, G. R., Velicer, W. F., & Prochaska, J. O. (1997). What makes a good staging algorithm: Examples from regular exercise. *American Journal of Health Promotion, 12*, 57–67.

Reeve, J., & Deci, E. L. (1996). Elements of the competitive situation that affect intrinsic motivation. *Personality and Social Psychology Bulletin, 22*, 24–33. doi:10.1177/0146167296221003

Reid, P. P., Compton, W. D., Grossman, J. H., & Fanjiang, G. (Eds.). (2005). *Building a Better Delivery System: A New Engineering/Health Care Partnership.* Committee on Engineering and the Health Care System. Geneva: IOM Press.

Reigeluth, C., & Schwartz, E. (1989). An instructional design theory for the design of computer-based simulations. *Journal of Computer-Based Instruction, 16*(1), 1–10.

Reis, J., Riley, W., Lokman, L., & Baer, J. (2000). Interactive multimedia preventive alcohol education: a technology application in higher education. *Journal of Drug Education, 30*, 399–421. doi:10.2190/LWMQ-9CQA-B78H-9MA7

Resnick, L. (Ed.). (1989). *Knowing, learning, and instruction: Essays in honor of Robert Glaser.* Hillsdale, NJ: Lawrence Erlbaum Associates.

Revere, D., & Dunbar, P. J. (2001). Review of computer-generated outpatient health behavior interventions: clinical encounters "in absentia." . *Journal of the American Medical Informatics Association, 8*, 62–79.

Reznick, R. K., & MacRae, H. (2006). Teaching surgical skills – changes in the wind. *The New England Journal of Medicine, 355*(25), 2665–2669. doi:10.1056/NEJMra054785

Rhodes, F., Fishbein, M., & Reis, J. (1997). Using behavioral theory in computer-based health promotion and appraisal. *Health Education & Behavior, 24*(1), 20–34. doi:10.1177/109019819702400105

Ricci, K. E., Salas, E., & Cannon-Bowers, J. A. (1996). Do computer based games facilitate knowledge acquisition and retention? *Military Psychology, 8*(4), 295–307. doi:10.1207/s15327876mp0804_3

Rideout, E., & Hamel, E. (Eds.). (2006). *The media family: Electronic media in the lives of infants, toddlers, preschoolers and their parents*. Menlo Park, CA: The Henry J. Kaiser Family Foundation.

Rieber, L. P. (1991a). Animation, incidental learning, and continuing motivation. *Journal of Educational Psychology, 83*(3), 318–328. doi:10.1037/0022-0663.83.3.318

Rieber, L. P. (1996). Seriously considering play: Designing interactive learning environments based on the blending of microworlds, simulations, and games. *Educational Technology Research and Development, 44*(2), 43–58. doi:10.1007/BF02300540

Rieber, L. P., & Kini, A. S. (1991b). Theoretical foundations of instructional applications of computer-generated animated visuals. *Journal of Computer-Based Instruction, 18*(3), 83–88.

Rizzo, A., & Kim, G. J. (2005). A SWOT analysis of the field of virtual reality rehabilitation and therapy. [Cambridge, MA: Massachusetts Institute of Technology Press.]. *Presence (Cambridge, Mass.), 14*(2), 119–146. doi:10.1162/1054746053967094

Rizzolati, G., Fadiga, L., Fogassi, L., & Gallese, V. (1999). Resonance behaviors and mirror neurons. *Archives Italiennes de Biologie, 137*, 83–99.

Roberts, B., Diller, D., & Schmitt, D. (2006). Factors affecting the adoption of a training game. In *Proceedings of the 2006 Interservice/Industry Training, Simulation, and Education Conference (I/ITSEC)*.

Rodenstein, M. (1988). *Mehr Licht, Mehr Luft - Gesundheitskonzepte im Städtebau seit 1750*. New York: Campus Verlag.

Rolnick, S. J. (1988). Self-management of pediatric asthma: Four programs being studied. *Journal of Pediatric Health Care, 2*(5), 264–266. doi:10.1016/0891-5245(88)90159-9

Rosen, R. (1985). *Anticipatory Systems. Philosophical, Mathematical and Methodological Foundations*. Oxford, UK: Pergamon Press.

Rothman, F., & Narum, J. L. (1999). *Then, Now, and In the Next Decade: A Commentary on Strengthening Undergraduate Science, Mathematics, Engineering and Technology Education*. Washington, DC: Project Kaleidoscope.

Rouse, R., III. (2005). *Game design theory and practice* (2nd ed.). Plano, TX: Wordware Publishing.

Royce, W. W. (1987). Managing the development of large software systems: concepts and techniques. In *ICSE '87: Proceedings of the 9th international conference on Software Engineering* (pp. 328-338). Los Alamitos, CA: IEEE Computer Society Press.

Rumelhart, D. E., & Ortony, A. (1977). The representation of knowledge in memory. In R.C. Anderson, R.J. Spiro, & W.E. Montague (Eds), *Schooling and the Acquisition of Knowledge*. Hillsdale, NJ: Lawrence Erlbaum Associates.

Ryan, R. M., & Deci, E. L. (2000). Self-determination theory and the facilitation of intrinsic motivation, social development, and well-being. *The American Psychologist, 55*, 68–78. doi:10.1037/0003-066X.55.1.68

Salas, E., & Cannon-Bowers, J. A. (2000). The anatomy of team training. In S. Tobias & J. D. Fletcher (Eds.), *Training and retraining: A handbook for business, industry, government, and the military* (pp. 312-335). New York: Macmillan.

Salen, K., & Zimmerman, E. (2004). *Rules of Play – Game Design Fundamentals*. Cambridge, MA: MIT Press.

Sandford, R., Ulicsak, M., Facer, K., & Rudd, T. (2006). *Teaching with games: Using commercial off-the-shelf computer games In formal education*. Retrieved from http://www.futurelab.org.uk/resources/documents/project_reports/teaching_with_games/TWG_report.pdf

Santora, M. (2006, January 12). Bad blood: East meets west, adding pounds and peril. *The New York Times*. Retrieved from http://www.nytimes.com/2006/01/12/nyregion/nyregionspecial5/12diabetes.html von Borries, F., Walz, S.P., & Böttger, M. (2006). Ausweitung der Schiesszone. *Archithese, April 2006*. (transl. by MK).

Satava, R. (2008). Competency, proficiency, and the next generation of skills training and assessment curricula using simulators. *Laparoscopy Today*, Aug. 13, 2008. Retrieved June 3, 2009, at http://www.laparoscopytoday.com/2008/08/competency-prof.html

Savery, J. R., & Duffy, T. M. (1996). Problem based learning: An instructional model and its constructivist framework. In B. G. Wilson (Ed.), *Constructivist learning environments: Case studies in instructional design*, (pp. 135-148). Hillsdale, NJ: Educational Technology Publications.

Sawyer, B., & Smith, P. (2008). *Serious games taxonomy*. Serious Games Initiative. Available from www.seriousgames.org/index2.html.

Sawyer, B., & Smith, P. A. (2008, February,). *Serious Games Taxonomy*. Paper presented at the Game Developers Conference, San Francisco, CA.

Schild, E. O. (1968). The shaping of strategies. In S.S. Broocock & E.O. Schild (Eds.), *Simulation Games in Learning*. Beverley Hills, CA: Sage Publications.

Schunk, D. H., & Ertmer, P. A. (1999). Self-regulatory processes during computer skill acquisition: Goal and self-evaluative influences. *Journal of Educational Psychology, 91*, 251–260. doi:10.1037/0022-0663.91.2.251

Schunk, D. H., & Zimmerman, B. J. (2003). Self-regulation in learning. In W.M. Reynolds, & G.E. Miller, (Eds.), *Handbook of psychology: Educational psychology,* (Vol. 7, pp. 59-78). New York: John Wiley.

Sechrest, R. C., & Henry, D. J. (1996). Computer-based patient education: observations on effective communication in the clinical setting. *The Journal of Biocommunication, 23*, 8–12.

Serious Games Forum. (2009). *What are Serious Games?* Retrieved June 02, 2009, from http://seriousgames.ning.com/

Shamir, M., & Sagiv-Schifter, T. (2006). Conflict, identity, and tolerance: Israel in the Al-Aqsa Intifada. *Political Psychology, 27*(4), 569–595. doi:10.1111/j.1467-9221.2006.00523.x

Shank, R. (1997). *Virtual learning: A revolutionary approach to building a highly skilled workforce*. New York: McGraw Hill.

Shaw, G. B. (n.d.). Retrieved on November 3, 2008 from http://thinkexist.com/quotes/george_bernard_shaw/

Shegog, R. (1997). *Computer-assisted instruction for self-management education in pediatric asthma*. Unpublished doctoral dissertation, University of Texas, Houston.

Shegog, R., & Bartholomew, L. K. (2004). Computer-assisted asthma education for children: Impact on self-management behavior. *International Review of Asthma, 6*(2), 70–86.

Shegog, R., & Sockrider, M. M. (2009). Computer-based applications for asthma education and management. In Harver, A. (Ed.), *Asthma, Health, and Society*. New York: Springer.

Shegog, R., Bartholomew, L. K., Craver, J., Sockrider, M. M., Mullen, P. D., & Pilney, S. (2004). Development of an expert system knowledge base: A novel approach to promote guideline congruent asthma care. *The Journal of Asthma, 41*(4), 385–402. doi:10.1081/JAS-120026098

Shegog, R., Bartholomew, L. K., Gold, R. S., Pierrel, E., Parcel, G. S., & Sockrider, M. M. (2006). Asthma management simulation for children: Translating theory, methods, and strategies to effect behavior change. *Simulation in Healthcare, 1*(3), 151–159.

Shegog, R., Bartholomew, L. K., Parcel, G. S., Sockrider, M., Czyzewski, D., & Masse, L. (2001). Impact of a computer-assisted education program on variables related to children's asthma self-management behavior. *Journal of the American Informatics Association, 8*(1), 49–61.

Shegog, R., Conroy, J., Murray, N. G., Agurcia, C., Kelder, S., & Prokorov, A. (2003). Process evaluation of ASPIRE: A CD_ROM based smoking curriculum for high school students. *American Public Health Association*, San Francisco, CA.

Shegog, R., McAlister, A., Hu, S., Ford, K., Meshak, A., & Peters, R. (2005). Using the web to impact smoking intentions in middle school students: A pilot test of the

"Headbutt" risk assessment program. *American Journal of Health Promotion, 19*(5), 334–338.

Sheldon, L. (2004). *Character development and storytelling for games*. Boston: Thomson Course Technology.

Sims, V. K., & Mayer, R. E. (2002). Domain specificity of spatial expertise: The case of video game players. *Applied Cognitive Psychology, 16*, 97–115. doi:10.1002/acp.759

Sinclair, J., Hingston, P., & Martin, M. (2007). Considerations for the design of exergames. In *Proceedings of Graphite 2007: 5th International Conference on Computer Graphics and Interactive Techniques*, (pp. 289-295).

Singh, V., & Mathew, A. P. (2007). WalkMSU: An intervention to motivate physical activity in university students. In *Proceedings of Conference on Human Factors in Computing Systems CHI '07*, San Jose, CA, (pp. 2657-2662).

Skinner, C. S., & Kreuter, M. W. (1997). Using theories in planning interactive computer programs. In: R. Street, W. Gold, and T. Manning (Eds.), *Health Promotion and interactive technology: Theoretical applications and future directions*. Mahwah, NJ: Lawrence Erlbaum Associates.

Slater, M. D. (2002). Entertainment education and the persuasive impact of narratives. In M.C. Green, J.F. Strange, & T.C. Brock (Eds), *Narrative Impact: Social and Cognitive Foundations*. Mahwah, NJ: Lawrence Erlbaum Associates.

Slater, M., Sadagic, A., Usoh, M., & Schroeder, R. (2000). Small group behaviour in a virtual and real environment: A comparative study. *Presence (Cambridge, Mass.), 9*, 37–15. doi:10.1162/105474600566600

Smith, P., Bowers, C., & Cannon-Bowers, J. (2008). Social Learning Aspects of MMOGs and Virtual Worlds. In R. Ferdig (ed.), *Handbook of Research on Effective Electronic Gaming in Education*. Hershey, PA: Idea Group Inc.

Smith-Jentsch, K. A., Zeisig, R. L., Acton, B., & McPherson, J. A. (1998). Team dimensional training: A strategy for guided team self-correction. In J.A. Cannon-Bowers & E. Salas (Eds.), *Making decisions under stress: Guidelines for individual and team training*, (pp. 299-312). Washington, DC: American Psychological Association.

SocialGames. (n.d.). Available at http://www.socialimpactgames.org

Stahl, S. (1983). *Vocabulary Instruction and the Nature of Word Meanings*. Paper presented at the 27th Annual Meeting of the College Reading Association, Alanta, GA.

Steinkuehler, C. (2005). *Cognition and learning in massively multiplayer online games: A critical approach*. Unpublished dissertation, University of Wisconsin-Madison.

Steinkuehler, C., & Williams, D. (2006). Where everybody knows your (screen) name: Online games as "third places." *Journal of Computer-Mediated Communication, 11*(4), Article 1.

Steinman, R. A., & Blastos, M. T. (2002). A trading-card game teaching about host defence. *Medical Education, 36*(12), 1201–1208. doi:10.1046/j.1365-2923.2002.01384.x

Stephenson, M. Y., & Southwell, B. (2006). Sensation-seeking, the activation model, and mass media health campaigns. *The Journal of Communication, 56*, S38–S56. doi:10.1111/j.1460-2466.2006.00282.x

Stetz, M., Long, C., Wiederhold, B. K., & Turner, D. (2008). Combat Scenarios and Relaxation Training to Harden Medics Against Stress. *Journal of CyberTherapy & Rehabilitation, 1*(3), 239–246.

Stewart, J., Cartier, J. L., & Passmore, C. M. (2005). Developing an understanding through model-based inquiry. In M.S. Donovan & J.D. Bransford (Eds.), *How Students Learn: History, Mathematics, and Science Inquiry in the Classroom*, (pp. 515-565). Washington, DC: National Academies Press.

Strecher, V. (1999a). The role of interactive strategies in cancer risk communication. *Journal of National Cancer Institute*.

Strecher, V. J. (1999b). Computer-tailored smoking cessation materials: A review and discussion. *Patient Education and Counseling, 36*, 107–117. doi:10.1016/S0738-3991(98)00128-1

Strecher, V. J., Greenwood, T., Wang, C., & Dumont, D. (1999b). Interactive multimedia and risk communication. *Journal of the National Cancer Institute. Monographs, 25*, 134–139.

Strecher, V. J., Seijts, G. H., Kok, G. J., Latham, G. P., Glasgow, R., & DeVellis, B. (1995). Goal setting as a strategy for health behavior change. *Health Education Quarterly, 22*(2), 190–200.

Street, R. L., & Rimal, R. N. (1997). Health promotion and interactive technology: A conceptual foundation. In R. Street, W. Gold, & T. Manning (Eds.), *Health Promotion and Interactive Technology: Theoretical Applications and Future Directions*. Mahwah, NJ: Lawrence Erlbaum Associates.

Sutton-Smith, B. (1997). *The Ambiguity of Play*. Cambridge, MA: Harvard University Press.

Swaffar, J., & Woodruff, M. (1978). Language for Comprehension: Focus on Reading a Report on the University of Texas German Program. *Modern Language Journal, 62*(1/2), 27–32. doi:10.2307/324112

Sweet, A. P., & Guthrie, J. T. (1996). How children's motivations relate to literacy development and instruction. *The Reading Teacher, 49*(8), 660–662.

Sweetser, P., & Johnson, D. (2004). Player-centered game environments: Assessing player opinions, experiences, and issues. In *Entertainment Computing – ICEC 2004: Third International Conference*, (pp. 321-332). New York: Springer-Verlag.

Sweetser, P., & Wyeth, P. (2005). GameFlow: A model for evaluating player enjoyment in games. *ACM Computers in Education, 3*(3), Article 3A.

Szulborski, D. (2005). *This is not a game: A guide to Alternate Reality Gaming*. Morrisville, NC: Lulu.com.

Tannenbaum, S. I., Mathieu, J. E., Salas, E., & Cannon-Bowers, J. A. (1991). Meeting trainees' expectations: The influence of training fulfillment on the development of commitment, self-efficacy, and motivation. *The Journal of Applied Psychology, 76*, 759–769. doi:10.1037/0021-9010.76.6.759

Taylor, R. (1980). *The Computer in the School: Tutor, Tool, Tutee*. New York: Teachers College Press.

Team Technology. (2008). Free personality test: MMDI™ questionnaire. *Team Technology*. Retrieved from http://www.teamtechnology.co.uk/mmdi-re/mmdi-re.htm

Thiele, A., & Schneibner-Herzig, G. (1983). Listening Comprehension Training in Teaching English to Beginners. *System, 11*, 277–286. doi:10.1016/0346-251X(83)90045-3

Thompson, C. (2006). Video games: Saving the world, one video at a time. *The New York Times*. Retrieved from http://query.nytimes.com/gst/fullpage.html?res=9901E3DB163FF930 A15754C0A9609C8B63&sec=&spon=&pagewanted=1

Thompson, J., Berbank-Green, B., & Cusworth, N. (2007). *The Computer Game Design Course. Principles, Practices, and Techniques for the Aspiring Game Designer*. London: Thames and Hudson Ltd.

Thoresen, C. E., & Kirmil-Gray, K. (1983). Self-management psychology and the treatment of childhood asthma. *The Journal of Allergy and Clinical Immunology, 72*(5), 596–610. doi:10.1016/0091-6749(83)90487-6

Triplett, J., & Bosworth, B. (2003). Productivity measurement issues in services industries: "Baumol's disease" has been cured. *Economic Policy review, Sept*, 23-33. Retrieved June 22, 2009 from http://econpapers.repec.org/article/fipfednep/y_3a2003_3ai_3asep_3ap_3a23-33_3an_3av.9no.3.htm

Tripp, M. K., Herrmann, N. B., Parcel, G. S., Chamberlain, R. M., & Gritz, E. R. (2000). Sun Protection is Fun! A skin cancer prevention program for preschools. *The Journal of School Health, 70*(10), 395–401. doi:10.1111/j.1746-1561.2000.tb07226.x

Tripp, S., & Bichelmeyer, B. (1990). Rapid prototyping: An alternative instructional design strategy. *Educational*

Technology Research and Development, 38(1), 31–44. doi:10.1007/BF02298246

Tully, T. (2005). *From Genes to Drugs for Cognitive Dysfunction*. Paper presented at the symposium: Reprogramming the Human Brain, April 2005, Dallas, TX.

Tuttle, W. (2008). Fallout 3 (PC). *Gamespy.com*. Retrieved February 3, 2009, from http://pc.gamespy.com/pc/fallout-3/924348p1.html

U.S. Department of Commerce, Economics and Statistics Administration, National Telecommunications and Information Administration. (2002). *A nation online: How Americans are expanding their use of the internet*, Washington, DC. Retrieved May 11, 2007, from http://www.ntia.doc.gov/ntiahome/dn/anationonline2.pdf

U.S. Department of Education, Institute of Educational Sciences. (2008). *Fast Facts*. Retrieved January 16, 2008 from http://nces.ed.gov/fastfacts/display.asp?id=72

United Nations Programme on Ageing. (2002). Retrieved on November 5, 2008, from http://www.un.org/ageing/.

USDHHS. (1997). National Asthma Education Program Expert Panel Report. Expert panel report 2. *Guidelines for the diagnosis and management of asthma*. Bethesda, MD: National Asthma Education Program, Office of Prevention, Education and Control, National Heart, Lung and Blood Institute, U.S. Department of Health and Human Services, Public Health Service; Report No.: 97-4051.

USDHHS. (1999). Science Panel on Interactive Communication and Health. In T.R. Eng & D.H. Gustafson (Eds.), *Wired for Health and Well-Being. The Emergence of Interactive Health Communication*. Washington, DC: US Department of Health and Human Services, US Government Printing Office.

USDHHS. (2007). National Asthma Education and Prevention Program, Expert panel report 3: *Guidelines for the Diagnosis and Management of Asthma*. Bethesda, MD: National Asthma Education Program, Office of Prevention, Education and Control, National Heart, Lung and Blood Institute, U.S. Department of Health and Human Services, Public Health Service.

USDHHS. (n.d.). *Healthy People 2010*. Office of Disease Prevention and Health Promotion. U.S. Department of Health and Human Service. Retrieved February 2008, from http://www.healthypeople.gov/

van Merriënboer, J. J. G., & Kirschner, P. (2007). *Ten steps to complex learning: A systematic approach to four-component Instructional Design*. London: Lawrence Erlbaum Associates.

VanPatten, B., Williams, J., Rott, S., & Overstreet, M. (Eds.). (2004). *Form-Meaning Connections in Second Language Acquisition*. Mahwah, NJ: Lawrence Erlbaum Associates, Inc.

Velicer, W. F., DiClemente, C. C., Prochaska, J. O., & Brandenburg, N. (1985). A decisional balance measure for assessing and predicting smoking status. *Journal of Personality and Social Psychology, 48*(5), 1279–1289. doi:10.1037/0022-3514.48.5.1279

Verrier, E. D. (2004). Who moved my heart? Adaptive responses to disruptive challenges. *Journal of Thoracic and Cardiovascular Surgery, 127*(5), 1235-1244. Retrieved May 4, 2007, from http://dx.doi.org/10.1016/j.jtcvs.2003.10.016

Vogel, E. (2005). Playgrounds swing away from merry-go-rounds to "composite structures." *Las Vegas Review-Journal, July 5, 2005*. Retrieved June 22, 2009 from http://www.reviewjournal.com/lvrj_home/2005/Jul-05-Tue-2005/news/26818011.html

Vogel, J. J., Vogel, D. S., Cannon-Bowers, J. A., Bowers, C. A., Muse, K., & Wright, M. (2006). Computer gaming and interactive simulations for learning: a meta-analysis. *Journal of Educational Computing Research, 34*(3), 229–243. doi:10.2190/FLHV-K4WA-WPVQ-H0YM

Vogel, J., Vogel, D., Cannon-Bowers, J., Bowers, C., Muse, K., & Wright, K. (in press). Computer and interactive simulations for learning: A meta-analysis. *Journal of Educational Computing Research*.

Vogler, C. (1992). *The writer's journey: Mythic structure for storytellers and screenwriters*. Studio City, CA: Michael Wiese Productions.

Volpe, C. E., Cannon-Bowers, J. A., Salas, E., & Spector, P. E. (1996). The impact of cross-training on team functioning: An empirical investigation. *Human Factors: The Journal of the Human Factors and Ergonomics Society*, 87-100.

Walters, S., Wright, J. A., & Shegog, R. (2006). A Review of Computer- and Internet-Based Interventions for Smoking Behavior. *Addictive Behaviors, 31*, 264–277. doi:10.1016/j.addbeh.2005.05.002

Wang, Y., & Beydoun, M. A. (2007). The obesity epidemic in the United States – gender, age, socioeconomic, racial/ethnic, and geographic characteristics: A systematic review and meta-regression analysis. *Epidemiologic Reviews, 29*(1), 6–28. doi:10.1093/epirev/mxm007

Warshauer, M., & Kern, R. G. (2000). Theory and practice of networked-based language teaching. In M. Warshauer & R.G. Kern (Eds.), *Networked-based Language Teaching: Concepts and practice* (pp. 1–19). Port Chester, NY: Cambridge University Press.

Watkins, S. A., Hoffman, A., Burrows, R., & Tasker, F. (1994). Colorectal cancer and cardiac risk reduction using computer-assisted dietary counseling in a low-income minority population. *Journal of the National Medical Association, 86*, 909–914.

Watters, C., & Duffy, J. (2004). Metalevel analysis of motivational factors for interface design. In K. Fisher, S. Erdelez, & E.F. McKechnie, (Eds.), *Theories of Information Behavior: A Researcher's Guide*. Medford, NJ: ASIST (Information Today, Inc.).

Watters, C., Oore, S., Shepherd, M., Azza, A., Cox, A., Kellar, M., et al. (2006). Extending the use of games in health care. In *Proceedings of the 39th Hawaii International Conference on System Sciences*, (pp. 1-8).

Weiner, B. (1982). An attribution theory of motivation and emotion. In H.W. Krohne and L. Laux, *Achievement, stress, and anxiety*. New York: McGraw-Hill.

Weiner, B. (1985). An attributional theory of achievement motivation and emotion. *Psychological Review, 92*(4), 548–573. doi:10.1037/0033-295X.92.4.548

Weiner, B. (Ed.). (1974). *Achievement motivation attribution theory*. Morristown, NJ: General Learning Press.

Weisman, S. (1983). Computer Games for the Frail Elderly. *The Gerontologist, 23*(4), 361–363.

Weiss, K. B., Gergen, P. J., & Wagener, D. K. (1993). Breathing better or worse? The changing epidemiology of asthma morbidity and mortality. *Annual Review of Public Health, 14*, 491–531.

Wexler, S., Corti, K., Derryberry, A., Quinn, C., & Van Barnveld, A. (2008). *Immersive Learning Simulations: A 360-Degree Report*, eLearning Guild.

Whitcomb, G. R. (1990). Computer Games for the Elderly. Symposium on Computers and the Quality of Life. In *Proceedings of the Conference on Computers and the Quality of Life*, George Washington University, Washington, DC, (pp. 112-115). New York: ACM SIGCAS.

White, R. W. (1959). Motivation reconsidered: The concept of competence. *Psychological Review, 66*, 297–333. doi:10.1037/h0040934

White, R. W. (1959). Motivation reconsidered: The concept of competence. *Psychological Review, 66*, 297–333. doi:10.1037/h0040934

Whitelock, D., Brna, P., & Holland, S. (1996). What is the value of virtual reality for conceptual learning? Towards a theoretical framework. In *Proceedings of EuroAIED*, Lisbon.

Wigal, J. K., Creer, T. L., Kotses, H., & Lewis, P. (1990). A critique of 19 self-management programs for childhood asthma: Part I. Development and evaluation of the programs. *Pediatric Asthma Allergy & Immunology, 4*(1), 17–39. doi:10.1089/pai.1990.4.17

Winter, S., Wagner, S., & Deissenboeck, F. (2008). A Comprehensive Model of Usability. *IFIP International Federation for Information Processing Proceedings* (pp. 106-122). EIS.

Wizards of the Coast. (2004). *Magic: The gathering: Noteworthy facts about the premier trading card game*. Available online at http://ww2.wizards.com/Company/Press/

Wolfe, D., & Jones, G. (1982). Integrating Total Physical Response Strategy in a Level I Spanish Class. *Foreign Language Annals*, *14*(4), 273–280. doi:10.1111/j.1944-9720.1982.tb00258.x

Wood, P. R., Hidalgo, H. A., Prihoda, T. J., & Kromer, M. E. (1993). Hispanic children with asthma: morbidity. *Pediatrics*, *91*, 62–69.

Wright, W. (2008). *Spore* [computer/video game]. Emeryville, CA: Maxis/EA Games.

Yamagishi, T. (1986). The provision of sanctioning as a public good. *Journal of Personality and Social Psychology*, *51*, 110–116. doi:10.1037/0022-3514.51.1.110

Yamagishi, T., & Sato, K. (1986). Motivational bases of the public goods problem. *Journal of Personality and Social Psychology*, *50*, 67–73. doi:10.1037/0022-3514.50.1.67

Yannakakis, G. N., & Hallam, J. (2007). Towards optimizing entertainment in computer games. *Applied Artificial Intelligence*, *21*(10), 933–971. doi:10.1080/08839510701527580

Yukawa, M. (1985). *Structural and psychological factors in social dilemmas*. Sapporo, Japan: Hokkaido University.

Zantow, K., Knowlton, D. S., & Sharp, D. C. (2005). More than fun and games: Reconsidering the virtues of strategic management simulations. *Academy of Management Learning & Education*, *4*(4), 451–458.

Zhao, Y. (2008). What Knowledge Has the Most Worth? *The School Administrator*. Retrieved June 22, 2009 from http://www.aasa.org/publications/saarticledetail.cfm?ItemNumber=9737

Zimet, E., Armstrong, R. E., Daniel, D. C., & Mait, J. N. (2003). Technology, transformation, and new operational concepts. *Defense Horizons*, *31*. Retrieved May 4, 2007, from http://www.ndu.edu/inss/DefHor/DH31/DH_31.htm

Zimmerman, B. J. (2000). Self-efficacy: An essential motive to learn. *Contemporary Educational Psychology*, *25*, 82–91. doi:10.1006/ceps.1999.1016

Zimmerman, C., Raghavan, K., & Sartoris, M. L. (2003). The impact of MARS curriculum on students' ability to coordinate theory and evidence. *International Journal of Science Education*, *25*, 1247–1271. doi:10.1080/0950069022000038303

Zimmerman, E. (2003). Play as research: The iterative design process. *Ericzimmerman.com*. Retrieved November 5, 2008 from http://www.ericzimmerman.com/texts/Iterative_Design.htm

Zuckerman, M. (1971). Dimensions of sensation seeking. *Journal of Consulting and Clinical Psychology*, *36*, 45–52. doi:10.1037/h0030478

About the Contributors

Jan Cannon-Bowers, Dr. Cannon-Bowers holds M.A. and Ph.D. degrees in Industrial/Organizational Psychology from the University of South Florida, Tampa, FL. She recently left her position as the Navy's Senior Scientist for Training Systems to join the School of Film and Digital Media at the University of Central Florida, Orlando, FL as an Associate Professor. As the team leader for Advanced Training Research for the Navy, she was involved in a number of research projects directed toward improving performance in complex environments. These included investigation of training needs and design for multi-operator training systems, tactical decision-making under stress, the impact of technology on learning and performance, training for knowledge-rich environments, human-centered design, human performance modeling and development of knowledge structures underlying higher order skills. At UCF, Dr. Cannon-Bowers is continuing her work in technology-enabled learning and human performance modeling. Her goal is to leverage and transition DoD's sizable investment in modeling, simulation and training to other areas such as entertainment, workforce development and life-long learning and education. To date, she has been awarded several grants to support this work, including a recent award by the National Science Foundation under their Science of Learning Center program. Dr. Cannon-Bowers has been an active researcher, with over 100 publications in scholarly journals, books and technical reports, and numerous professional presentations. She is a Fellow of the Society of Industrial and Organizational Psychology (Division 14 of the American Psychological Association).

Clint Bowers, Ph.D. currently resides on the University of Central Florida's Psychology Faculty as well as being Co-Director of the RETRO (Research and Emerging Technology Research Organization) Lab at UCF's Institute for Simulation and Training. Clint's current research interest is in the use of technology to facilitate teamwork. This takes several forms. The first includes basic research on the nature of effective teamwork and the factors that influence it. A second research thrust involves the training of teams. Clint is especially interested in the use of training technologies and simulation in training team skills. Finally, Clint is interested in the use of technology to assist teams in their day-to-day tasks. This includes research in information visualization, groupware, and other hardware/software systems. Besides his interest in team performance, Clint is also involved in several student projects. These include the use of virtual reality to teach deaf children, the neuropsychology of spatial abilities, and the use of warnings to manage predictable non-compliance.

* * *

About the Contributors

Dr. Anya Andrews earned her Ph.D. in Instructional Technology from the University of Central Florida (UCF) and has worked as a learning strategist for a number of training and education organizations located in the Central Florida Hi-Tech Corridor. Anya is a true expert in the areas of e-learning, simulation and training, and performance support solutions. As the lead instructional systems architect of the RETRO (Recent and Emerging Technologies Research Organization) Lab, Dr. Andrews provides pedagogical and research expertise to the training development efforts focused on leveraging advanced gaming technology towards effective instructional solutions. Dr. Andrews has presented extensively on the subject of serious games both nationally and internationally.

Christopher Ault is an Assistant Professor in the Interactive Multimedia Program a The College of New Jersey. His teaching and research span a range of subjects, from video games to user experience design to digital media production and literacy. He is also a consultant to major corporations in the area of web accessibility. Chris was a researcher and adjunct professor in NYU's Interactive Telecommunications Program. He holds a bachelor's degree from the Plan II Honors Program at the University of Texas, and a master's from ITP at NYU.

Holly Blasko-Drabik graduated in 2004 from Penn State Erie, the Behrand College with a B.S. in Psychology and is currently a Research Assistant at the University of Central Florida, pursuing a Masters in Modeling and Simulation and PhD in the Applied Experimental Human Factors program. She is currently working at the RETRO Laboratory on the design and testing of serious games. Her research interests include working memory and recall, usability testing, software and game development, and software interface design.

Susan L. Coleman, PhD, CPT, is the Chief Performance Officer for IDSI. Dr. Coleman oversees all phases of human performance analyses, while specializing in instructional systems design processes. She earned a Ph.D. in Instructional Technology and Design and has analyzed performance and designed and developed performance solutions since 1983. She spent the last 18 years analyzing and designing training systems for the military. Dr. Coleman conducts instructional design research, training effectiveness evaluations, design analyses, technology integration front-end analyses, and human performance improvement analyses.

Ms. Lisa Czlonka is a Research Associate in the Dynamic Decision Making Laboratory in the Department of Social and Decision Sciences at Carnegie Mellon University. She earned her MSW at the University of Hawaii and is currently pursuing an MBA degree from Carnegie Melon's Tepper School of Business.

Wallace Feurzeig, BBN Principal Scientist, is a computer scientist with extensive experience in mathematics education, artificial intelligence, and programming languages. The central focus of his research is the creation and application of advanced technology to improve learning and teaching in mathematics and science. His early work included the design and implementation of the first context-sensitive, mixed-initiative instructional system and the first versions of the Logo educational programming language. He has developed computer-based instructional tools and applications in clinical medicine, flight training, and computational linguistics. He has designed instructional microworlds for engaging students in scientific inquiry, including a visual programming language for facilitating mathematical

About the Contributors

exploration and experiment (Function Machines). His current research is on evolutionary techniques for the construction and development of software agent systems with capabilities for adaptation and learning.

Stephen M. Fiore, Ph.D., holds a joint appointment with the *Cognitive Sciences* program in the Department of Philosophy and the Institute of Simulation and Training at UCF. He earned his Ph.D. in Cognitive Psychology from the University of Pittsburgh (2000) working primarily in the Learning Research and Development Center. He is Director of the *Cognitive Sciences Laboratory* (CSL) and is co-editor of recent volumes on Distributed Learning and on Team Cognition; he has published extensively in the area of learning, memory, and problem solving at the individual and group level. Through numerous collaborative efforts, Dr. Fiore has helped to manage over $10 million in research funding from organizations such as the National Science Foundation, the Transportation Security Administration, the Office of Naval Research, and the Air Force Office of Scientific Research.

Dr. Cleotilde Gonzalez is an Associate Research Professor and Director of the Dynamic Decision Making Laboratory in the department of Social and Decision Sciences at Carnegi Mellon University. She earned a Ph.D. in Management Information Systems from Texas Tech University in 1995. Her research lies at the intersection of Behavioral Decision Making, Cognitive Psychology, and Computer Science. She employs a wide range of research methods including laboratory experiments with interactive simulations, computational and cognitive modeling to investigate how people make decisions in dynamic environments.

Clarissa Graffeo began working at the University of Central Florida's Institute for Simulation and Training (UCF IST) during her undergraduate studies. After receiving her bachelor's degree in Digital Media from UCF, she has continued to work for IST as a full time Research Assistant and is preparing to begin a Masters degree. Clarissa has developed a love for games, both tabletop—including card, board and pen-and-paper role playing games—and video games. She has also studied many creative pursuits, largely in the areas of writing and theater. During her undergraduate studies, she decided to combine these creative pursuits with her interest in computers by studying digital media. During this time, Clarissa developed an interest in game design, and works both studying and designing games for learning at IST. She also works as a freelance writer, and is interested in writing and narrative in games, particularly Alternate Reality Games.

Dr. Talib Hussain is Senior Scientist at BBN with a broad interest in learning and training for both machines and humans. He is currently co-Principal Investigator on the ONR-sponsored Tools for Games-Based Training and Assessment of Human Performance project, which is investigating advances in authoring technology for pedagogically strong game-based training systems and is developing game-based training solutions to support Navy recruit training. He was recently the development lead on the DARPA-sponsored Plan Order Induction by Reasoning from One Trial (POIROT) project, which is applying a broad range of machine learning and artificial intelligence techniques to learn procedural knowledge based on a single observation of a human performing a task. He was the principal investigator on the DARPA-sponsored Gorman's Gambit project, part of the DARWARS program, which studied the design issues involved in training military teamwork skills using modern multi-player game technology.

Jeanne D. Johnston (PhD Human Performance, Indiana University) is an Assistant Professor of Exercise Physiology in Indiana University's School of Health, Physical Education, and Recreation (HPER). Professor Johnston's research centers on examining the relationship between physical activity, health and quality of life in diverse populations including the aging population and worksites. Recently she has been focusing on promoting physical activity and healthy lifestyle habits within the college student population. She is currently leading technological and educational focused research projects to assess the impact of interventions on physiological, anthropometric, and health outcomes within the college population. She is the principal investigator for a Robert Wood Johnson Foundation Grant titled "BloomingLife: The Skeleton Chase" which is an alternate reality game designed to increase physical activity within the college population. Her work has been published in outlets including *American College of Sports Medicine* and *International Journal of Sport Nutrition and Exercise Metabolism*.

Rachel Joyce: After earning her MA in Interactive Media from UCF, Rachel became a Research Associate Faculty Member at the Institute for Simulation and Training specializing in management in of both game production and learning science research efforts for the RETRO (Recent and Emerging Technology Research Organization). Major accomplishments include completing a National Science Foundation funded physics serious game in 2008 called Lunar Quest. Lunar Quest, a 3D MMOG (Massive Multiplayer Online Game) hybrid with integrated 2D flash minigames designed for high school physics students and college freshman, is the first ever grant from NSF funding a informal learning science tool in the form of a game. By late 2009 Rachel and her development team will complete the Robert Wood Johnson funded game for health, Guardian Angel, designed to assist in Relapse Prevention Therapy. Future projects include a RTS (Real-Time Strategy) serious game redesign for the US Army's Commander Training Course as well as minigames for the Office of Naval Research.

Henry Kelly is President of the Federation of American Scientists (FAS). He received a PhD in Physics from Harvard University and has worked for the US Arms Control and Disarmament Agency, the Congressional Office of Technology Assessment, the US Department of Energy, and the Office of Science and Technology Policy. His work has included information technology policy including policy for information technology research. He convened the President's Information Technology Advisory Committee (PITAC) and facilitated the creation of an interagency program to support education technology research. At FAS his work has included designing and building several computer based games and simulations in collaboration with FAS staff and universities and businesses collaborators. He has also worked with the Congress to increase attention on the benefits of research focused on use of technology to improve learning.

Dr. Alan Koenig is a Senior Research Associate at UCLA/CRESST where he specializes in the application of innovative uses of technology for delivering and assessing instruction. His research focuses on the design and implementation of computer-based games and simulations designed for classroom and/or military training environments. His current work centers on the development of automated assessment systems for high fidelity games and simulations used in the military. Prior to joining CRESST, Dr. Koenig spent 10 years working in the technology sector as both a software developer and mechanical design engineer. Dr. Koenig holds a PhD in Educational Technology from Arizona State University, a BS in Mechanical Engineering from the University of Hartford, and a BA in Economics from the University of Connecticut.

About the Contributors

Martin Knöll: Born in 1981 in Darmstadt, Germany. Martin Knöll studied architecture and urban planning at the University of Stuttgart, Germany. He worked as an architect in Berlin and Stuttgart and is currently a doctoral candidate at the institute of modern architecture and design (IGMA) at Stuttgart University (funded by the State of Baden-Württemberg). He has worked on the serious game concept "DiabetesCity" for children and young people with diabetes (funded by Karl-Steinbuch-Stipendium). Since he is concerned himself, Knöll became interested in the correlation between computer-assisted health care, serious gaming and urban design. He has recently contributed "DiabetesCity - How Urban Designers can help Diabetics" to the first "Conference on Electronic Health Care 2008" in London, UK and has held lectures about urban theory in London, Berlin and Stuttgart. From May 2009 to April 2010 he is doing a research year at the Lansdown Centre for Electronic Arts at Middlesex University in London. He is currently working on his doctorate on "pervasive health games", dealing with the development of health and urbanism within in a perspective of partizipative planning. He will be a research fellow to the Akademie Schloss Solitude in Stuttgart, Germany in 2011.

Dr. John Lee's current research is related to technology-based assessments in a variety of Navy/Marine Corps contexts. He is currently working on the development of a computer-based assessment tool for assessment of Tactical Action Officers (TAO) in a simulated CIC (Combat Information Center) onboard Navy ships called the Multi-Mission Team Trainer (MMTT). He is also working on a simulation-based re-certification assessment of marksmanship coaches' fault checking abillity that delivers just-in-time, individualized instruction that utilizes Bayesian nets for diagnosis and remediation. A third project he is also working on is a game based assessment project for the Navy related to assessment of complex skills (starting with damage control) also using Bayesian nets for real time and after action assessment of skills including situation awareness, decision making and communication. His research interests include data-informed decision making, knowledge mapping, simulation-based assessments, and distance learning.

Anne Massey (PhD Decision Sciences and Engineering Systems, Rensselaer Polytechnic Institute), is the Dean's Research Professor at the Kelley School of Business at Indiana University. Professor Massey's research focuses on focuses on how IS and technology can be used to support individual, group, and organizational performance. She is particularly interested in the extent to which technology features and contextual factors affect performance, and under what circumstances. Current areas of research interest include knowledge-intensive processes, collaborative work to support innovation, and health information technologies. She has been the recipient of competitive research grants from the *Robert Wood Johnson Foundation, Advanced Practices Council of the Society of Information Management, the Center for Innovation Management Studies*, as well as foundation grants from several firms. Her research has garnered industry-based support and funding and field work with several companies including Eli Lilly, IBM, and Nortel, among others.

Rudy McDaniel, Ph.D., is an Assistant Professor of Digital Media at the University of Central Florida (UCF). His research interests include XML, narrative theory, video game technologies, and knowledge management frameworks. He received his doctorate from the University of Central Florida's Texts and Technology program after building an online software application for the narrative classification and analysis of organizational knowledge. He holds additional degrees in Psychology, Technical Writing, and Computer Science. Rudy is co-author of *The Rhetorical Nature of XML: Constructing Knowledge* *in Networked Environments* (Routledge, 2009) and is technical editor for *Emotion Notions: Modeling Personality in Game Character AI (*forthcoming from Cengage Learning). Rudy is currently producing ethical learning games for multiple clients (including EthicsGame.com in Denver, CO) out of the Partnership for Research on Storytelling Environments (PROSE) Lab at UCF.

Dr. Claudia L. McDonald is responsible for leadership and management of the Center for Virtual Medical Education at Texas A&M University-Corpus Christi, a federally designated and funded center of excellence. She has designed and implemented the Center's signature project Pulse!! The Virtual Clinical Learning Lab and overseen commercialization of technology developed at the Center. Dr. McDonald is principal investigator for Pulse!!, a federally-funded research and development project exploring whether high-level learning occurs in virtual space created by state-of-the-art videogame technologies. Dr. McDonald is a board member for the Lone Star Research & Education Network, a collaboration of Texas higher education institutions building a next generation optical network to support research requiring access to global resources and related services. She represents A&M-Corpus Christi and the Texas A&M University System for distance learning initiatives at the Texas Higher Education Coordinating Board.

Ellen S. Menaker, PhD, CPT, is the Chief of Research and Evaluation for Intelligent Decision Systems, Inc. Dr. Menaker has over 30 years of experience in the fields of research and evaluation, cognitive development, and human performance. Dr. Menaker oversees the design, data collection, and analysis phases of research; and various types of analyses and evaluations. She specializes in learning theory, measurement, and instructional systems design. Her academic and industry experiences include conducting research for various military, governmental, and educational entities. Recent studies have focused on implementation of learning strategies, including use of massive multiplayer online games (MMOGs) and the integration of SCORM modules into gaming environments. Recent literature reviews have focused on the application of learning theory, strategies for developing situational awareness, and identification of foundational skills for lifelong learning. Dr. Menaker is an active member of AERA, ISPI, and she serves on the Education Committee for IITSEC.

David Metcalf: With a 15-year history in Web-based and mobile learning, Senior Researcher Dr. David Metcalf combines an academic grounding and continued University involvement with a strong history of industry-centered training and simulation, providing learning innovations for 3Com, Fujitsu, FedEx Ground, Tyco and many others. As a research faculty member with the University of Central Florida's Institute for Simulation and Training (UCF IST), Dr. Metcalf continues to bridge the gap between corporate learning and simulation techniques and non-profit and social entrepreneurship with the foundation of the Mixed Emerging Technology Integration Lab (METIL). Simulation, mobilization, outsourcing, visualization systems, and operational excellence are current research topics. Dr. Metcalf frequently presents at industry and research events shaping business strategy and use of technology to improve learning and human performance.

Kerry Moffitt is a Scientist at BBN Technologies. He works primarily in the lightweight immersive training group, applying his entertainment game development experience to the design and implementation of serious games. He has also developed software for voice communication, conversational analysis, learning management, and speech-based training systems. Before coming to BBN, Mr. Moffitt served in

About the Contributors

a number of engineering and management roles developing 3D interactive entertainment games over the course of ten years. He led and managed diverse teams of engineers, designers, and artists; developed a steady stream of game, library, graphics, and system modules; optimized code down to the assembly level; and shipped seven titles on the Opera, PC, and PlayStation 2 platforms. Mr. Moffitt obtained his BA in Computer Science and Minor in Philosophy from UC Berkley in 1994. He works in Java, C++, and various scripting languages.

Carrie E. Murphy is a Graduate Teaching Associate and Research Assistant in the AEHF program at the University of Central Florida. She graduated in 2004 from Northern Kentucky University with a B.S. in Psychology. Her research interests include metacognition and learning, in particular the active process of navigating a physical space and our memory for three dimensional space. She teaches Perception and works as a research assistant on projects involving online learning, internationalization, educational technology and professional development.

Curtiss Murphy is a Project Engineer in the AMSTO Operation of Alion Science and Technology. He manages the game-based training and 3D visualization development efforts for a variety of Marine, Navy, and Joint DoD customers. He is a frequent speaker and author and specializes in open source technologies such as the Delta3D Game Engine (www.delta3d.org). He has been developing and managing software projects for 17 years and currently works in Norfolk, VA. Curtiss holds a BS in Computer Science from Virginia Polytechnic University.

Mihai Nadin holds advanced degrees in Electrical Engineering and Computer Science and a post-doctoral degree in Philosophy, Logic and the Theory of Science. He has written extensively and lectured around the world on mind, cognition, visualization, various aspects of human-computer and human-technology interaction, and anticipation. He has provided consulting to major corporations and state agencies (in the USA, Europe, and Asia). His education, interests, and professional life combine engineering, digital technology, semiotics, cognition, and anticipatory systems. He founded the antÉ–Institute for Research in Anticipatory Systems in 2002. Since 2004, Dr. Nadin has served as Endowed Professor at the University of Texas at Dallas. His research focuses on theoretical and applied aspects of Anticipatory Systems. With Project *Seneludens*, he introduced Serious Games and Games for Health to UTD in 2004. He was involved in computer graphics since the late 1960s. In 1978, he created the computer game *FrogLeap*.

Teresa Marrin Nakra is an Assistant Professor of Music at The College of New Jersey, where she teaches Music Technology, Composition and Music Theory. Her research areas include Conducting Technologies, Musical Robotics, and Interactive Music Systems. Prior to her appointment at TCNJ, she was the Technical Director of the Computer Music Studio at MIT and a Clifton Visiting Artist at Harvard University. Teresa founded and runs Immersion Music Inc., a non-profit organization that builds technical solutions for musicians, artists, educators and museums. Teresa holds an A.B. in Music, magna cum laude, from Harvard University, and an M.S. and Ph.D. degrees from the MIT Media Laboratory, where she worked with Tod Machover, Rosalind Picard, John Harbison, and Marvin Minsky. She has received numerous distinctions for her academic work, including Research Fellowships from IBM, Motorola, and Interval Research Corporation.

Denise Nicholson, Ph.D., is a Senior Research Associate at UCF's Institute for Simulation and Training, a faculty member of the Modeling and Simulation graduate program and an affiliated Research Scientist at the College of Optics and Photonics: CREOL. Prior to joining UCF in 2005, she was the Deputy Director of the Science and Technology Program Office at NAVAIR Training Systems Division in Orlando, FL. She led basic, exploratory and advanced R&D programs for ONR, DARPA, DMSO and NAVAIR in the areas of Virtual Technologies and Environments, Augmented Cognition, and Human Systems Modeling and Integration. Denise received her B.S. in Electrical Computer Engineering from Clarkson University and holds both M.S. and Ph.D. degrees in Optical Sciences from the University of Arizona. She has over 17 years of experience in research and development for DOD applications and is a Certified Modeling and Simulation Professional (CMSP).

Kelly Pounds' multi-dimensional career began as a schoolteacher in Orange County (FL) Public Schools where she spent twelve years as a classroom teacher. Kelly next took her passion for learning and instructional design into the corporate environment as a Technology Designer for the Disney University where she developed computer-based and classroom training products for leaders at the Walt Disney World® Resort. Her next corporate role was at Hard Rock Cafe International as Manager of Organization Development where, as an innovative strategist, she helped the corporate leadership collaboratively develop strategic planning processes and products. Kelly's enthusiasm for learning is currently shared through her role as Vice President of i.d.e.a.s. Learning where she puts her past work experience and her master's degree in Educational Technology to work everyday. i.d.e.a.s. is an innovation studio employing its core competency of storytelling to create entertainment, marketing, and learning products.

Sara Raasch is a designer working at 42 Entertainment. She has a Masters degree in Digital Media from the Georgia Institute of Technology. Prior to attending Georgia Tech, Sara received a Bachelors of Science degree from the University of Central Florida. Sara has worked on several game design projects, including leading a team of five Central Florida Digital Media students to win the first annual Red Bull BUILD game design competition. After Red Bull BUILD Sara continued designing as a research associate at the Institute for Simulation and Training, where she worked on the DAU CardSim design team. She continued her research into serious game design at the Experimental Games Lab at Georgia Tech, where she designed and developed Terra Viva, a green alternate reality game. Sara currently works in Alternate Reality Game Development at 42 Entertainment.

Yolanda A. Rankin, Ph.D., is a Research Scientist at IBM Almaden Research Center. Her primary research interests involve the design and evaluation of virtual worlds as collaborative spaces that promote knowledge acquisition and simulate business applications. Yolanda first gained industry experience in telecommunications at Lucent Technologies, developing IS41 wireless features, providing first tier customer technical support for wireless service providers, and managing Y2K deployment and TDMA Overlay. As a senior program manager at Luxcore Networks, she managed the product development cycle of optical networking subsystems. Awards include the National Science Foundation Graduate Research Fellow, the Northwestern University Graduate School Fellow, the Alliance of Graduate Education and the Professoriate Scholar, the Patricia Roberts Harris Fellow, and the Tougaloo College Presidential Scholar. Yolanda completed her Ph.D. in Computer Science at Northwestern University, a M.A. in Computer Science at Kent State University, and a B.S. in Mathematics at Tougaloo College.

About the Contributors

Bruce Roberts is a Lead Scientist at BBN Technologies in Cambridge, MA. Over the years, he has developed simulation-based intelligent tutoring systems in several domains, including avionics troubleshooting, shiphandling, and air-to-air combat operations. His interest in automated instruction includes some of the earliest work integrating simulation, computer graphics, and expert systems for training; namely, STEAMER, which taught principles of operation for Navy ship steam propulsion plants. More recently, he led the rapid development and successful deployment of DARWARS Ambush!, a widely adopted multi-player game-based training system. As one of the Principal Investigators on the ONR VESSEL project, he is currently developing game-based training for Navy recruits.

Dr. Alicia Sanchez is the game czar for Defense Acquisition University (DAU), where she coordinates overall strategy for the use of games and simulations to meet learning and performance objectives for DAU and their military and government partners.

Dr. Alicia Sanchez specializes in the implementation of games and simulations into a variety of learning environments. Leveraging decades of research in Education and Simulations, Alicia's focus lies in the appropriate use of games within curriculum and emerging technologies that continuously redefine the potential of games based learning options. Alicia completed her doctoral work with the University of Central Florida, at the Institute for Simulation and Training's Modeling and Simulation program. Since completing her degree, she has served as a Research Scientist at the Virginia Modeling, Analysis and Simulation Center prior to being named Defense Acquisition University's Games Czar.

Jason Seip has been creating real-time game art assets for over ten years. His professional experience includes 3D modeling, character animation, and texture generation using applications such as 3D Studio Max and Photoshop. He has modeled and animated characters to be displayed in the user interface for the commercially released game Kohan II: Kings of War. Over the past five years, he has worked as an Art Lead in the serious games industry, working on training games for the Army, Marines, Navy, and Air Force through clients such as the Army Research Institute. Said games have explored domains such as cultural training, rank/insignia recognition, and IED threat awareness. His work has ranged from 2D 'casual' game play mechanics to fully immersive 3D environments with simulated humans capable of expressing their different emotional states. Work presented in this paper was conducted while he was a graphic artist at CHI Systems, Inc.

Dr. Ross Shegog, Ph.D. is an Assistant Professor of Behavioral Science at the University of Texas School of Public Health and Associate Director of Communication in the University of Texas Prevention Research Center. His research focuses on the application of instructional technology in health promotion and disease prevention. Clinic-based projects have included education and decision-support to enhance asthma management by children, families, and community physicians, for self-management training for HIV+ youth, and for smoking cessation decision support for dentists. School-based projects have included the promotion of smoking cessation and prevention for middle and high school students; promotion of HIV/STD and pregnancy prevention for middle school children; and parental support to impact television viewing behaviors of their children. Dr. Shegog has also contributed to the development of a novel exercise program for worksites and an Internet-based violence prevention intervention. His interests include family, travel, and developing stories and visual arts.

Lee Sheldon has written and designed 20 commercial video games and MMOs, is the author of the book *Character Development and Storytelling for Games,* and is a contributor to several books on game design including *Game Design: An Interactive Experience and Second Person*. Before his career in video games, he wrote and produced over 200 popular television shows, including Star Trek: The Next Generation, Charlie's Angels, and Cagney and Lacey. As head writer of the daytime serial Edge of Night he received a nomination for best writing from the Writers Guild of America. He has been twice nominated for Edgar awards by the Mystery Writers of America. While at IU, he has served as the lead designer on the narrative-driven MMO Londontown and worked on the serious games Quest Atlantis, Virtual Congress, and most recently The Skeleton Chase.

Marcus W. Shute is the Vice President of Research and Sponsored Programs at Clark Atlanta University. A widely published subject matter expert in optical communications, polarization phenomena, wireless communications, materials science, electronic materials, and recently gaming, Marcus holds several patents in the areas of optical fiber design and characterization, polarization phenomena, optical fiber amplifiers, and optical networking. A registered Professional Engineer, he earned the M.S. and Ph.D. degrees in Mechanical Engineering from the Georgia Institute of Technology, the S.M. degree in Materials Science and Engineering from the Massachusetts Institute of Technology, and a B.S. in Mechanical Engineering from Tennessee State University. Awards: 1999 Golden Torch Award for Engineering Excellence - National Society of Black Engineers, the Council of Outstanding Young Engineering Alumni Award - Georgia Institute of Technology, 2005 National Technical Association Achiever of the Year Award - Engineering, and Senior Member of the Institute of Electrical and Electronics Engineers.

Peter Smith is the Director of the Training Technologies Center at the Joint ADL Co-Lab. In this position he works primarily in the Games and Virtual Worlds Domains. Peter has participated in the design and production of numerous Serious Games in the Academic, Government, and Industrial domains. Peter has also participated to the operations of the Serious Games Showcase & Challenge and the Serious Games Initiative.

Timothy J. Smoker is a human performance researcher and consultant based in Orlando, Florida. Mr. Smoker has extensive professional experience in usability and human-machine/human-system integration. Mr. Smoker is currently active in both University research and private industry; he currently serves, and has served, as key research personnel on a number of projects for a variety of sponsors. He holds a Bachelor's degree in Psychology from DePaul University (2003), and a Master's in Human Systems in Modeling and Simulation from the University of Central Florida (2009). Timothy Smoker and his wife Alexandra have lived in the Central Florida area since 2004.

Vance Souders has over 10 years of software engineering experience and is proficient in multiple languages and APIs including C++, C#, .NET, .NET micro, WPF, WCF, DirectX and the XNA framework. In addition to his software engineering experience, Mr. Souders has extensive experience with all parts of the game and simulation development pipeline including task analysis, game design, management, production and marketing. He is active in the development community and is a founding member of the Philadelphia chapter of the International Game Developer Association. Mr. Souders leverages his diverse knowledge at Firewater Games to design and create innovative solutions for multiple platforms including cellular, custom embedded systems, PC and XBOX 360. Work presented in this paper was conducted while he was a senior engineer at CHI Systems, Inc.

Dr. Richard Wainess is a senior researcher with the National Center for Research on Evaluation, Standards, and Student Testing (CRESST) in the University of California Los Angeles' Graduate School of Education and an adjunct associate professor in the University of Southern California's Rossier School of Education. Dr. Wainess' research interests center on the use of games and simulations for training and assessment of adult learners. His most recent work is focused primarily on assessment of problem solving and decision making using computer-based interactive tools. He has authored and co-authored numerous reports, articles, and book chapters and has presented at many conferences on the topic of games and simulations for learning, with a particular emphasis on instructional methods, cognitive load theory, motivation, and learning outcomes. Richard has a B.A. in Radio-Television Production, an MS.Ed in Instructional Technology, and a Ph.D. in Educational Psychology and Technology.

Ann Warner-Ault received her Ph.D. from Columbia University in 2007 and currently teaches Spanish language, literature and culture at The College of New Jersey. In addition to her research in the fields of Spanish and Spanish-American literature, Ann pursues potential applications for interactive and digital media in the foreign language classroom. As Columbia's first New Media Fellow in Spanish, she introduced professors and instructors to online and interactive teaching materials. Ann also worked with a colleague to create STOR, the Spanish Teaching Online Resource, an archive of original Spanish-teaching materials that engage students with popular songs, culture and current events. Most recently, her research has focused on the use of kinesthetic game interfaces as a vehicle for natural language acquisition.

Ursula Wolz, Associate Professor of Computer Science and Interactive Multimedia, is the Principal Investigator for the NSF "Broadening Participation in Computing via Community Journalism for Middle Schoolers" program, and was Principal Investigator of "Multidisciplinary Game Development" (Microsoft Research). She is a nationally recognized computer science educator who has taught students including disabled children, urban teachers, and elite undergraduates. She is a co-founder of the Interactive Multimedia Program at TCNJ. She has a background in computational linguistics, with a Ph.D. in Education from Columbia Teachers College, and a bachelor's degree from MIT, where she was an undergraduate researcher on constructivist computing environments in Seymour Papert's Logo Lab.

Index

Symbols

3D game 106
3D version 111
19th century picture 261

A

accelerometers 276, 278, 279, 281
Accreditation Council on Graduate Medical Education (ACGME) 235
acquisition strategies 81
Acute Stress Disorder (ASD) 248
adult seniors 162, 171
adventure-style gaming strategy 57
aggressive agile development 65
Alcoholics Anonymous (AA) 253
Alternate Reality Game (ARG) 270, 271
American College of Surgeons (ACS) 241
American Medical Colleges (AAMC) 236, 243
a multiplayer scenario-based card game 81
analog computer-based game 150
analytic thinking 236
AnticipationScope 150, 163, 165, 166, 168, 170, 171
anticipatory performance 150
anticipatory perspective 161, 162
anticipatory processes 150, 168
Anticipatory Profile 150, 165, 166, 173
anxiety level 110
A one-way Analysis of Variance (ANOVA) 186
apague el despertador 112
a psychological origin 257
ARCS model 196, 207, 208, 226
ARG play 283
Artificial Intelligence 102
asthma-specific behaviors 213, 214
Attribution Theory 196, 202, 206, 208
auditory learners 107, 109
auditory stimulus 199
autism 151, 171

B

baby boomers 235
baby-boom retirements 233
Balance of Power 106
basic knowledge 122
behavioral skills 249, 250
behavioral theory 197, 201, 221, 222, 229
behavioral therapy 150, 158, 159, 160, 161, 163, 165, 174
behavioral training purposes 256
body language 254
Brain Age 250
Brotherhood of Steel 27
bureaucracy 5
Business, Contracting and Finance (BCF) 9
business financial management 9

C

CALL tools 180, 184
cancer treatment 151
card game system 81
CardSim 81, 82, 84, 87, 88, 89, 93, 94, 95, 96, 97, 98
case-authoring system 242
case-based development 235
case-development 240
Centre for Pervasive Healthcare (CfPH) 262

chief executive officer (CEO) 238
chronic disease 205, 210, 214, 221, 235, 271
chronological manner 51
CISD methods 249
Civilization IV 34, 38, 39, 41, 42, 44, 45
climax 17, 18
clinical education 263
clinical training opportunities 233
Cognitive Behavioral Therapy (CBT) 249
cognitive constructivist work 3
cognitive faculties 175
cognitive framework 214, 218
Cognitive Learning Theory 3
cognitive task analysis (CTA) 54
collaboration 237, 238, 243, 244
collaborative effort 48
collaborative model 190
collaborative social dialogue 190
collaborators 239, 240, 242
college-age population. 271
color-coding 88, 89, 91
Combat-Related Post-Traumatic Stress Disorder (PTSD) 251
commercial game development 50, 51
commercial games 48, 50, 57, 65
commercial tools 129
communication methods 262
communicative performance skills 182
community-based interventions 270, 271
compelling task environment 49
complex cognitive-behavioral self-regulatory skills 213
complex environments 134
complex learning concept 2
complex situation 134
complex story 104
Computer Assisted Language Learning (CALL) 180
Computer Assisted Language Learning (CALL) tools 180
computer-based games 181
computer-based lessons 215
computer-based technology 198
computer-based training 50, 51
Computer Based Training (CBT) 1
computer-empowered learning 104

computer games 48
computer gaming technology 47, 48
computer science 113
computer scientists 157, 159
conception of games 159
conceptualization 13, 28
Concord Consortium 128
conflict resolution 135
constructivist model of learning 106
content design 207
context of validation 159
contextual relevance 256
continuous source of feedback 50
Contracting Officer 81, 83
Controlled Breathing (CB) 249
control symptoms 213
conversational fluency 178, 180, 188, 189, 192
conversational skills 182
Coping Training (CT) 249
craving-management techniques 252
Critical Incident Stress Debriefing (CISD) 248
cross-training 87
CT tools 249
cynicism 111

D

damage control central (DCC) 56, 61
Dance Dance Revolution (DDR) 270
data-driven approach 65, 74, 75
data-driven design 65
DAU CardSim 81, 82, 84, 87, 88, 89, 93, 94, 95, 96, 97, 98
DDM research 135
decision-making 134, 136, 139, 144, 148, 203, 208, 214, 215
decision-making objectives 54
decision-making simulator 252
decision-making skills 250
Defense Acquisition University (DAU) 1, 9
Department of Defense (DoD) 124
development efforts 307, 309
development model 208
Diabetes 263, 265, 267, 268
didactic curricula 233
digital documentation tools 263

digital games 261
digital system 81
digital version 81, 95, 104
documentation tool 263
DoD simulations 126
dramatic immersion 13
dramatis personae 19
dynamic content sequencing 255
Dynamic decision making (DDM) 134

E

economic development 136
educational computer games 104
educational games 103, 104, 111, 112, 113
Educational Mini-Games 1
educational potential 45
educational science 48, 51, 52
educational system 118, 120
education solutions 246
edutaining concept 267
effective game-based learning interventions 247
effective learning paradigms 271
empathy-infused content 250
empirical research 293, 301
empirical work 290
end-users 41
Engagement and immersion 50
English as Second Language (ESL) 179, 185
English proficiency 178, 179, 180, 187, 192
enhance self-efficacy 50
environment 13, 14, 16, 20, 21, 22, 23, 25, 26, 27, 28
environmental model 129
environmental triggers 211, 212, 214, 218
epidemiology 198, 232
equivalent achievement 209
EverQuest® 178, 182, 185, 189
evidence-based health education 196, 223
evidence-based interventions 201
evidence-based programs 196
exergames 154, 166
existing schemata 3
extrinsic motivation 4, 5, 7, 11

F

face-to-face 54, 73
fact-based content 102
Fallout 3 13, 15, 22, 24, 25, 26, 27, 28, 29, 30
fantasy-based 21, 23, 27
fantasy-based environments 27
Federations of American Scientists (FAS) 129
fictional narrative centered 273
fidelity 298, 302
Final Fantasy 152
Fitness & Wellness LLC (FWLLC) 272
Flooding Ontology 60, 61

G

game-based learning 247, 250, 254
game-based learning method 250
Game-based tasks 49
game-based technologies 237
game-based training 47, 48, 50, 51, 52, 69, 70, 73, 75, 78, 290
game-based training design 50, 73
game-based training prototype 51
game-based training systems 47, 48, 50
game controller 106, 109, 113
game designers 48, 57, 63, 68, 69, 71, 73, 263, 267
game developer 198, 203, 209
game dynamics 199, 200
game engines 48
game environments 45
game genres 22
game-informed practices 182
game mechanics 59, 70, 73
game-oriented culture 151
game-play 102, 103, 104, 109, 110, 112, 113
game-related blogs 270, 271
game's context 103
game world 14, 15, 22
gaming approach 57
gaming environments 249
gaming for learning 200
gaming technology 246, 247, 249, 256
genuine learning 103

Geometry Tutor 128
gesture-based mechanics 113
global economy 117
goblin antagonist 14
graphical environment 103
graphical representation 89
graphic design 48, 51
group communication 282
guidance system 9

H

hand-held controller 106
hard-coded game 65
Health Belief Model 196, 202, 205
health-care delivery systems 235
health-care education 236
health education 196, 197, 198, 202, 203, 223, 224, 225, 227
health game developer 198
health game intervention 196
Health promotion 198, 231
health promotion programs 196, 197, 210, 223
HealthyHome 262, 268
highly active learning 49
Higinbotham 150, 151
hippocampal volume 173, 175
HIV infection 205
Holy Trinity 16
homo ludens 160
host cells 123
human behavior 197, 201, 203
human performance assessment 48, 51
human physiology 240
human-simulation trainers 234
human user 103
hypertension 257

I

idiomatic sophistication 113
IHC 198, 199
immersion 13, 14, 15, 16, 20, 25, 26
Immersive Learning Simulations (ILS) 249
immersive training systems 47
Immune Attack 121, 123, 124, 125, 126, 127, 128, 129, 130

Indiana University (IU) 272
indicated intervention 248
Individual Resources (IRs) 254
in-game dictionary 179, 189, 192
in-game dictionary module 189, 192
in-game social interactions 178, 189, 190, 191
insouciant 104
Institute for Creative Technologies (ICT) 251
Institute for Simulation and Training (IST) 251
Institute of Modern Architecture and Design (IGMA) 267
instructional design 48, 50, 51, 58, 60, 61, 64, 67, 68, 69, 70, 71, 73, 74, 75, 76
instructional design models 209
instructional design practitioners 255
instructional environments 49
instructional material 207
Instructional Systems Design (ISD) 254
Intel Corporation 85
Intelligent Decision Systems 47, 77
intelligent tutoring systems 48, 50, 51
intentionality dimension 206
interaction design 106, 113
Interaction frequency 49
interactive digital 182
Interactive digital media 179
interactive game-based system 251
interactive problem solving tools 135
interactive story 15
interactive storytelling 13, 14, 15, 20, 21, 28, 29
interchangeable atomic units 28
Intervention Mapping framework 210
Israeli-Palestinian conflict 134, 136, 137, 139, 146, 147

K

kinesthetic 102, 103, 104, 107, 108, 109, 110, 111, 113, 114
kinesthetic interface 103
kinesthetic learners 104, 107, 108
kinesthetic learning 102, 110, 111, 113, 114
knowledge of XML 189
knowledge, skills, and abilities (KSAs) 81

knowledge, skills and attitudes (KSAs) 87
KSA cards 82, 83, 89, 90, 94, 95

L

language acquisition 102, 107, 108, 109, 110
language learning 103, 108, 109, 110, 113
language-learning game 103, 104
language pedagogical supports 180
language socialization 188
learner challenge 200
learner mastery 200
learners 291, 294, 295, 296, 298, 299, 300, 303
learning curve 185
learning environment 179, 181, 183
learning experience 70, 207, 297, 298, 300
learning experience active 250
learning games 291, 295
learning goals 200
learning objectives 2, 8, 10, 14, 15, 16, 17, 22, 28
learning outcomes 240, 291, 292, 293, 294, 295, 296, 297, 298, 299, 300, 303
learning simulations 246
learning styles 292
Learning Styles Index (LSI) 107
lifestyle diseases 260, 261
Likelihood Model 207
logical groupings 190
logistics 83, 92, 93
Logo programming environment 104
long-term literacy 7
long-term memory 3, 4
ludologists 20
ludology 113
Lunar Colonization Academy 5
Lunar Quest 5, 6

M

management-related goals 215
management training 85
Massively Multiplayer Online Game (MMOG) 5
massively multi-player online role playing game (MMORPG) 151, 178, 185
medical applications 151

medical community 233, 234, 235
medical education 233, 235, 237, 239, 240, 241, 242, 243
Medical education 233
medical-education community 243
medical-education experience 242
medical learners 233, 234
medical personnel 157
medical simulation 234
medication taking 211, 212, 213, 214, 218
mental health care 247, 248, 257
mental health-related serious game 246
mental health training 246, 247, 248, 250
METIL 89, 91
METIL team 89, 91
microphone 106
Mii avatars 109
military health system 247
Mini-Game 1, 8
mini-game paradigm 2
mini-games 1, 2, 3, 5, 6, 7, 8, 9, 10
mini-programs 33
Mixed Reality Game 262
MMOG environment 6
model-based world 49
monitoring symptoms 212, 213
Morphology of the Folktale 18
motion-based controller 102
motion-capture data 150
motion-sensitive controllers 103
motivational theories 196, 197, 221
movie production 48
multi-faceted approach 222
multimedia 103
multiplayer card game 81
multiple user dungeons (MUDs) 184
multiple user-skill levels 34
Myers-Briggs Type Indicator (MBTI) 138

N

narrative taxonomy 13, 16, 22, 24, 28
narratology 13, 15, 16, 17, 19, 20, 21, 28, 29, 30
National Academy of Sciences (NAS) 121
national health objectives 196
National Science Foundation (NSF) 1, 128

Native English Speakers (NES) 185
Naval Service Training Command (NSTC) 52
near-endless iterations 44
Near Field Communication (NFC) 264
Near Field Communication (NFC)-tag 264
Neuroscience to Science 151
No Child Left Behind 123
non-commercial training programs 48
non-diabetics 266
non-fantasy-based environments 27
non-narratologists 15
non-playable characters (NPCs) 26, 57, 180, 181, 182
non-social interactions 191
novel strategy 196
NPCs 16, 26, 27

O

observation-based focus groups 254
online courses 9
ontology 60, 61, 63, 69
Oregon Trail 104
oscilloscope 150
Outcome expectation 204

P

paramedical practitioners 242, 243
paramilitary organization 27
PeaceMaker 134, 135, 136, 137, 138, 139, 144, 145, 146, 148
pedagogy 48, 49, 68, 125, 130
pedometers 263
Perceived behavioral control 204
personality, religious affiliation 136
personal protective equipment (PPE) 63, 64
Pervasive Health Games (PHGs) 260, 262
PFC Ratner 9, 10
physical activity 158, 166, 260, 261, 263, 264, 270, 271, 272, 273, 278, 279, 280, 281, 282, 283, 284, 285, 286
Physical inactivity 270, 271
physical interactions 107
physical manifestation 109
physiological effects 174
physiological sensors 150
Piaget's theory 107

platform 47, 53, 68, 150, 151, 165, 170, 171
Player Characters (PCs) 181, 182
plot 13, 14, 16, 17, 18, 19, 21, 22, 23, 24, 25, 26, 27, 28
Poetics 17
political philosophy 261
Post-Traumatic Stress Disorder (PTSD) 247, 248, 251
potential psychological health issues 248
PRECEDE model 196, 202, 203
Prevention Therapy (RP) 251
Preventive health behaviors 197
problem-based case 240, 242
problem-solving 250
problem-solving tasks 179
product prototype 51
Progressive Muscle Relaxation (PMR) 249
Project management tools 129
Project Manager 81
proof-of-concept 102, 112, 113, 115
protagonist 14, 21, 23
prototype training systems 49
psychobiological treatments 249
psychoeducation methods 251
psychological attractiveness 279, 280, 281, 282, 283, 284
psychological component 257
psychological health care 257
Psychological Health (PH) 251
psychology 198, 228, 231, 290, 291, 293, 301, 302, 304
PTSD 151, 171, 173
Pulse!! 233, 234, 236, 237, 238, 239, 240, 241, 242, 243, 244
Pythagorean Theorem 14, 28

Q

Quality in Use Integrated Map (QUIM) 36
Questionnaire for User Interface Satisfaction (QUIS) 42, 32
quiz-module 266

R

Rapid Iterative Testing Evaluation (RITE) 37
R&D community 300

realistic rendition 162
real-life 52, 53, 54
real-life situations 155, 160, 249
real world 178, 179, 191
real-world concepts 6
real-world experiences 104, 107, 109
real-world interactions 109
real-world possibilities 240
reprobes 105
Retro-Future Moon Colony 5
rhythmic foot patterns 181
rigid taxonomy 125
Risky behaviors 197
RITE method 37, 38, 39
role-playing games (RPG) 14, 33, 153, 184, 194

S

schemata 3, 4
schizophrenia 155
scientific foundation 155
scientific literature 160
screen-based version 103
Second Language Acquisition (SLA) 178, 179
second language pedagogical methods 192
second language teaching methodology 184
selective intervention 248
Self-efficacy 204, 216, 217, 223, 294, 295, 301
self-judgment 214
self-management 205, 219
self-management behaviors 197, 211, 213, 215, 218, 220
self-monitoring 203, 214, 295
self-regulatory 211, 213, 214, 215, 217, 218, 219
self-regulatory behaviors 214
self-regulatory skills 213, 215, 218
Semantic memory 4
Seneludens 156, 160, 161, 162, 163, 166, 168, 171, 174, 176
sensation-seeking 208
sensory-motor experience 103, 107, 108, 109
sensory-motor period 107
sensory stimuli 256
serious educators 290

Serious Game applications 306
Serious Game developers 309
serious games 13, 14, 15, 16, 17, 19, 20, 21, 22, 24, 25, 27, 28, 34, 40, 44, 45, 178, 179, 181, 184, 193, 229, 310
serious games research 222
serious game story 22, 28
Serious Pervasive Games (SPGs) 261
sexually risky behavior 205
short-term memory (STM) 4
short-term treatment models 247
shovelware 104
Sick-role behaviors 197
simulation-based education 234
simulation-based training 47, 48, 50, 51
simulation-based training system 48
situated learning 3
Situation Awareness, Communications, and Decision Making 60
Skeleton Chase 270, 271, 273, 277, 278, 279, 280, 281, 282, 283, 284, 287, 288
social activity 110
Social Cognitive Theory 3, 4, 196, 202, 203, 214, 217, 219, 223
social context 112
social dynamic 110, 115
social interaction 86, 110, 153, 174, 184, 265, 266
social interconnectedness 160
social learning 271
social network 174
social support 248
Social Values Orientation (SVO) 139
Social Values Orientation (SVO) scale 139
society organization 34
sociology 198
socio-psychological aspects 135
socio-psychological theories 135
socio-psychological variables 135
software developers 48
software development 50, 51, 65, 73, 75
Sony Online Entertainment (SOE) 189
Spanish-learning game 103
Spanish-learning video game 102
split-second reflex 106

stability dimension 206
stages of change 205, 224, 228
state-of-the-art gaming technologies 249, 256
stigma 247, 248, 257
stimulus control, 206
stimulus-control 253
story-based training 48, 51
storytelling 13, 14, 15, 17, 19, 20, 21, 22, 24, 25, 27, 28, 29, 30
stress-inducing events 247
stress prevention interventions 249
stress recognition 251
subject-matter experts (SMEs) 238
supply chain management 134
symbolic domain 85
symbolic representations 203
syntactical structure 183
systematic design process 207
systematic health promotion 210
system designers 48
Systems Engineering lead 81
Systems Usability Scale (SUS) 32, 40
Systron analog computer 150

T

tactical combat casualty care (TCCC) 242
Tactical Language Training System (TLTS) 180
tactual learners 107, 109
taxonomy 199, 229, 306, 307
Telemedicine Advanced Technology Research Center (TATRC) 251
Tetris 1
text-based computer games 106
textual information 182
text/voice messages 270, 271
theoretical construct 15
theoretical dimensions 28
theoretical methods 215, 217
theory-based principles 166
Theory of Planned Behavior 202, 204
Theory of Reasoned Action 196, 202, 204
The Zuckerschloss 264, 266
three-dimensional virtual simulation 235
toolbox 15
Total Physical Response 103, 107, 114, 115

Total Physical Response (TPR) 103
traditional classroom instruction 178, 185, 186, 187, 188
traditional interventions 270, 271
traditional learning objectives 179
training development effort 256
Training Technology Lab (TTL) 246, 251
Transtheoretical Model 196, 202, 205, 229
traumatic event-related memories 249
Triadic reciprocity 203
trust attitude 134, 137, 139
typology 260

U

UI Builder 189
universal intervention 248
usability methods 35, 45
User Interface (UI) 189
U.S. workforce 117, 118, 120, 132

V

video games 15, 19, 20, 27, 102, 104, 105, 109, 110, 111, 112, 134, 135, 139, 143, 145, 146, 179, 180, 181, 182, 185, 192, 193, 249, 250, 290, 293, 300
video games medium 102, 104
violent manifestation 135
virtual characters 181, 183
Virtual Clinical Learning Lab 233, 234
virtual court 150
virtual environments 249, 251
Virtual Field Trips (VFTs) 6
virtual heroine 14
virtual landscapes 128
virtual learning space 235
virtual reality experts 157
Virtual-reality learning 236
Virtual Reality Stress Inoculation Training (VR-SIT) 249
virtual-reality training 234, 235
virtual reality (VR) 234, 237, 251
Virtual Reality (VR) Assessment 251
virtual realm 178, 179
Virtual simulation 235
virtual space 233, 235, 237, 240

virtual spaces 190
virtual world 181, 182, 183, 184, 191
virtual-world medical facility 240
virtual worlds 13
Virtual-world technologies 233, 234
visible goal indicators 234
visual representation 60
vocabulary acquisition 178, 179, 185, 186, 187, 188, 189, 190, 192
vocabulary memorization 9
voice-over-IP (VoIP) 37
VR health-education systems 241
VR learning platforms 243
VR technologies 241, 243
Vygotsky's theory 294

W

Watch, Discover, Think, and Act (WDTA) 211
WDTA simulations 218
web-based course 10
web-based games 270
Well-designed games 50
white-coat effect 262
Wii balance-board 114
Wii console 109
Wii-Mote 103, 104, 109, 110, 111, 112, 113, 115
working memory (WM) 4
World Health Organization (WHO) 263
World of Warcraft (WoW) 33

X

XML code 189